The Geology of **Switzerland**

The Geology of Switzerland

AN INTRODUCTION TO TECTONIC FACIES

Kenneth J. Hsü

PRINCETON UNIVERSITY PRESS · PRINCETON, NEW JERSEY

Copyright © 1995
by Princeton University Press

Published by Princeton University Press, 41 William Street,
Princeton, New Jersey 08540
In the United Kingdom: Princeton University Press,
Chichester, West Sussex

Library of Congress Cataloging-in-Publication Data

Hsü, Kenneth J. (Kenneth Jinghwa), 1929–
[Geologie der Schweiz. English]
The geology of Switzerland : an introduction to Tectonic Facies / Kenneth J. Hsü
    p.   cm.
Includes bibliographical references and index.
ISBN 0-691-08787-3
1. Geology—Switzerland.   I. Title.
QE285.H7813   1994
554.94—dc20   94-18000 CIP

This book has been composed in Times Roman

Princeton University Press books are printed on acid-free paper and meet the guidelines for
permanence and durability of the Committee on Production Guidelines for Book Longevity of the
Council on Library Resources

Printed in the United States of America

10 9 8 7 6 5 4 3 2 1

To Hermann Eugster, Rudolf Trümpy, their teachers, colleagues, and students, who have given us the geology of Switzerland

Contents

List of Figures

Foreword

by
Alfred G.
Fischer

Among earth scientists, Kenneth Hsü stands out for the wide range of fields to which he has brought not only competence but new light. Never content with the present level of understanding, he has ever reached out to seek and find new horizons. His scholarship rests on a pervading interest in the development of scientific concepts, extending far beyond the earth sciences. At the same time, his books for the lay public have been extraordinarily effective in conveying the excitement in exploring the workings and history of the Earth.

Much of his life has been spent at the Eidgenössische Technische Hochschule in Zürich, Switzerland, in contact with Alpine problems. Here he eventually came to teach the "flagship" course on the geology of Switzerland—a course taught, over a century, by a chain of the grand masters of Alpine geology: Heim, Staub, and Trümpy.

With advancing insight, each of these masters had come to see the development of the Alps in a somewhat different light, and Hsü was no exception. To this task he brought incision, imagination, and two special advantages. Whereas his predecessors had grown up with Alpine geology from the beginning, Hsü came to it from the outside, with geological experience gained elsewhere, and free of the load of dogma that invariably surrounds any scientific problem. Also, he came at the time of a major revolution in geological thought: the conception of plate tectonics.

The Alps are a complex jumble of displaced slices of crust (nappes or thrust sheets), variously deformed and partly metamorphosed, and displayed in spectacular scenery. No other mountain belt has been studied in such detail, by generations of geologists with ever-growing heaps of petrographic, paleontologic, geochemical, and geophysical information; and nowhere else has such a complex nomenclature been developed. From the beginning it was realized that most of the sediments had been formed on the sea floor, and had been deformed. As models evolved it became clear that this mountain belt, less than 100 km wide, must represent remnants of an ancient sea, Tethys, closer to 1000 km wide, driven together in a long and complex history full of puzzles and seeming contradictions.

Plate tectonics and the advances associated with it provided explanations for previously enigmatic rocks and structures, and some understanding of the dynamics. An oceanic terrane had grown, separating Europe and Africa, and had then been squashed; the crust was subducted and its cover piled up in thrust slices. Hsü discusses each of the traditional slices but relegates some to the status of mélanges—puddings of mixed rocks that provide a kind of matrix for the more coherent underthrust slices dragged into subduction zones. Other thrust sheets developed as overthrusts at the surface. Whereas most structures verged northward, northward underthrusting provided some antithetic structures. Deformation was not paroxysmic but continuous, though punctuated by episodic collisions.

Finally, the overall development seems not to have been that of a simple "Pennine Ocean" or Tethyan axis that drove Europe and Africa apart in Jurassic time and collapsed in the Cretaceous-Eocene. Rather, the complexity of terranes suggests that this portion of Tethys grew as a series of oceanic "back-arc" basins surrounded by tectonic arcs, each with its own sets of sediments, subduction zones, and history—basins of which the Eastern Mediterranean is a remnant. Thus Alpine history was not one of Africa crushing a seaway against Europe, but rather a progressive growth and subduction of oceanic basins and surrounding arcs against each other, and the progressive welding of their remnants against the south-European margin. Indeed, one comes to wonder if Italy is just another such arc, being rotated toward the Alpine belt by the growth of the Tyrrhenian basin. Africa has not collided with Europe—at least not yet.

The spotlight thus comes to rest on the origin and history of these back-arc basins, whose very transience implies a large role in the construction of the crust.

Other compressed mountain belts have also seemed simple in the beginning and complex on deeper study. While each is an "individual," all share some general features, and for the purpose of comparison Hsü advocates classification of tectonic units into three *tectonic facies*: the *raetic*, for thrust sheets involving rigid basement; the *celtic*, for nappes involving mobilized basement; and the *alemannic*, for the foreland fold and thrust belts of décollement type. Some tentative comparisons provide excursions to the Appalachians, to China, and to the Western Cordilleras of the Americas.

Thus Hsü proceeds from basic stratigraphic concepts and from discussions of rock sequences to imaginative megatectonic models. Interspersed are bits of personal recollection and philosophy, and the warning that creativity is being discouraged in an increasingly "professionalized" science.

Hsü makes no pretense of presenting the ultimate truth about the Alps. Like its predecessors, his is a model that attempts to explain, as simply as possible, the structure of the Alps as now known, in terms of the state of the art in our understanding of the formative processes. The reader may disagree with some of the interpretations and conclusions, but this is a great forward leap in the interpretation of the Alps and of mountain belts in general.

June 22, 1994

Preface

to the American Edition

I had a reputation as being a bad teacher, and I asked a teaching assistant for an explanation. He told me that students go to class to prepare for examinations. A good teacher gives orderly presentations, but I burst onto the podium and tell the students that I come to inspire, not to inform. Students may be fascinated by my stories, but they feel that they do not learn anything that helps them prepare for the examinations. Adding to the problem is the fact that German is my second foreign language, and a student who is convinced that he cannot understand will not understand. To prevent a catastrophic confrontation between me and the undergraduates, I was spared, for almost two decades, the obligation to teach beginning courses. However, when my colleague Rudy Trümpy retired, his duty of teaching the course "The Geology of Switzerland" to geographers, foresters, engineers, and beginning geology students fell onto my shoulders, thanks to the trust of the young colleagues who had joined the Institute.

The catastrophe came sooner rather than later. I was still congratulating myself on my success as a lecturer, thinking that I had inspired young beginning students to new insights in science, when I had to receive a delegation that represented the 200-plus students in my class. Nobody could understand what I said. Facing the rebelling students, I asked them what they wanted. "*Skriptum!*" they all roared—they wanted a set of lecture notes.

I have always considered myself a public servant, paid to serve the students. I would give them the *Skriptum* if that was what they wanted, even though I had never written any lecture notes in my 30 years of teaching. It was "no big deal," because I had already started writing a book on the geology of Switzerland in order to introduce the tectonic facies concept. The students in my class got lecture notes for the rest of the semester. The class next year had a whole set in English. The class after that had a set translated into German by my colleague Ueli Briegel. Last year's class had the *Skriptum* in book form, *Geologie der Schweiz: Ein Lehrbuch für den Einstieg und eine Auseinandersetzung mit den Experten* (Birkhäuser, Basel).

I was going back to work on my manuscript on tectonic facies, thinking that I had to change the elementary textbook into a "scholarly treatise," phrased in technical terms, that would sound sufficiently foreign to geologists who are not experts in Alpine geology. But I changed my mind about how to revise after I read numerous reviews of the opus. My critics may or may not agree with me on the geological interpretations presented in this textbook, but they all praised my pedagogical approach and my philosophy of the making of science. I was particularly encouraged to leave the text as it was by my young friend Celal Sengör, who told me that he read the book in a few hours while sitting in a Zürich police station waiting to register the theft of his camera. "It would be a shame to sacrifice the readability for the formality of unintelligibility," he told me.

I was finally convinced that Sengör was right after I talked to the students in my class. I asked them if they could understand my lectures in German. I was encouraged when I received an affirmative answer: "We understand you because of your enthusiastic intonations; we probably would not understand if you talked in a monotone." Reading a book on the geology of Switzerland is difficult for someone who does not know much about Swiss geography, and who is not familiar with the technical terms of Alpine geology. I recall my frustration when I tried to learn the geology of Switzerland in 1957. I knew enough German then, but I was stopped by the many unfamiliar proper nouns. To describe, for example, the cover picture of the Birkhäuser opus, one would have to write that "the *Mytilus Dogger* of the *Brianconnais facies* of the *Médianes rigides* (*Klippen Nappe*) and the *Zoophycus Dogger* of the *Subbrianconnais facies* of the *Médianes plastiques* on the *Mythen* are thrust above the *North Helvetic Priabonian Nummulitenkalk* and the *Ultrahelvetic Wildflysch* near *Seewen* in *Canton Schwyz.*" In this one sentence, one finds a total of sixteen lithostratigraphical (Mytilus Dogger, Zoophycus Dogger, Nummulitenkalk, Wildflysch), paleogeographical (Brianconnais, Subbrianconnais), tectonic (*Médianes rigides*, *Médianes plastiques*, Klippen Nappe, North Helvetic, Ultrahelvetic), chronostratigraphical (Priabonian, Dogger), and geographical (Mythen, Seewen, Schwyz) names. These names are known and even well known to Swiss undergraduates in geology, but are "Greek" to foreigners who are not experts in Alpine geology. I did not get very far in 1957 when I plodded through Heim, Collet, and Cadisch, day after day. All those beautiful details were useless to me, as they could not be stored up in my memory as active knowledge. The resistance to learning was overcome only when I picked up Bailey's *Tectonic Essays, Mainly Alpine*. In one afternoon of browsing through its pages, I caught the essence of the Alpine magic. No, we do not need another Heim, or an updated Cadisch. I have to learn from Bailey, and the title of the textbook for students was originally *Tectonic Facies, Mainly Alpine*.

I have, therefore, left the first twelve chapters of this book on the geology of Switzerland essentially as they were written, except for minor textual changes. Knowing that the book was originally intended as a textbook for nonmajors, the reader will not, I hope, misconstrue my efforts to provide simple explanations of fundamental concepts in geology as condescension. I should also be excused for being less than completely comprehensive, because the first twelve chapters of this book are not intended to be a geological guide, but rather to provide the background knowledge for the introduction of the tectonic facies concept.

The subtitle of the Birkhäuser opus is *A Textbook for Beginners and a Discourse for Experts*. In this English edition, I have added, for experts, a discourse on comparative tectonics and on the applicability of the tectonic-facies concept to interpret the geology of mountain ranges outside Switzerland. Using comparative anatomy to understand mountain-building is, of course, not a new idea. The expression *vergleichende Tektonik* is essentially the title of Stille's classic monograph, and *The Anatomy of Mountains* is a recent book edited by Schaer and Rodgers that narrates the life stories of several mountain belts. This approach searches for the basic body-plan (*Bauplan*). Each mountain chains may be different, but they all share some common characteristics, having been built according to the same blueprints. Mountains formed by the collision of two continental plates belong to a different class than those that owe their genesis to the subduction of ocean lithosphere under continent. These are the two major kinds of mountain on the earth, the Tethyan and the Circum-Pacific, as proposed by Eduard Suess more than a hundred years ago. Now that we recognize that collision is preceded by subduction, we can conclude that all Tethyan mountains must have undergone a Circum-Pacific stage of tectonic evolution before the collision.

The *Bauplan* concept of anatomy is derived by analogy with architecture. The essential elements of buildings are the same: all buildings have a foundation, floor, walls, ceiling, and roof. If one knows the style of architecture, one could reconstruct the appearance of the whole edifice on the basis of a few stone or brick fragments lying around the site of the ruin. With the presence of a foundation and a few tile fragments, one could conclude that a house once stood on that spot, even if all the building stones, for one reason or another, had disappeared. Likewise with plants—a tree with root, stem, and crown had to be there before the pollen or leaves could come into existence. The presence of pollen grains or leaf fragments is sufficient evidence that a tree once existed. The comparison is applicable to geology, after one recognizes that mountains have a *Bauplan* too. This knowledge dictates that we cannot reach a conclusion merely on the basis of what we can see; we also have to consider what must have been there. Collision-type mountains have a suture zone, and a tectonic facies representing that suture must be (or have been) present in each such orogenic belt, even if it cannot be found now, having been buried, eroded, or faulted away. The methodology of science is not only induction on the basis of observations, but also deduction on the basis of theories.

The geology of the Swiss Alps has given us not only a geometrical *Bauplan*, but also an understanding of the processes of building—mountain-building. The history of the tectonic evolution of the Alps is not only kinematic, but also dynamic. Adopting this, the best-studied mountain chain of the world, as a model, the task of resurrecting a "temple" (an ancient orogenic belt) from its ruins (scattered outcrops) becomes workable.

It is perhaps no coincidence that a book introducing the concept of tectonic facies should be based upon the geology of Switzerland. Only when I was writing the final chapter, on non-Smithian stratigraphy, did I realize that the great contributions to Alpine geology were possible because the masters of the pioneer days had the concept, even though they did not use the term. Swiss geologists have long recognized that the mapping methodology introduced by William Smith cannot easily be adopted for studying metamorphic terranes. The map units such as Combin zone, Barrhorn series, and Monte Rosa nappe are not rock-stratigraphic units; they are units of mélanges or of tectonic subfacies.

Five years ago, when this work was first conceived, I thought that *The Geology of Switzerland* would be Part 1, and the application Part 2 of an ambitious undertaking. As I reach the twilight of my career, however, I see the limitations, and I do not think that I shall ever have the opportunity of doing justice to a "Part 2." I have thus decided to add only four chapters, to introduce the tectonic facies concept, to my original manuscript for Birkhäuser's *Geologie der Schweiz*, namely

Chapter 13. The Tectonic Facies Concept
Chapter 14. Global Tectonic Facies
Chapter 15. A Tectonic Facies Map of China
Chapter 16. Theoretical Geology

Chapter 13 gives a historical perspective: the concept evolved after the invention was found useful. Chapter 14 is an effort to introduce a common language for communication. The geological histories of the classic mountain belts of the world, namely the Appalachians, the Caledonides, and the American Cordilleras, have been worked out by generations of geologists. This short chapter uses the terminology of the tectonic facies concept to summarize my understanding. Chapter 15 is a long abstract of a book I may yet write on the geology of China, if there is enough incentive left after my retirement. The interpretations of the orogenic histories of Chinese mountains are not conclusions, but working hypotheses to be tested by the younger generation of Chinese geologists, many of whom I have worked with in the field during the last decade. Chapter 16 may turn out to be my swan song in geology, in which I have my last words on the stamp-collector's mentality of some of my colleagues. I have been sufficiently discouraged during the last few years that I have vowed that I shall never again write for a refereed journal. The reviewers of this opus may yet convince me that I should not write any more books, either. I feel more and more lonesome, and am saddened to find that I seem to be the odd man out in my scientific philosophy. Have I been walking too fast, or am I just a Don Quixote no longer adaptable to modern times?

The work turned out to be far more demanding than I had anticipated. I would like to express my gratitude to the many persons who helped make this opus possible. Ueli Briegel redrafted numerous illustrations when we coauthored the German edition of *The Geology of Switzerland*. Eva Pour proofread the manuscript and made many corrections; she and David Brunn helped in many other ways. Albert Uhr, scientific illustrator, and Urs Gerber, photographer, did a great service in adapting for this English-language edition the many illustrations that were originally prepared for the German text. I am indebted to Birkhäuser Verlag Basel, for permission to reproduce copyrighted materials from the first twelve chapters of the German edition of the book. I am grateful to many colleagues (Clark Burchfiel, T. A. Cross, J. C. Crowell, Warren Hamilton, R. D. Hatcher, and Celal Sengör) for permission to reproduce their published figures as illustrations for the new chapters of this book, and I acknowledge with thanks the following publishers for permission to reproduce several copyrighted figures: the Annual Review of Earth and Planetary Sciences (vol. 15, © 1987 by Annual Reviewers Inc.) for Figs. 14.1 and 14.2; the American Journal of Science for Fig. 14.13; the Geological Society of London for Figs. 14.4 and 14.7; and John Wiley & Sons for Fig. 14.8.

Hanspeter Funk, Helmut Weissert, Gregor Eberli, and Paul Felber read the German edition of *The Geology of Switzerland*; their comments led to improvements in the manuscript. Celal Sengör has always been an inspiration and a source of encouragment whenever I have ventured into unchartered waters. My Chinese colleagues, Wang Qincheng, Chen Haihong, Li Jiliang, Sun Shu, and the late Yao Yongyun, contributed much to the evolution of my thinking on the concept of tectonic facies, while we were working together in China. Christine and Peter Hsü were very patient during the years when I was sitting at the dinner table and thinking about the opus. Finally, I feel a deep indebtedness to Switzerland and to Swiss geology, and I would like to dedicate this opus to my father-in-law, Hermann Eugster, my friend Rudolf Trümpy, and to their teachers, colleagues, and pupils who have given us the wonderful geology of Switzerland.

Kenneth J. Hsü
Zürich, October 30, 1992

Preface

to the
Swiss Edition

I was a student once, and I often thought that lectures were a waste of time. Some lectures probably are a waste of time, when students could get more out of using the time to study on their own. Yet lecturing is still the *modus operandi* of instruction, from primary schools to universities. It has been time-tested universally, and its value should not be underestimated. Furthermore, we teachers are asked to give lectures; we have little choice. I have, therefore, had to give much thought to the possibility of optimization. Why would you want to attend my lectures?

I would like to see you here not because you have to be here. As university students, you have freedom of choice. You come because you can learn from my lectures what you cannot by reading a book. Students go to lectures often because textbooks are not readable. Few authors are used to writing in a conversational style, and hardly any science textbooks get published that way. Textbook authors are required to be explicit, succinct, and, worst of all, comprehensive.

Comprehensive books serve as ready references for teachers; they can choose what they want to teach from the encyclopedic coverage of comprehensive books. But textbooks are often nightmares for students, because they do not know where to start (and where to end) in a big, thick textbook. Lectures have an advantage because they can never be truly comprehensive. A course has to be taught in so many hours. Constrained by the boundary condition of time, lecturers have to choose their materials. Attending lectures, students are told what is important, what is less important, in a discipline where the knowledge is unlimited. Students in our university, therefore, demand lecture notes from their instructors. Those notes serve to define a more limited area for candidates to prepare for their examinations. This book, like many other textbooks, is an outgrowth of my lecture notes.

Books are seldom written with regard to a reader's current qualifications. A reader who does not understand now can come back tomorrow, or next year. Lecturers are, however, not supposed to ignore the students' capacity to learn. If they do, they will hear complaints, and I have heard many complaints in my teaching career. After 25 years of teaching and a reputation for being a poor lecturer, I have finally appreciated the difference between what I should teach and what students can learn. I can assure you that it is not my purpose to teach you all that I know about the geology of Switzerland. I give lectures because I hope you will learn something that you should and that you can.

You attend lectures, as I said, because you do not always understand what you read. Often you may have difficulty fathoming the reasoning behind an author's statement. I remember my own frustrations as a student. I was often stopped by a strange-sounding technical term in a textbook, or because I came across an elegant equation that was well known, or that could be proved. Well, the term was probably defined in a technical dictionary I did not have. The equation might be well known to an expert or easily proven by the knowledgeable, but it was not known to me nor could it not be proved by average users of the textbook.

When you face such a problem, you either waste many hours puzzling over a small point, or you give up and throw the book aside. You cannot face the author and expect an immediate answer to your question. Lecturing is different. Any teacher with pride and sensitivity is prepared to answer questions on the subject he or she is teaching, and has to know what he or she is talking about. Ultimately, no teacher can hide behind semantics in a lecture hall. Ultimately, no student need be too shy to ask questions and get to the bottom of things. This is the way learning should be.

Lecturing may be a monologue, but thoughtful lecturers do try to interact with their audience. They go back and repeat if they see an abundance of puzzled expressions in the audience, or they make a long story short when the audience seems bored. One communicates with just as much repetition or omission as is necessary. In contrast, authors pound away on their typewriters. They try to anticipate their readers, but can never be sure.

We have to realize that lectures are tailor-made for students. Authors of textbooks do not, however, labor under such constraints. If tailor-made clothes do not fit, the misfit is irrevocable. Generations of my students have complained about my lectures, which were specially designed for them. Yes, a designer often misjudges a customer's size, and I have been changing the content and methodology of my teaching. Textbooks are, however, not tailor-made for a particular group of readers. They have excesses and deficiencies, but they serve useful purposes: Lecturers can use a textbook as the basis for their lectures; they can subtract from or add to what is written. Students can use a textbook to teach themselves, if they miss a lecture or a crucial explanation during a lecture. They cannot expect to learn all they need to know from one textbook. But if they learn at least something from it, the book has justified its existence.

Lectures are transient yet final. Books are permanent, yet transient. Those may sound like inscrutable quotes from Lao-tzu's *Tao-te Ching*, but the fact remains that mistakes in a book can be corrected, but lectures and lecture notes cannot. I realized that when I went over some notes taken of my lectures by students; errors I made while writing on the blackboard were copied down and, in many cases, "immortalized." There were also mistakes that were made either because I had not mastered the subject matter or because the students did not understand my explanation. Thus errors and mistakes made during a transient lecture are mostly irrevocable. Now in this textbook, I have to write down what I said, but my colleagues and I have an opportunity to correct errors and remedy mistakes; I can also think over again some of the problems that I did not explain very well during my lectures and try to do better. If this printing or this edition cannot do the job, there can be a second printing or a revised edition. What is said is said, but what is written can always be changed.

This book is written as a history of inquiry, especially my history of inquiry. Geological interpretations are not stated as self-evident truth, but as approximations of understanding, deduced from incomplete observational data and imperfect knowledge of natural processes. The purpose is not so much to inform, but to invite inquiries to reduce the element of falsehood in our understanding.

I have often said that one can teach science like one teaches a foreign language. Students learn a foreign language by studying its grammar and acquiring enough of a vocabulary to read, speak, and write. The grammar of the language of science is scientific logic, and the vocabulary consists of the technical expressions, which are commonly abbreviations of scientific concepts. You learn to read, speak, and write science by knowing the principles and terminology; you do not need to memorize all the information contained in an encyclopedia of science.

Geologists study the history of the Earth, and rocks are an archive of the Earth's history. The books in the archive are, for the uninitiated, books written in a foreign language. I have no intention of making a summary of all the books in the archive. Teaching, in my viewpoint, is not primarily aimed at transmittal of information. Information is transmitted only to illustrate analytical methodology, logic, and principles.

I would like to make it clear that our curriculum for students is not designed to make lecturing like reading an encyclopedia. My lectures are not an oral summary of the *Geological Guide of Switzerland*; I do not even attempt to tell you where all the formations are, and what the age, stratigraphy, and tectonics are of those rocks cropping out in different cantons of Switzerland. The purpose of my lectures is to teach you to "read" the rocks as if they were foreign-language books in a library, so that you can acquire knowledge on the geology of any part of Switzerland when you need such information during the practice of your future profession. The purpose of this opus is to record this effort in written words.

During my discussions with students concerning my lectures on the geology of Switzerland, I heard complaints, often after one or two lectures, that they did not see the big picture in individual lectures. I had to dismiss the criticism as a manifestation of impatience. Good science books read like a detective story. Many incidents are recounted and many leads are suggested, but they do not seem to have much relevance to the main theme. That method of presentation is what makes the subject interesting, with all the confusion and suspense. Eventually, the loose ends are tied up. I have to ask your patience, if you feel somewhat lost in the details during the first nine lectures. I assure you that those details are given only because they are the necessary building blocks for the final conclusion, to be presented at the end of the opus.

Writing for the students in the beginning class, some of whom have had no introductory geology, I have to assume that some of the readers do not have more knowledge about the earth than a middle-school student. We start with the Jura Mountains with an exposition of the origin of stratigraphy as a science, and the evolution of ideas on tectonics. This introduction is followed by a chapter on the geology of the Swiss Midland. Interpretation of the Recent and late Pleistocene sediments in terms of observable processes is the foundation of actualism in geology. The Molasse stratigraphy is an illustration of classic stratigraphy, relating temporal changes of facies to the interplay of subsidence and sedimentation. At the same time, the science of isotope geology is introduced to illustrate the application of chemistry and physics in order to understand geological processes.

The rest of the volume is devoted to an exposition of the geology of the Swiss Alps on the basis of modern plate-tectonic theory. Here my discussions with the expert begin. Departing from the classical compartmentalization, which separates stratigraphy from structural geology, I formalize the concept of tectonic facies. The idea is not new, but the terminology is, and it is published for the first time in this elementary textbook. The basic motivation is to focus on the interrelation of sedimentation, deformation, and paleogeography. Each tectonic facies is characterized not only by its style of deformation, but also by its sedimentary association, and its paleogeographic position in the framework of the plate tectonics. The term Helvetic refers not only to the pile of cover thrusts, but also to the mainly shallow marine carbonate deposits on a passive continental margin. The evolution of the geological structures is influenced not only by the kinematics of plate displacement; the structures have become what they are because of their sedimentary history and paleogeographical evolution. Using the tectonic facies concept, numerous units have been recognized, and discussions of the geology of those units constitute the next chapters (4–11) of the book, namely:

> The Helvetic stratigraphy, as a model of sedimentation on a passive continental margin and as an illustration of the methodology for paleogeographical reconstructions (Chapter 4).
>
> The Flysch, their stratigraphy, sedimentology, tectonics, and paleogeography, as a source of information on the geological history on and beyond the margin of a disappeared ("consumed") ocean (Chapter 5).
>
> The Prealpine Klippes, their stratigraphy, tectonics, and paleogeography, to illustrate the principle of identifying nappes by recognizing the telescoping of stratigraphic facies (Chapter 6).
>
> The Pennine core nappes, their petrography and the inferred deformational and metamorphic histories, as an illustration of the fate of underthrust passive continental margin (Chapter 7).
>
> The Bündnerschiefer, schistes lustrés, North Penninic Flysch and other sediments involved in subduction and collision processes of deformation (Chapter 8).

The ophiolites, vestiges of a former ocean, to illustrate the principles of mélanges and non-Smithian stratigraphy (Chapter 9).

The Schams Paradox, as a model of the nonlinearity of geological evolution (Chapter 10).

The Austroalpine nappes, their stratigraphy and tectonics, to portray the evolution of a passive margin that became an active margin at the beginning of the Alpine orogenesis (Chapter 11).

The general conclusions for this elementary textbook are, of course, the classical stratigraphy and structural geology; the main conclusions of the Swiss masters have stood the test of time. However, we would be irresponsible if we did not introduce the new concepts of geology that have led to more correct interpretations of some well-known phenomena. In Chapter 4, for example, the classical interpretation of the so-called "Wang Transgression" is discarded in favor of a new theory of hemipelagic sedimentation on passive margins. Similarly, the new idea of flysch as part of accretionary prism is introduced to beginning students, instead of the once popular idea of gravity-sliding.

The Antigorio recumbent fold is almost as valid a concept today as when Gerlach first made the postulate, more than 100 years ago. The tectonic superposition of the six Penninic nappes is still considered the same as that proposed by Argand at the turn of the century. Yet no one could deny the great revolution in our understanding of the geology of mobilized basement nappes and of ophiolites. Even the concept of the core nappes has to be revised in the case of its application to interpret the Monte Leone and Lebendun in the west, or the Adula in the east. Instead of accepting one great Mischabel nappe, as advocated by Master Staub, we find less contradiction in adopting a modern interpretation that places the Bernard and Monte Rosa on two sides of the Piedmont Ocean. The "Platta nappe" has disintegrated into a unit of ophiolite mélange, and the Margna has become an Austroalpine exotic in such a mélange.

Although both belong to the Brianconnais/Subbrianconnais facies, the interpretation of the Median Prealps has not changed greatly since Schardt wrote his brilliant analysis in 1893. On the other hand, the tectonic evolution of the Schams Nappes has been a center of controversy for more than half a century. I was accused of talking over the heads of my beginning students when I presented an exposition of the geology of Schams to my class. In fact, the idea was as obvious as the egg of Columbus once it was proposed. I can understand that most of my Swiss colleagues would find it difficult to accept the simple solution of a century-old paradox, but modern popes can no longer prevent Galileos from teaching scientific truth.

There is really not much dispute about the tectonics of the Austroalpine, except that between the Austrian and Swiss schools on the relative superposition of two groups of nappes. The controversy whether the rocks under the Engadine Window belong to the Austroalpine or Penninic can be resolved, if one accepts the postulate of mélange. The paradox of the Bergell Granite is not all that mysterious, because highly mobilized continental basement may have been partially melted to form granite magma and discordant intrusion.

The twelfth chapter is a résumé of the geologic history of Switzerland, and we have a chance to throw a glance at the European Hercynides. A model of the Mesozoic evolution of the Tethys, as presented in my 1987 Fermor lecture, is described. The model for Neo-Alpine orogenesis is the one first proposed by me in 1979.

The infusion of original ideas into an elementary textbook can introduce new insight as well as falsifiable opinions. We take the risk of writing such elementary textbooks because scientists have learned more modesty during the second half of the twentieth century. We always search for truth. Once upon a time, we thought we had found it, and the "truth" is taught to beginning students in elementary textbooks. Karl Popper told us, however, that truth is elusive. We can never find the truth, but it exists because there is falsehood. We approach the truth a step further each time we are proven wrong. The duty of teachers of science is thus not to impress on the uninitiated that the professors know best; our duty is rather to tell the students that we do not know for sure, but we think it could be so and so, and we could be proven wrong. This elementary textbook is written in this spirit.

Kenneth J. Hsü
Zürich, May 1991

The Geology of **Switzerland**

1.

The Jura Mountains

Driving from Basel to Bern, one crosses the Jura Mountains. Exposed on roadcuts are layered rocks. These are pages of a history book on the earth's history. Geology is a translation of the earth's history into our language, and geologists are interpreters, expressing in words or figures the history stored in the rocks.

Principles of Stratigraphy

Layered rocks are beds or strata, and the study of strata is called stratigraphy. Unlike the natural phenomena described in physics or chemistry, the earth's history is too complicated to be expressed by formulae or digital models. Stratigraphy is an analogue model of the earth's history, and "laws" in stratigraphy are expressions of common sense. Nicolaus Steno (1638–1686), a bishop of Florence, stated the obvious when he formulated the "law of superposition": the oldest sedimentary layer was the first to be laid down, and it is thus found at the bottom of a pile.

Roadcuts are rarely more than a dozen meters high, and even the Grand Canyon in Arizona is only a few thousand meters deep. Nowhere do we find a sequence of strata, or a stratigraphic section, which encompasses the oldest and the youngest ever deposited on earth. The second common-sense "law of stratigraphy" is the observation that a sedimentary layer has a certain horizontal extent, also known as the "law of lateral stratal continuity." A limestone layer could, for example, be a key stratum: a limestone layer containing many mollusk remains may be the highest (or youngest) bed at one roadcut, but the lowest at the next. Stratigraphical correlation is the methodology of certifying that the limestone exposures at the two roadcuts belong to one and the same bed. Through stratigraphical correlation, we can use a columnar section of stratigraphy to express our knowledge of the relative ages of strata in a region.

One of the earliest stratigraphical sections was presented by Johann Gottlob Lehmann and George Christian Füchsel in the eighteenth century to describe the sequence of a group of sedimentary rocks in Thüringen. They noted three rock units, always in the same descending order at different places in central Germany, namely:

Variegated, shales, marls, limestone, and gypsum
Limestone with abundant mollusk remains
Red sandstone

The sandstone was always at the bottom and the gypsum was always near the top. *Buntsandstein*, *Muschelkalk*, and *Keuper* were the names eventually given to designate those units, and the whole group was called the Triassic, referring to its threefold division.

Lehmann and Füchsel recognized the value of rocks as a record of the ancient history of the earth. Füchsel was one of the first to propose that continuous series of strata of the same composition constitute a formation, which is now defined as a mappable rock unit. The three formations of the Triassic constitute a system, the Triassic System: Buntsandstein is the lowest, or the Lower Triassic, and is also the oldest, or Early Triassic, in age. The Muschelkalk and Keuper are Middle and Upper (Late) Triassic, respectively. The modifiers Lower, Middle, and Upper refer to the superposition of the rock unit, and Early, Middle, and Late refer to the time when those Triassic rock units were laid down. We see that the term Triassic has acquired a double significance: we speak of Triassic rocks when we see this group with a threefold division at a roadcut or in the mountains, or we can write about a time, the Triassic Period in the earth's history, or the period of time when the sedimentation of the Buntsandstein, Muschelkalk, and Keuper took place in northern Germany.

The same threefold division of Triassic rocks is found in the Jura Mountains of Switzerland. In the eighteenth century, naturalists thought in terms of universal formations—that similar sequences of rock formations have been precipitated from water everywhere on the surface of the earth. Advocates of this school of thought were called "Neptunists," and their master was Abraham Gottlob Werner. It seemed reasonable to assume that the Buntsandstein, Muschelkalk, and Keuper of the Jura Mountains were laid down at the same time as the German Triassic. This correlation of the two Triassic sequences is, as we now know, correct, but the empirical verification of the interpretation had to wait until the principles of biostratigraphy were worked out by an English surveyor, William Smith.

As a young man just finished with his apprenticeship, Smith was hired to do surveying for a network of inland waterways to be built by the British government for the transport of coal. He joined the project in 1794 and for the next six years surveyed every segment of its course and supervised construction. Excavation laid bare outcrops on both sides of the newly dug channel, but almost no two sections showed exactly identical sequences of strata. Limestones, shales, sandstones, chalks, and clays, like strange faces in a crowd, seemed to possess no distinctive features by which they could be identified. Yet, after six years of growing familiarity, Smith realized that each stratum contained its own peculiar fossil remains, and that the sequential relation of the fossil faunas along the excavated channel was everywhere the same. Using the fossils as guide, Smith published in 1815 a geological map of England and Wales. Thus was born the science of biostratigraphy, on which the geological chronology was to be based. We now know that the Triassic of the Jura Mountains are strata that are the same age as the Triassic of central Germany, not only because the same rock sequences were found in those two different regions, but also because similar faunal groups were found in both places. In the framework of the theory of evolution, similar faunal groups can be postulated to have lived during the same time period. The word Triassic acquires thus a third meaning; the phrase Triassic *faunas* refers to the fossil remains that are found in Triassic *formations* of the Triassic *age*.

Geological time determined by stratigraphic superposition was relative. The Triassic strata of Germany overlie a group of twofold division (Rothliegend and Zechstein), which was known as Dyassic (twofold) but is now named Permian (after the town of Perm in the Urals). The Triassic rocks are overlain by limestone formations called Jurassic. The Triassic is younger than the Permian, and older than the Jurassic. Since the development of radiometric dating, the Triassic can be expressed in terms of Ma, or millions of years before present; the latest results suggest the Triassic Period spanned some 50 million years, between 250 and 200 Ma.

Biostratigraphy permits the correlation of strata. Rocks of very different kinds could be correlative, and determined to have been deposited during the same time interval in the past. The result is the recognition that there are no "universal formations," and this conclusion is verified by radiometric dating. The eighteenth-century teaching of the Neptunist master Werner is wrong; the lateral continuity of any rock unit is not global. Rocks of Triassic age may have the same threefold subdivision of Buntsandstein, Muschelkalk and Keuper in northern Germany and in the Jura Mountains, but the Triassic of the Eastern or Southern Alps consists of different kinds of rock units.

The stratigraphical sequence of the Alpine Triassic was worked out by Austrian geologists during the second half of the last century; they proposed a sixfold division of the Triassic in the Alps, and the units in descending order are designated as follows:

Rhätic
Noric
Carnic
Ladinic
Anisic
Scythic

Unlike the German Triassic of sandstone, limestone, and variegated beds, the Alpine Triassic rocks are mainly carbonates—limestone ($CaCO_3$) or dolomite ($CaMg(CO_3)_2$). The geological age of the units was defined on the basis of the faunas contained in the rocks of those units. Among the faunas, an important element is ammonites, or extinct floating organisms similar to the nautiluses of today. Many ammonite species are cosmopolitan. Faunas with affinity to the Triassic faunas of the Alps have been found in Triassic rocks on other continents: Anisic or Ladinic have been identified in faraway places such as China and North America, and such fauna have been found in rocks quite different from those of the Alpine Triassic. Scythic, Anisic, Ladinic, Carnic, Noric, and Rhätic do not constitute globally correlative rock units, or formations, but they designate globally correlative faunas living at about the same stage of the earth's history. The term stage is introduced in stratigraphy to denote a certain interval of geological time. The time of the Triassic Period is thus divided into six stages: Scythic Stage, Anisic Stage, Ladinic Stage, Carnic Stage, Noric Stage, and Rhätic Stage. By correlating faunas that lived at the same time, geologists have found out that the Lower Triassic Buntsandstein Formation was laid down during the Scythic Stage, the Middle Triassic Muschelkalk Formation during the Anisic and Ladinic stages, and the Upper Triassic Keuper Formation during the Carnic, Noric, and Rhätic stages.

Stratigraphical Taxonomy

Studying the faunal sequences of sedimentary strata in different parts of the world, geologists during the last century have come up with a division of geological time into eras, periods, and epochs, as shown in Table 1.1.

The oldest fossiliferous strata in the Jura Mountains are Carboniferous. They are overlain by Permian, Triassic, Jurassic, Cretaceous, Tertiary, and Quaternary sediments. The oldest rocks underlie the Carboniferous strata; they are thus older than Carboniferous, or Paleozoic in age. The term Paleozoic refers to the era of ancient life

Table 1.1. Geological Time Scale

Period	Epoch	Date Began (millions of years ago)
Cenozoic Era		
Quaternary	Holocene	0.01
	Pleistocene	1.8
Tertiary	Pliocene	5.3
	Miocene	23.7
	Oligocene	36.6
	Eocene	57.8
	Paleocene	66.4
Mesozoic Era		
Cretaceous		144
Jurassic		206
Triassic		245
Paleozoic Era		
Permian		286
Carboniferous		360
Devonian		408
Silurian		438
Ordovician		505
Cambrian		570
Precambrian Eras		

forms, which lived some 600 to 250 Ma ago. No sedimentary formations older than the Carboniferous are known in the Juras, or elsewhere in Switzerland. Rocks of those older Paleozoic rocks are Devonian, Silurian, Ordovician, and Cambrian, named after places in England or tribes in Wales. Still older rocks are called Precambrian; fossils with hard skeletons are hardly ever found in those oldest rocks.

The oldest rocks of the Jura Mountains are igneous rocks emplaced kilometers deep below the earth's surface and crystallized under conditions of relatively high temperature and pressure. They are called crystalline rocks and form the basement of the Jura Mountains. After the consolidation of the igneous rocks from magmas, the region was uplifted and subjected to erosion. The upper Paleozoic Carboniferous and Permian rocks, mainly terrestrial sediments, were then deposited on river banks or flood plains, underlain by the basement rocks.

The Triassic is the Germanic Triassic discussed previously. The Buntsandstein finds its recent analogue in the red beds of deserts. The red color is derived from the mineral hematite, Fe_2O_3, which is formed by the reaction of hydrated iron minerals ($Fe_2O_3 \cdot xH_2O$) with ground water flowing through the pores of rocks, an alteration process after sedimentation called diagenesis. When the Triassic desert became submerged under marine waters, abundant molluscs populated a shallow sea-bottom and the Muschelkalk was deposited. The Keuper includes sediments deposited under a variety of marine, lagoon, or tidal-flat conditions, including residues of evaporated sea-water, forming rocks called evaporites. Gypsum, a hydrated calcium sulfate ($CaSO_4 \cdot 2H_2O$), is a common evaporite mineral, whereas anhydrite is a dehydrated sulfate; they are soluble in ground water, but are found in some outcrops. The more soluble salts, such as halite ($NaCl$) or potash salts, are not exposed; they are encountered only below ground in boreholes.

The bulk of the rocks in the Jura Mountains are Jurassic formations of the Jurassic System or Period. The name Jurassic is derived from the fact that those rocks were first described in the Jura region by Leopold von Buch, a well-known pioneer, who spent several years in the Neuenburger Jura in the early nineteenth century. Von Buch noted that the region is underlain mainly by limestones and referred to them as the *Jurakalk* or Jurassic limestone. He used the term later to describe similar rocks in Poland (von Buch 1805), and eventually the term Jurassic was used to designate not only the rocks but also the period of geological history during which those rocks were deposited.

Not all the sedimentary strata in the Jura region are Jurassic. Underlain by the Triassic mentioned above, the Jurassic formations of the region are overlain by strata of the Cretaceous System. How do we distinguish a Jurassic formation from a Triassic or Cretaceous formation?

The placement of a boundary between geological systems is a matter of opinion and is often controversial, because each group of paleontologists may have different opinions. There were endless debates during the last century where the top or bottom of a system should be. Eventually geologists came to the realization that geological boundaries are arbitrary or man-made. The practice now is to leave the decision to international agreement: different subcommissions of the Commission on Stratigraphy of the International Union of Geological Sciences are to reach a majority opinion to "hammer a golden spike" between two beds at the type locality of a boundary section. In fact, of course, no golden spike will be hammered into the ground; "golden spike" refers to the horizon, or the contact between two beds, that is designated by "legislation" to represent the instant of time separating two geological periods (or epochs or stages). The determination of the boundary between two periods elsewhere has to be based upon correlation to the horizon marked by the imaginary "golden spike." Although the golden spikes at the top and bottom of the Jurassic System have not yet been hammered in, there is general agreement that the base should be defined by the oldest fossil zone in a biostratigraphical stage called Hettanginian, and the top by the youngest fossil zone in a stage called Tithonian.

The Jurassic System of the Jura region, like the Triassic, also has a threefold division: Lias (Lower Jurassic), Dogger (Middle Jurassic), and Malm (Upper Jurassic). Leopold von Buch used to talk about the Black Jura, the Brown Jura, and the White Jura, because the Liassic marls of southern Germany are dark gray, the Dogger iron oolites there give the rocks a brownish stain, and the Malm limestones are nearly pure calcium carbonate and give the whole formation of the region a white appearance. The Lias, Dogger, and Malm are further subdivided into numerous formations.

The Jurassic Period is also divided into the Early, Middle, and Late Jurassic epochs, and further subdivided into various stages: the Liassic stages are the Hettanginian, Sinemurian, Pliensbachian, Domerian, and Toarcian; the Dogger stages are the Bajocian, Bathonian, and Callovian; and the Malm stages are the Oxfordian, Kimmeridgian, and Tithonian. The names of the stages are derived from the place where the rocks of that stage are best developed; the rocks of the Bathonian Stage are named, for example, after a sequence of rocks near the town of Bath, and those of the Oxfordian are named after Oxford in England.

The Cretaceous rocks are the youngest of the Mesozoic, the era of the middle life forms. They are mainly shallow marine limestones like the Jurassic, but they are found only at scattered localities in the Juras; much has been removed by erosion.

The Cenozoic derives its name from the fact that it is the era of the new life forms. Fossil molluscs from the Cenozoic may be identical or very similar to those found on modern beaches, or they may belong largely to extinct species. An English geologist, Charles Lyell, suggested early in the last century the terms Eocene, Miocene, and Pliocene to designate the epochs of the dawn of, the less, and the more recent life forms, respectively. Later, the Paleocene, Pleistocene, and Holocene epochs were introduced to designate the ancient (more ancient than Eocene), the most recent, and the wholly recent life forms (see Table 1.1).

Although the Cenozoic strata may once have covered much of the Jura region, only remnants here and there are preserved from erosion. The Eocene sandstone and bauxitic clay were deposited at a time when the climate was warm and wet in Switzerland, before the sedimentary strata of the Alps were piled up to form mountains. The Oligocene strata bear evidence of fluvial connection between the Molasse basin in the Swiss Midland and the Rhein-Graben. The Miocene are also mostly deposits of stream gravels, although there is evidence of occasional incursions of sea water, which inundated at least part of the Jura region. The Pliocene and Quaternary sediments are mainly stream gravels, deposited when the Jura Mountains were being formed.

The Facies Concept

It is now common knowledge that sedimentary rocks are not universal formations. Sedimentary strata of any given age may be a limestone at one place, a marl at another, and a shale at a third locality. A Swiss geologist, Amand Gressly (1814–1865), was the first to take note of this fact, and he introduced the concept of facies to denote the different manifestations of sediments.

Gressly (1838) developed the idea while he was working in the Jura. He wrote (pp. 10–11): "In the regions that I have studied, perhaps more than anywhere else, very variable modifications, both petrographic and paleontologic, everywhere interrupt the universal uniformity that, up until now, has been maintained for different stratigraphic units in different countries."

He innovated the concepts of lithofacies and biofacies. The two major lithofacies identified by Gressly in the Jura are (1) a thin-bedded marly limestone facies, and (2) a massive reef-limestone facies. Geological studies during the last century have verified Gressly's conclusion and have added much detail to his observations. I shall illustrate this lateral facies change with a discussion of the different types of sedimentary strata deposited during one stage, the Oxfordian, of the Jurassic Period in the Solothurn Jura, as described by a former student at ETH, R. Gygi (1969).

Oxfordian is a stage of the Late Jurassic Epoch, but there was much dispute as to the exact interval of time that should be called Oxfordian. In older geological literature, we see terms such as Argovian and Sequanian (e.g., Ziegler 1956). Stratigraphers now use the term Oxfordian in a broader sense, so that the Oxfordian, *sensu stricto*, in older literature is now considered Early Oxfordian, and the Argovian and Sequanian have become more or less equivalent to the Middle and Late Oxfordian. The rocks deposited during the different stages can be identified on the basis of ammonite faunas found in those rocks.

The Lower Oxfordian strata are a ferruginous oolitic limestone (*eisenoolitischer Kalk*), encountered everywhere in the Jura Mountains of Canton Solothurn. The basal middle Oxfordian is a thin unit of limestone and marl, called Birmenstorf Beds (*Birmenstorfer Schichten*). Both of these units show little lateral facies changes in Solothurn.

The bulk of the Middle Oxfordian consists of Effingen Beds (*Effinger Schichten*) and Günsberg Beds (*Günsberger Schichten*). The former are clay, marl, and argillaceous limestone, whereas the latter are mainly thin-bedded limestone. As indicated by the stratigraphical profile (Figure 1.1), near Olten the Effingen Beds are 200 m thick, while the Günsberg Beds are only about 20 m thick. But the marly Effingen strata become thinner and the Günsberg strata thicker in a northwesterly direction, because some of the marly beds are changed into limestones.

The Günsberg Beds include coralline limestone, quartz-rich and glauconitic limestone, cross-bedded oolitic limestone, pelletoid limestone, bioclastic limestone, oncolitic limestone, and argillaceous limestone, as well as some dolomite. The thickness of the formations ranges from about 30 to 90 m in Canton Solothurn. From its type locality at Gschlief near Günsberg eastward toward Olten, the coralline limestone becomes the dominant lithology, whereas the other beds pinch out.

The overlying upper Oxfordian Holzflue Beds are mainly oolitic limestone, which

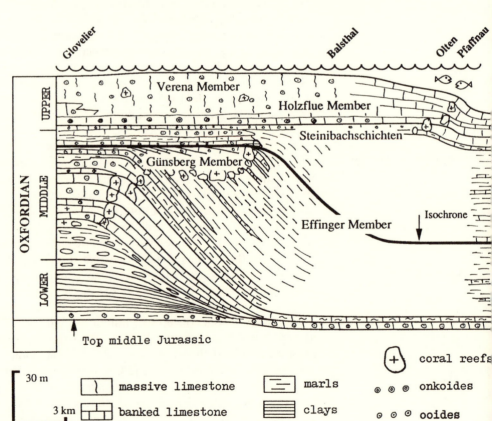

Fig. 1.1. Lateral facies variation of strata of Oxfordian age in the Solothurn Jura (after Gygi 1969).

used to be known as white Sequanian oolite (*oolithe blanche séquanienne*). This formation is separated from the underlying Günsberg Beds by a prominent pelletoid limestone bank. The oolite is commonly cross-bedded. Lenticular masses of massive coral-algal limestone are encountered locally. Also intercalated are thin layers of white limestone of Ste-Vérène (*calcaire blanc de Ste-Vérène*); these are porous pelletoid and locally oolitic limestones, with sparry calcite cement. They are easily weathered, forming niches on a limestone cliff.

The Holzflue Beds range in thickness from 20 to 90 m. Like the underlying Günsberg Beds, the Holzflue are changed, in an easterly direction, into a massive coralline limestone near Olten. Farther east are mainly skeletal limestone beds of the same age (Figure 1.1).

Comparative Sedimentology

The facies change is easy to comprehend when geology is based on actual observations. In the eighteenth century, geologists envisaged universal formations. Abraham Gottlob Werner described a succession of rock formations in the vicinity of Freiburg in Saxony, and he taught his students that sedimentary successions elsewhere on the surface of the earth should be the same. Werner seldom traveled; he relied upon the Bible, and thought that sedimentary strata were all deposited during the different stages of Noah's Flood when the whole surface of the earth was submerged under the Deluge. Geologists of the nineteenth century departed from that tradition, and modern geology is founded upon the assumption of actualism. We now believe that the world millions or billions of years ago was, in principle, not much different from what it is today: there were oceans, shelf seas, and lagoons, and there were continents with rivers, lakes, marshes, tidal flats, etc. Now, as then, different kinds of sediments, or strata of different facies, are laid down in various environments: gravels and sands are and were the deposits of streams; sands, silts, and muds are and were laid down in coastal waters; clays and calcareous oozes settle and settled on deep-sea floor, etc. The interbedding of terrestrial deposits and marine sediments is a testimony to changing paleogeography, when land was inundated by sea and when continent again became dry land.

Geologists use the expressions "continent" and "ocean" in a broad sense. A continent, in the jargon of geologists, is not necessarily that part of the earth lying above sea level, but is characterized by having an underground structure which is different from that under the ocean. The underground, or the earth's outermost shell, is called the earth's crust, or simply crust. The crust overlies a heavier substratum, called mantle. A continent is different from the ocean because the crust under a continent is thicker, ranging from 30 to 70 km, than that under the ocean, which is commonly 5 to 10 km thick. Such a crustal structure causes a continent to stand thousands of meters above the ocean floor. The upper surface of a continent is not everywhere subaerial: the exposed part of a continent is land, and the submerged parts of a continent are called continental shelf, continental slope, shelf sea, marine platform, carbonate bank, etc. Sediments on continents are thus not necessarily terrestrial; they are marine when the continent is partly or wholly submerged below a shelf sea, such as the North Sea of Europe.

The actualistic principle of geology assumes that sedimentary rocks are deposits of past eras, when there were mountains, deserts, rivers, lakes, flood plains, swamps, beaches, and tidal flats on land, and shelves, slopes, shelf seas, platforms, and banks submerged under shallow seas. The fossil organisms in the Jurassic formations of the Jura Mountains belong mostly to the kinds that lived in shallow marine waters. This fact suggests that the Jurassic sediments were laid down in shallow sea-bottom, i.e., on a submerged part of the European continent.

Recalling that the Permian/Triassic red beds were laid down in deserts or on flood plains, the deposition of marine limestones in the Jurassic is evidence of a great geo-

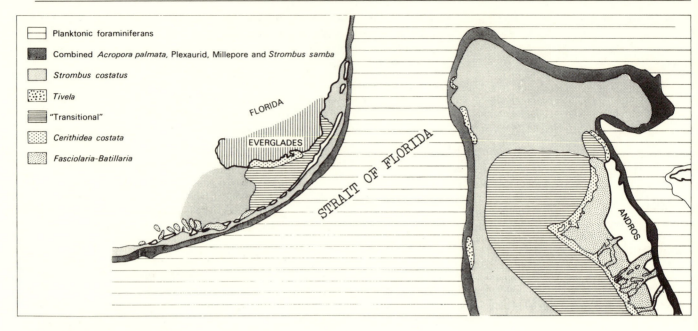

Fig. 1.2. Biofacies distribution of Recent marine sediments of the Florida-Bahama region (after Reading 1986).

graphical change that took place since the Triassic. The European continent of the Jurassic Era no longer stood high above sea level, having been gradually inundated by the sea. This process of flooding a land is called marine transgression. Late in the Jurassic Period, at the time of the Oxfordian Stage, large parts of Europe were submerged under a shelf sea. Switzerland was then part of a marine platform, or a carbonate bank. There were coral reefs, shelf lagoons, fore-reef slopes, oolite bars, tidal channels, and so on. Different kinds of deposits were laid down in different environments. This is the reason why geologists were able to learn to reconstruct the geography of a bygone world—paleogeography—on the basis of looking at rocks.

One of the easiest ways to reconstruct paleoenvironments is to compare the sedimentary rocks with similar sedimentary deposits in modern environments. After sediments undergo diagenesis, they are lithified to make rocks. The changes are, however, mainly chemical and mineralogical; the appearances of the rocks commonly do not change much. Fossil skeletons are preserved, and the shapes of the particles, be they oolitic (rounded concentric shells of calcite enclosing a nucleus) or pelletoid (elliptical concentric shells), are not much changed. Geologists use the term texture to describe the size, shape, and origin of particles in a sediment. Another term, sedimentary structures, refers to the larger features of a sedimentary layer, such as whether it is laminated, thin-bedded, cross-bedded, massive, etc. Textures and structures are little modified by lithification. We can thus try to determine the depositional environment of ancient rocks through a comparison of their texture and sedimentary structures with those of modern deposits in known environments.

The Oxfordian rocks of the Jura Mountains have their modern analogues in the sediments of the Bahama Platform. This shallow marine bank is separated from North America by the Florida Strait (Figure 1.2); thus not much terrigenous detritus could reach the Bahamas, where only carbonate sediments, derived mainly from debris of fossil skeletons, are laid down. Coral reefs grow on the windward side of the platform, whereas skeletal sands are laid down on the platform edge, especially on the foreslope of coral reefs (Figure 1.3). Oolites are deposited on tidal bars in active tidal channels. The sediments in the shallow-water interior of the Great Bahama Bank are mainly mud and pellet. Not only are different types of sediments found in various environments of the Florida-Bahama region, different types of fauna are also known (Figure 1.2). The lithofacies distribution of the upper Oxfordian strata of the central Jura is shown

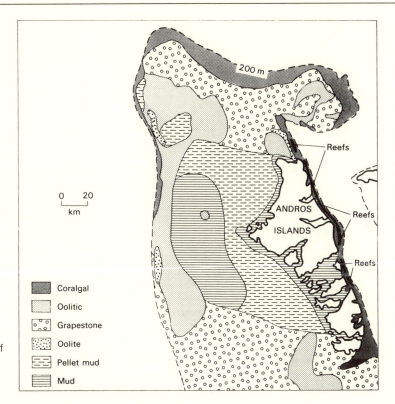

Fig. 1.3. Lithofacies distribution of Recent marine sediments of the Bahama region (after Reading 1986).

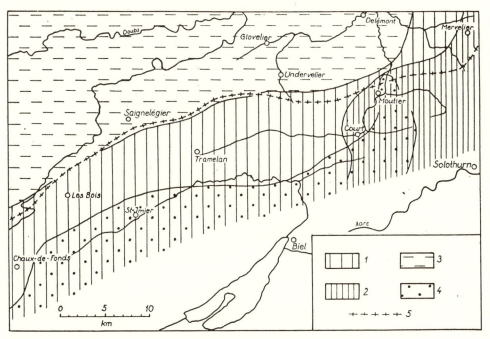

Fig. 1.4. Paleogeography of the Central Jura region in late Oxfordian time (after Ziegler 1956): (1) marly beds with coral limestone at base; (2) coral limestone and oolite; (3) marls and limestones; (4) green pelletoid limestone; (5) southern boundary of the so-called Rauracian facies.

in Figure 1.4. The map shows that coral reef and oolite banks were present in a WSW-ENE trending zone between La Chaux-de-Fonds and Solothurn, whereas marls and lime muds were deposited in the interior of a carbonate bank northwest of the platform edge.

Comparative sedimentology enables us to conclude that the loose sediments of the Bahamas, when lithified, form the massive coralline, the bioclastic, the oolitic, the pelletoid, and the very fine-grained (micritic) limestones of Oxfordian age in the Jura

Mountains. There is, however, one essential difference: the area of the Jura carbonate deposition was not completely separated from European land, and terrigenous debris is thus common in the Oxfordian rocks. The contamination of such clastic components has resulted in the formation of quartz-rich limestone, marl, or argillaceous limestone in the Jura. In contrast, the sediments of the Bahamas are more than 99% calcium carbonate.

Décollement Tectonics

A hiker entering a valley in the Jura Mountains cannot help but see vertically dipping beds exposed on the walls of stream-cuts. These sedimentary layers were originally laid down on flat or nearly flat sea-bottom, but they are now found standing on their ends. The originally horizontal layers have been deformed (Figure 1.5).

The folding of sedimentary formations in the Jura region is a mountain-building process. As late as the eighteenth century, naturalists did not think mountains were made; it was thought that they had always been there. Sedimentary formations are present on top of some of the highest mountains. Werner and his Neptunist pupils considered this fact evidence that there was once a global ocean, covering the whole earth; mountains to Werner were simply the places where the greatest quantity of precipitates had been heaped up. Werner refused to be bothered by the problems of why the primordial ocean disappeared, and where all the water went. "When you meet with an insuperable difficulty, look it steadfastly in the face—and pass on," was the advice of the great teacher (see Geikie 1905).

Water-laid deposits should be horizontal, but inclined beds are common in the mountains. Horace Bénédict de Saussure noted this in the Alps. Again, the question was evaded. Jean Baptiste Lamarck, the great French naturalist, assumed that the inclination of the strata was a manifestation of the natural slope of the surface on which the sediments were originally deposited, like the talus slope of mountains. This is, of course, wrong, if we recognize, on the basis of elementary physics, that the angle of repose of loose sediments is less than 35 degrees. De Saussure knew that; it was manifestly absurd, he told us, to think that pebbles in vertical beds of conglomerate in

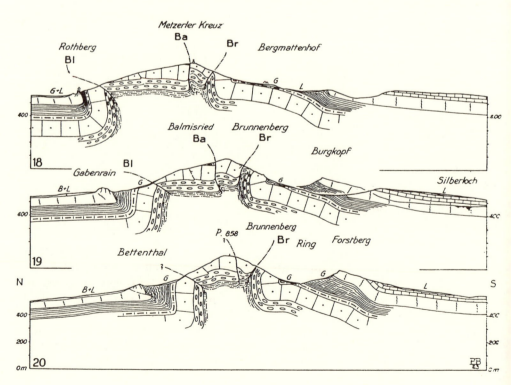

Fig. 1.5. Jurassic folding south of Basel (after Bitterli 1945).

Switzerland could ever have been deposited in such a position. Nor could loose skeletal debris in Jura limestones, I might add, have been glued onto vertical walls during their deposition.

Mountain-building is an essential element of James Hutton's "theory of the earth," first proposed in 1795. In formulating his theory, Hutton relied heavily upon the observations by de Saussure of the "many remarkable instances of the bending of the strata, particularly where the small stream of Nant d'Arpenaz (near Geneva) forms a cascade, by falling over the face of a perpendicular limestone rock. The strata of this rock are bent into circular arches, extremely regular, and with their concavity turned to the left" (Playfair 1802, p. 222). Hutton was, of course, describing the folding of the Jura Mountains. He dismissed the absurd notion of his contemporaries that the bending of strata took place because with "certain great caverns or vacuities, having been opened in the interior of the globe, a great part of the waters which formerly covered its surface, retired into them, and much of the solid rock also sunk down at the same time " (Playfair 1802, p. 220). Hutton proposed the modern theory that "the chain of Jura is secondary, and the beds which compose it . . . are bent in such a manner, that in a transverse section of the mountain, each layer would have the figure of a parabola" (Playfair 1802, p. 223). Hutton envisioned some great convulsions that, from time to time, have shaken the very foundations of the earth, and caused lateral and oblique thrust that gave rise to the contortions of strata such as those of the Jura.

Although the vertical layering seems to point right to the center of the earth, even de Saussure noted that the layers are bent back to form an arch. Two centuries of geological work have revealed that the "arches," or what we now call folds, are "skin deep"; only the thin sedimentary cover of the Jura has been folded, not the oldest sedimentary layers nor the crystalline basement of the Jura (Figure 1.6). This pattern of folding has been called Jura-type or "thin-skinned" folding. During such a deformation process, the cover is sheared off or detached from its underground, like a tablecloth from the table top, and this process of separation is called *décollement*.

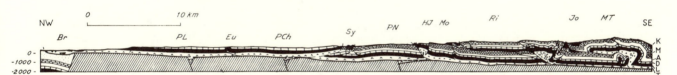

Fig. 1.6. Structural cross section across the Central Jura Mountains (after Laubscher 1965). K: Cretaceous; M: Malm (Upper Jurassic); A: Argovian (Upper Jurassic); D: upper Dogger (Middle Jurassic); L: Lias (Lower Jurassic) and lower Dogger; T: Triassic.

Why should the sediments of the Jura Mountains be folded, while those of the Jura Plateau are still flat-lying? When did the deformation take place? How fast? What were the forces, or how were the stresses induced, to cause the deformation? Where did those forces come from?

These questions on making mountains are the central questions in a theory of the earth. When Hutton wrote that there were subterranean forces causing convulsions from time to time, he implied "violent" and "episodic" deformations. Georges Cuvier, studying the faunas of the Paris Basin, discovered remarkable differences between faunas of different geological periods. Dinosaurs died out at the end of the Cretaceous Period, and the Cenozoic Era was distinguished by several distinct faunas in succession. A citizen of the French Revolution, Cuvier postulated that earth revolutions caused the mass extinction of the earth's inhabitants. His countryman, L. Elie de Beaumont (1830), made a synthesis of Hutton's violent deformations and Cuvier's revolutions, and proposed that revolutionary convulsions causing mountain-making were the revolutions that caused the mass extinctions at the end of successive periods. Elie de Beaumont listed nine such revolutions: prior to Carboniferous, prior to Muschel-

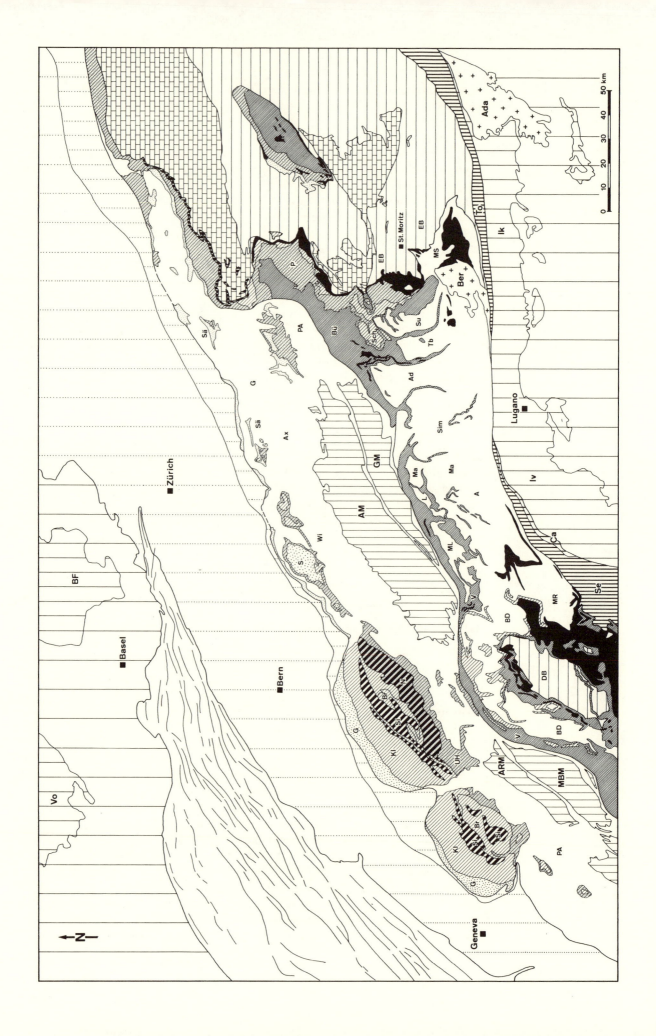

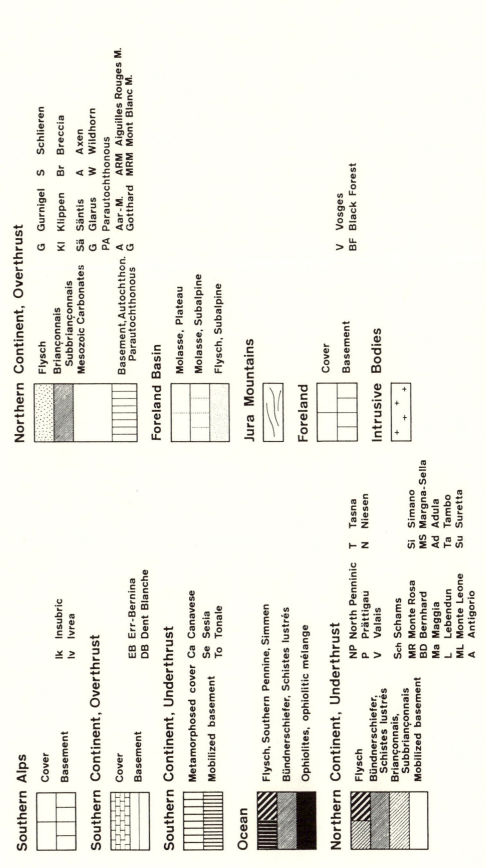

Fig. 1.7. Tectonic units of Switzerland. The three major tectonic units of Switzerland are the Jura Mountains, the Swiss Midland (Molasse Basin), and the Alps. The subdivision of the Alpine tectonic units will be discussed in Chapter 3.

kalk, post-Triassic and pre-Jurassic, post-Jurassic and pre-Cretaceous, etc. Between each episode of mountain-making, stability prevailed and conformable sequences of sedimentary strata were deposited. The Jura Mountains, in Elie de Beaumont's view, were formed during one of the last two revolutions, which took place shortly before the sedimentation of Recent alluvium. Refinement of this somewhat antiquated view has persisted into modern days. Modern geological literature uses the word "phase" in place of Elie de Beaumont's "revolution," and the Alps are postulated to have been formed by a "Cretaceous Eo-Alpine," an "Eocene Meso-Alpine," and a "Miocene Neo-Alpine phase," whereas the Jura Mountains were folded during a "Pliocene phase."

A revolution in earth science has given us the theory of plate tectonics. We now postulate that the earth's crust and upper mantle constitute a lithosphere and the lithosphere is broken up into numerous plates. The plates move at rates of centimeters per year in the same direction for millions of years. First ocean plates were dragged beneath continental margins, in a process called subduction, forming mountains such as the circum-Pacific mountain chains. The eventual consumption of ocean plates caused the continental plates to collide. Plate collision causes compressive stresses in the earth's crust, and the Alpine type of mountain-building is a consequence of such collision.

In the new theory, it is postulated that Africa and Europe once belonged to one continent, called Pangea. The two were rifted apart, so that an ocean came into existence early during the Jurassic Period. This ocean between Africa and Europe, called the Tethys, continued to expand until the Early Cretaceous. The sedimentary strata of the Jura region were deposited on or near the north shore of the Tethys. From the beginning of the Cretaceous Period, the African plate marched northward continuously until Africa collided with Europe; the last remnant of the ocean was consumed and the final collision took place during the Late Eocene. Deformation under compression is the mountain-building process by which the Alps were formed (Hsü 1989).

Figure 1.7 is a schematic tectonic map of Switzerland. The Jura Mountains in northwestern Switzerland are characterized by a series of southwest-northeast trending folds, separated from the Alps by the Swiss Midland. The deformation of the Jura was, however, a continuation of the Alpine deformation, because the European and African plates continued to press together even after the collision, causing the continued folding and thrusting of the Alps and the genesis of a foreland basin in the Swiss Midland, while the Jura lay, at first, outside of the realm of deformation. Eventually, however, in the Late Miocene or Early Pliocene, the basement under the Jura Mountains (and the Midland) was thrust under the basement of the Alps, causing the rise of the Alps. At the same time, the sedimentary cover above a Triassic evaporite horizon was detached and pushed northward to form the thin-skinned folding that is typical of the Jura Mountains (Laubscher 1965; Hsü 1979).

2.

The Swiss Midland

The Swiss Midland is the country of rolling hills between the Alps and the Jura Mountains. The outcropping rock formations belong exclusively to the Tertiary or Quaternary periods; rocks of the Mesozoic Era are encountered only in boreholes.

The names Tertiary and Quaternary are relics from a bygone era. An Italian naturalist, Giovanni Arduino, writing in the early eighteenth century about the mountains in the vicinity of Padua, Vicenza and Verona, first spoke of Primary, Secondary, and Tertiary formations. This usage was adopted by Johann Gottlob Lehmann. The oldest were called Primitive or Primary, including granites, gneisses, schists, and other rocks that have neither bedding nor fossils. The next were termed Secondary, for cemented sedimentary rocks that commonly contain fossils; Lehmann's Triassic belonged to the Secondary. The third class, the Tertiary, included still younger and weakly consolidated sediments. Eventually a fourth, the Quaternary, was added to denote alluvium and terrace gravels.

Confusion began to arise when fossils were found in the so-called Primary rocks. Furthermore, the Secondary also proved to be a rather arbitrary subdivision when a major break in faunal successions was found between the Permian and the Triassic. The presently used terms Paleozoic, Mesozoic, and Cenozoic were introduced by John Phillips in 1840, and the division, as we have mentioned previously, was based upon the fossil record. According to this scheme, the Cenozoic Era includes the Tertiary and Quaternary periods of the previous nomenclature, and these two names were thus retained. The Mesozoic includes only the younger of the Secondary periods, namely the Triassic, Jurassic, and Cretaceous. The Paleozoic rocks include both the older Secondary formations and what have also been called Transitional Formations between the Primary and the Secondary. Phillips's scheme has stood the test of time, because his three eras turned out to be rather natural divisions: the beginning of each is defined by remarkably sudden faunal changes, heralding the starts of the ancient, middle, and recent forms of life on earth. With the establishment of this nomenclature, we have a chronological classification of strata that includes the names Tertiary and Quaternary, but not Primary or Secondary.

The Tertiary strata of the Swiss Midland are Molasse formations, deposited on land or in shallow marine waters. The Quaternary are the deposits of the Ice Age (Pleistocene) and the Recent (Holocene) alluvial and lacustrine deposits. We shall start this chapter by looking back from the Recent into the history of the past.

The Quaternary Ice Age

Every schoolchild in Switzerland learns that there was an Ice Age, during which the whole country was buried under an ice cap like the continent of Antarctica is today. They are even taught that there were four glacial stages, namely Günz, Mindel, Riss, and Würm, which were separated by interglacial stages of warm climate. But how did we find out about the Great Ice Age?

The discovery is a Swiss story. In the valleys and on the plains of the Swiss Midland, farmers are used to digging up stones and rocks of various sizes, some of which weigh many tons. The boulders are made up of rocks that are quite different from those cropping out in the nearby countryside; they are exotic, or *Findlings* (Figure 2.1). Their nature is so alien, diluvialists of the eighteenth century thought they had found some relics of Noah's Flood; the exotics were believed to have been carried down from mountains by huge floods at the time of the Deluge.

Charles Lyell had a more realistic estimate of the stream power; he postulated instead that there might have been ice-rafting during the time of Noah—stones and boul-

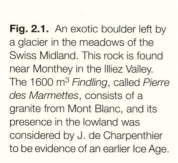

Fig. 2.1. An exotic boulder left by a glacier in the meadows of the Swiss Midland. This rock is found near Monthey in the Illiez Valley. The 1600 m³ *Findling*, called *Pierre des Marmettes*, consists of a granite from Mont Blanc, and its presence in the lowland was considered by J. de Charpenthier to be evidence of an earlier Ice Age.

ders frozen in drifting icebergs were dispersed by the Flood and deposited as exotics when the ice melted away.

Lyell had not seen glaciers in his youth, and he had no idea of the power of glacial transport. Furthermore, he did not believe that conditions on earth during the past had been so different that glacial ice could have advanced to the Swiss Midland. Swiss farmers were not bound by such philosophical dogmas, and they were used to seeing the advance and retreat of mountain glaciers. They took for granted that the erratic blocks were brought down at a time when Switzerland was much colder; they were brought down by glaciers coming from the mountains to the plains. Jean-Pierre Perraudin, a mountain guide from the Valais, communicated this opinion to a number of his academic acquaintances. Since some of the findlings (an anglicized technical term in geological literature) were found on the sides of valleys, hundreds of meters above the valley floor, Perraudin had to advance the bold idea that those Alpine valleys were once filled completely with glacial ice! Such an imaginative postulate from a layman did not sit well with the academics; they dismissed the idea lightly as the tall tale of an illiterate.

Ignace Venetz, a Swiss civil engineer, was the first to listen attentively to Perraudin, and was the first to publish, in 1821, an article advocating the theory of glacial transport of findlings. In less than a decade, Venetz had made enough observations to venture an even more daring hypothesis to an audience gathered at the 1829 meeting of the Swiss Society of Natural Sciences (SNFG) at the Hospice of the Great St. Bernard—he proposed the existence of an Ice Age in the not too distant past. Immense glaciers coming down mountain valleys coalesced and formed a great ice sheet that covered practically all of Switzerland, as well as other parts of central Europe. This seemingly outrageous hypothesis was rejected by all except Jean de Charpenthier. He and his young protégé, Louis Agassiz, who had just become professor of geology at the University of Neuchâtel, devoted much of their time and energy during the ensuing decade to verifying the idea and to formulating a theory of the Ice Age.

The proponents of the new theory claimed that the climate on earth was not always the same as it is now. Historical records showed that Swiss mountain glaciers grew at times when the climate was colder. During the "Little Ice Age" of the seventeenth century, for example, mountain glaciers expanded to such an extent that their snouts came down and deposited moraine in valleys where grass and wild flowers now flourish. At some more remote time during the Ice Age, it was even colder, and all of Switzerland could have been covered by an ice sheet. Large and small boulders were pushed to the valleys and plains by the glaciers of the Ice Age, and they were left standing in the meadows as findlings long after the ice had melted and the glaciers had retreated back to the mountains.

The theory was very sensible, but the geological profession was dominated during the nineteenth century by the English, especially by Lyell, honored by many as the father of modern geology. The postulate of an Ice Age stood in gross contradiction to the Lyellian dogma that conditions on earth have remained the same since the very beginning of time. When Agassiz brought the new theory to England, Lyell was slow to change his mind. His teacher, William Buckland of Oxford, had seen the glaciers and moraines in Switzerland and was convinced by the young man. Buckland and Agassiz visited Lyell at Kinnordy, Lyell's manor in northern England, and they showed him a beautiful cluster of glacial moraines within two miles of his house. Lyell finally had to change his mind and abandoned the ice-rafting theory of the origin of exotic blocks. The theory of the Ice Age was accepted by the English establishment, and Agassiz went on to North America, where he discovered that a large part of that continent had also once been covered by glacial ice.

Germany was slow to accept the new theory. Findlings are not uncommon in northern Germany, but professors in Berlin who had never seen glaciers clung to the Lyellian postulate of iceberg transport, long after Lyell himself had discarded the mistaken notion. The German resistance lasted almost four decades, until 1875, when a Swedish Arctic researcher, Otto Martin Troell, came to Berlin and convinced some of the young Germans that the striations on the findlings had been scratched by glacial action. Toward 1880, the "old soldiers" faded away, and Albrecht Penck, who had in his youth spent many seasons in the Eastern Alps, where glaciers abound, swept away the last agnostics with his masterful analysis of the history of Alpine glaciation.

Penck was a geographer and a student of landforms. Working in southern Germany, he noticed that modern streams meander weakly in valleys underlain by gravel deposits. He reasoned that modern streams did not have enough power to transport gravels, and that this coarse debris was brought down by rivers coming out of glaciers during an ice age. He further found that stream terraces on both sides of a valley are underlain by the same type of gravel deposit as that in the modern valley, and reasoned that those gravels must have been laid down during earlier ice ages. Since he was able to distinguish four levels of terraces, he postulated that there must have been four ice ages. Using the names of the streams where he had made his observations, Penck and his associate Brückner came up with the names Günz, Mindel, Riss, and Würm, names known to every European schoolchild (see Penck and Brückner 1909). In between were the interglacial stages, when the climate was as warm as, or even warmer than, that of the present.

Geologists in North America used a different approach. They pointed out that glaciers push rock and mud debris forward, aside, and under as they advance, forming end, side, and ground moraines. The moraines were weathered under the prevailing warm climate of the interglacial stages, so moraines of different ice ages are separated by soil horizons. T. C. Chamberlin of the University of Chicago and his associates counted the number of ice ages on the basis of the number of ground moraines, or tills, that have been deposited on the plains of North America. Like Penck, they found four ice ages, which were named after the states in which the moraines are best exposed, namely Nebraskan (oldest), Kansan, Illinoian, and Wisconsinan. It was generally assumed that the four American stages corresponded to the four European stages.

Neither the American nor the European classic approach promised a complete record of glaciation history. Moraine deposits or tills laid down during an earlier glaciation may have been eroded away by a later glacier. In Switzerland, for example, the Riss glaciers seemed to have been the most powerful, and they managed to remove much of the earlier record. Also, the recognition of river terraces and their correlation from one stream valley to another presented problems and controversies. In fact, Penck at first thought that there were only three terraces, corresponding to three glacial stages, before he adopted his fourfold division. Later on, other scientists found more than four river terraces, and thought that there were perhaps more than four stages of continental glaciation. Nevertheless, theories tend to become "incontroverti-ble truth" once they are written in schoolchildren's textbooks. We have to preserve our Günz, Mindel, Riss, and Würm, even if we have to invent "substages" Riss I, Riss II, Würm I, Würm II, Würm III, and so on to retain our magic number.

The ocean is a receptacle of sediments, and a complete record of the history of the Ice Age should be found in ocean sediments. A twenty-meter piston core can yield a climatic record of the last two million years, and oxygen-isotope geochemistry is the key with which to decipher the record. The principle is based upon the consideration that oxygen isotopes in water are fractionated during evaporation: the lighter mole-cules (with ^{16}O) are preferentially concentrated in the vapor phase, leaving the liquid phase preferentially enriched with molecules containing the heavier ^{18}O atoms. Nor-mally, water vapor in the atmosphere is condensed and precipitated as rain and snow, to be carried back by rivers to the oceans. The oxygen-isotope composition of the ocean water could thus remain more or less constant. During the glacial stages, how-ever, much of the water vapor, enriched with oxygen-16, was condensed to form the snow and ice in the ice caps of the polar regions and in continental glaciers. In fact, so much water was stored in glaciers that the worldwide sea level dropped some 100 m below the present level. The sea water of glacial stages was enriched in O-18, the O-16 having been preferentially stored in glacial ice. The skeletons of fossil organisms liv-ing then were likewise enriched in oxygen-18, compared with those of organisms liv-ing during warmer interglacial stages. This difference in isotope composition of fossil skeletons is further enhanced because the partition of oxygen isotopes in water mole-cules of ocean water and those in skeletons of marine organisms is temperature depen-dent; a calcite crystal precipitated from a warmer ocean contains relatively more oxy-gen-16 molecules than one precipitated out of a colder ocean. Marine organisms living in the warmer surface waters of interglacial stages should thus have an excess of ^{16}O in their skeletons compared with those living during glacial stages. The alternating changes in the oxygen-18 content of marine microfossils is thus a document of the coming and going of the glacial stages (Figure 2.2).

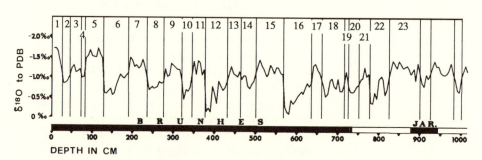

Fig. 2.2. Oxygen-isotope record (below), and paleomagnetic record in a core from the Pacific Ocean (after Shackleton and Opdyke 1976). The glacial stages have even numbers and the interglacial have odd numbers. Note the more negative $\delta^{18}O$ values of ocean sediments during the interglacial stages, when the surface sea-water temperature was warmer. Brunhes is the name of the present magnetic-polarity epoch, which started about 700,000 years ago. Jaramillo (Jar.) is the name given to a so-called magnetic event, when the earth's magnetic pole was situated at the North Pole as it is today. In between was an interval during which the north magnetic pole was situated somewhere in Antarctica, an interval of negative polarity. The ocean sediments at this site were deposited at a rate of about 1 cm per 1000 years, so the depth in centimeters corresponds more or less to the age in thousands of years.

Cesare Emiliani, using cores from the Caribbean Sea, was a pioneer in applying isotope geochemistry to paleoclimatology. In 1955 he found that there might be far more glacial stages than the four suggested by Penck. This conclusion has since been verified by Shackleton and Opdyke (1976), who counted 19 alternate cold and warm stages during the last 0.7 million years, and more than 40 such stages during the Qua-ternary. Isotope-stage 1 is the Holocene, which started 10,000 years ago. The classical

Würm stage is now believed to designate the glaciations that started some 80,000 years ago; it corresponds to two cold glacial stages (isotope-stages 2 and 4) interrupted by a warmer, or interstadial, stage 3. The Würm-Riss interglacial stage is correlative isotope-stage 5, which consists of three warm (5a, 5c, 5e) and two cold (5b, 5d) intervals. Back in time beyond that, Penck's Riss, Mindel, and Günz cannot be correlated exactly with the isotope-stages defined on the basis of the ocean record. Radiometric and other dating methods indicate, however, that continental glaciation in the northern hemisphere started some 2.5 million years ago, long before the Günz stage.

The Lake Zürich Record

Much of the Swiss Midland is covered by moraines—ground moraines, side moraines, and end moraines. The positions of end moraines give an indication of the former advances of glaciers. In the Zürich and Limmat Valley region, Hug (1917) distinguished three glacial stands during the Würm glaciation: Killwangen Stadium, Schlieren Stadium, and Zürich Stadium.

Gravel deposits are common on valley bottoms and on terraces on the sides of valleys, as Penck found in streams called Würm, Riss, Mindel, and Günz. A very thick sequence of lake deposits, consisting of lignite (*Schieferkohle*) and terrace gravels, covered by a glacial moraine, is exposed in a quarry near Buchberg, near Wangen in Canton Schwyz. The intercalation of lignite indicates deposition under warm conditions. The question is its age: Is the warm climate that of a Würm interstadial (isotope-stage 3) or of the interglacial (isotope-stage 5)?

The Buchberg lignite deposit was once considered a Riss-Würm interglacial on the basis of pollen analysis, but the carbon-14 date of about 40,000 years BP (before present) suggested an interstadial origin. Studies done with an acceleration mass spectrometer indicate, however, that slight contamination of carbon samples is almost unavoidable: Precambrian graphite often gives ages of 38,000–40,000 years BP. The question of the lignite age must thus be left open. An investigation of the Quaternary history of Lake Zürich by lake drilling led to the conclusion that the Buchberg lignite deposits are Riss-Würm interglacial (Hsü and Kelts 1984).

From the ocean record and the study of moraines, it can now be concluded that the coldest climate prevailed globally some 18,000 years ago. The glaciers retreated rapidly for several millennia, from 15,000 to 11,000 BP. A temporary reverse, during the so-called Young Dryas time (11,000–10,000 BP), was marked by a cold climate, to be followed by a sudden global warming, which heralded the beginning of the Holocene Epoch.

During the summer of 1980, we made an attempt to find a continuous Quaternary sequence in the sediments of Lake Zürich. A borehole was drilled through the sediments under the deepest bottom of the lake into the Molasse bedrock (Hsü and Kelts 1984).

The youngest sediments of the lake are lacustrine chalk, which consists mainly of $CaCO_3$; detrital silts and clay are present in varying amounts, commonly constituting less than 10% of the bulk. The chalk is thinly laminated near the top. Each lamina is a varve, or an annual sediment, consisting of a light chalk and a dark detritus-rich layer, deposited during the summer (by chemical precipitation) and winter (by settling from suspension), respectively. Counting the laminae, we determined that the varve sedimentation began around 1890. The laminations deposited during the late nineteenth century are barely discernible, and they are not visible at all in even older sediments. We believe, however, that those older sediments were also varves, but have been homogenized; the original laminated structure has been destroyed by organisms that lived, plowed through, or ate the bottom sediments, a process called bioturbation. The condition for lacustrine-chalk deposition has persisted since the beginning of the Holocene Epoch 10,000 years ago.

The last Pleistocene sediments were deposited when the Linth glacier retreated and the lake basin was a deep-water lake. The uppermost are glacial-lacustrine varves, laid down during a time, more than 10,000 years ago, when the climate of the Zürich region was similar to that of the high Alpine valleys (above 1600 m) today. During the summer season, silt and clay were deposited by underwater suspension currents, also called turbidity currents; these sediments have a typical graded-bedding structure, showing a decrease in grain size from the sharp bottom contact to the top. During the winter season, a paper-thin lamina (some 0.1 mm thick) of very fine clay was deposited when the lake was frozen; clay particles less than 2 micrometers in size still suspended in the water column were the only source material for the winter deposit. Still older sediments of Lake Zürich, deposited 13,000–15,000 years ago, when the Linth glacier first retreated but still stood nearby, are centimeter-thick graded beds of silt or silty sand. These are turbidity-current deposits or turbidites, laid down during the occasional thawing of the lake, which must have been almost permanently frozen.

Still older sediments of the Lake Zürich borehole are those deposited during a glacial stage. Two kinds are recognized, till and dropstone-mud. The till was the bottom-moraine deposit of a glacier, when the Lake Zürich basin was frozen down to the bottom. The dropstone-mud was deposited in a subglacial lake; this sediment was formed when a stone originally lodged in the ice of the frozen lake dropped down to a muddy bottom, after the melting of the surrounding ice.

The oldest sediments of Lake Zürich, at 150 m below the lake bottom, or about 100 m above the present sea level, are gravels and coarse sands, similar to those found nowadays in the valley downstream from the Morteratsch glacier in the Engadine. Geophysical investigations indicated that the Linth glacier cut deeply into the interglacial deposits, such as those exposed at Buchberg to form the Lake Zürich basin (Hsü and Kelts 1971). The river gravels were probably deposited at the beginning of the Würm glaciation; the Linth had to cut deeper when the worldwide sea level dropped some 100 m. below that of the present.

We have no sediments older than those of the Würm glacial stage in Lake Zürich. In other valleys of the Swiss Midland, Riss-Würm interglacial and Riss glacial sediments have been encountered. Still older glacial sediments are terrace gravels and loess deposits cropped out here and there in the Midland, but nowhere can we find a record of the Great Ice Age in Switzerland as complete as that preserved in the oceans.

Tertiary Molasse Formations

The Quaternary sediments of the Swiss Midland overlie unconformably the Molasse strata. The Molasse basin, a site of terrestrial and shallow marine sedimentation from Early Oligocene to Late Miocene, is situated between the Jura Mountains and the Alps, or, more exactly, in front of the Alps (Figure 2.3). This type of basin has been called a foreland basin. The Swiss Molasse formations are typical foreland-basin deposits. Thick terrestrial clastics and shallow marine interbeds deposited in similar tectonic settings elsewhere have been called sediments of the molasse facies. Molasse, in that case, is used as a name for a sedimentary facies that recurs again and again in different parts of the world during different periods of geological history.

The Molasse deposits of the Midland are Tertiary sedimentary rocks, ranging from several kilometers thick just north of the Alpine front to several hundred meters thick toward the Jura Mountains (Figure 2.4). Conglomerates, also known as Nagelfluh, are common; also present are sandstones, siltstones, and shales. The Molasse deposits have been divided, since the time of the pioneer Bernhard Studer (1853), into four formations, which are, from top to bottom,

Upper Freshwater Molasse: Upper/Middle Miocene
Upper Marine Molasse: Middle/Lower Miocene
Lower Freshwater Molasse: Lower Miocene/Upper Oligocene
Lower Marine Molasse: mainly Lower Oligocene

The youngest Molasse strata are about 10 million years old. Between them and the Pleistocene lake sediments is a gap in the record. The foreland basin of Molasse sedi-

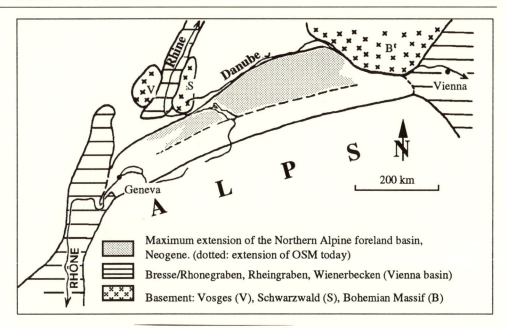

Fig. 2.3. Geological framework of the Molasse basin (stippled) (after Bürgisser 1980).

Maximum extension of the Northern Alpine foreland basin, Neogene. (dotted: extension of OSM today)

Bresse/Rhonegraben, Rheingraben, Wienerbecken (Vienna basin)

Basement: Vosges (V), Schwarzwald (S), Bohemian Massif (B)

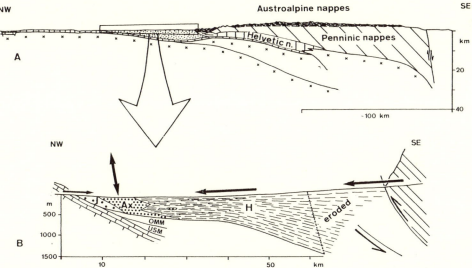

Fig. 2.4. Asymmetry of the foreland basin (after Bürgisser 1980): (A) structural cross section across central Switzerland, showing the location of the Molasse basin (stippled); (B) structural cross section of the Molasse basin across northeastern Switzerland (tenfold vertical exaggeration). Ax: quartz sandstone; H: Hörnli deposits; J: Molasse conglomerate of the Jura Mountains; OMM: Upper Marine Molasse; USM: Lower Freshwater Molasse.

mentation was changed into hilly countryside land during the Late Miocene. Erosion of the Molasse formations then supplied detritus to the Ur-Rhine and Ur-Rhône; terrestrial gravels of the Pliocene are found only here and there in Switzerland.

Freshwater Molasse

The Freshwater Molasse formations consist mainly of sandstone and conglomerate beds (Figure 2.5). The Nagelfluh conglomerates of the Freshwater Molasse formations have pebbles and cobbles ranging in size up to dozens of centimeters across. The clasts of each formation decrease in size toward the north, away from the source terranes. The Nagelfluh were deposits of gravels, dumped out of raging floods of mountain streams coming out of a rising mountain chain.

The lithology of the Nagelfluh is a source of information on paleogeography. The clasts in the Lower Freshwater Molasse are derived mainly from the Austroalpine nappes, and from the ophiolite mélanges. Those in the Upper Freshwater Molasse contain abundant components from the Austroalpine, Penninic, and Helvetic elements. This sedimentation history records the unroofing of the nappes in the Alps, as work by many Swiss geologists indicated.

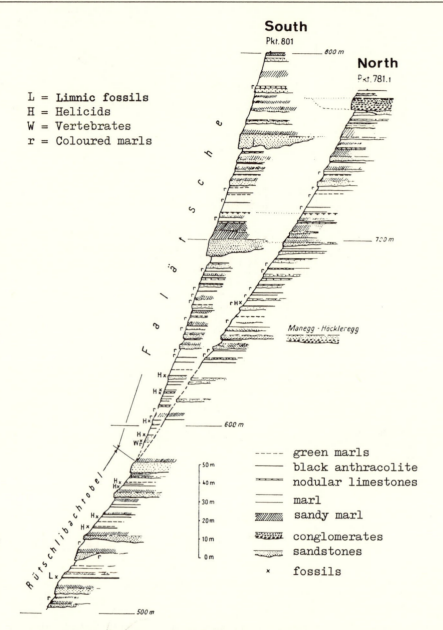

Fig. 2.5. Stratigraphical section of the Upper Freshwater Molasse, Rütschlibach-Falätsche (after Pavoni 1967). L: lacustrine fossils; H: snails; W: small vertebrates; r: red marl.

Molasse sediments are very thick, and correlation of individual units is difficult because index fossils are not abundant. A former student, H. Bürgisser (1980), made a special study of an index horizon—the "Appenzellergranit"—in the Upper Freshwater Molasse, and his study helps us to visualize the geography of the Swiss Midland some 15 million years ago. The "Appenzellergranit" is not a granite, but a hard conglomerate. The outcrops of this conglomerate unit have been found to extend from St. Gall to the shore of Lake Zürich, and are encountered in subsurface west of the Reuss River in Canton Zug. Underlying the conglomerate is a lacustrine limestone, an exceptional carbonate bed in the thick Molasse formations. The "Appenzellergranit" is hard because the gravels have been cemented by calcite carbonate.

The presence of the lacustrine chalk is the key. There must once have been a large lake in eastern Switzerland, extending from the Liechtenstein border to Zürich and Zug at least, and probably farther west. The lake probably came into existence when a huge landslide blocked the drainage system of the Molasse basin. The mountain streams of eastern Switzerland could then no longer reach the Ur-Rhône (or the Ur-

Rhine); instead, they were drained into this huge lake. Eventually, the hindrance was eroded away, and the streams of the east were again integrated into the Ur-Rhône (or Ur-Rhine), when the uppermost Nagelfluh of the Upper Freshwater Molasse were deposited.

The "Appenzellergranit" is only an index horizon. The history of its sedimentation gives a cinematic view of Switzerland during a geological instant, which lasted, however, tens or hundreds of thousands of years. The bulk of the Freshwater Molasse formations are not lacustrine, but river deposits. Swiss geologists used the term "alluvial fans" to describe the depositional environment of the Nagelfluh, and spoke of the Rigi Fan, Hörnli Fan, Napf Fan, etc. The word fan is, however a misnomer, because alluvial fans are generally desert features. Fan deposits are gravels and sands laid down by flash floods as they issue from the mouth of an arroyo (a normally dry creek).

The Oligocene and Miocene climates in the Swiss Midland have been reconstructed on the basis of fossil evidence. Plant remains, such as palm fossils (*Apeibopsis, Chamaerops*, etc.), are especially abundant in the Upper Freshwater Molasse. Rene Hantke (1954) was able to conclude from paleobotanical evidence that the Swiss Midland in Late and Middle Miocene time had a climate similar to that of the southeastern United States today. Forests typical of a humid climate, with the tree species *Tsuga* (*Hemlockstanne*), *Sciadopitys* (*Schirmtanne*), or *Abies* (*Tanne*), were widespread in the Swiss Midland during the Early Miocene and Oligocene, when the Lower Freshwater Molasse sediments were being deposited, although some more arid bushes and grasses, such as *Ephedra*, grew on sandy or gravelly river banks.

A more adequate modern analogue for the Molasse can be found in northeast India. The Oligocene and Miocene rivers emptying into the Molasse basin were comparable to the tributaries of the Ganges, the Kosi, the Bramaputra, and other large and small rivers coming out of the Himalaya Mountains (Figure 2.6). The rivers had broad valleys and braided channels. The sediments were deposited on gravel banks or bars during high-water seasons. Finer sediments remained largely in suspension and were carried away by the Ur-Rhône and Ur-Rhine.

Marine Molasse Formations

The marine Molasse occur closer to the Alpine front in the Midland. Some of the best outcrops have been found in the vicinity of Lucerne (Roessli 1967). The Upper Marine Molasse strata, ranging from late Early Miocene (Burdigalian) to Middle Miocene (Helvetian) in age, are well exposed, for example, in the Renggbach section (Figure 2.7). The well-bedded sandstone is interbedded with marl and siltstone. The Lower Marine Molasse crops out near Horw, south of Lucerne. Gray marl is the dominant lithology, and thin sandstone beds are intercalated. Marine shelly fossils, including *Cardium, Cyrena* spp., and other marine fauna, have been found in these and other Marine Molasse formations.

These hemipelagic sediments of the Lower Marine Molasse are practically identical in lithology, and they were deposited in an environment similar to that of the *flysch du Val d'Illiez* of western Switzerland (see Hsü 1970). Lower Oligocene turbidite sequences, similar sedimentologically to Flysch but cropping out north of the Alpine front, have also been considered a Lower Marine Molasse unit (Schere 1966).

Tectonics of the Molasse Foreland Basin

There is still the widespread impression that the Molasse was deposited after the Alpine orogenesis. This idea of "synorogenic" Flysch "and post-orogenic" Molasse was first proposed by Arbenz in 1919. Incorporated into a geosynclinal theory of orogenesis, the synorogenic Flysch was considered the sediments of foredeeps in front of advancing nappes, and the post-orogenic Molasse the deposits in a basin formed (by extension?) after the Late Eocene paroxysm of compressional deformation. In fact, the

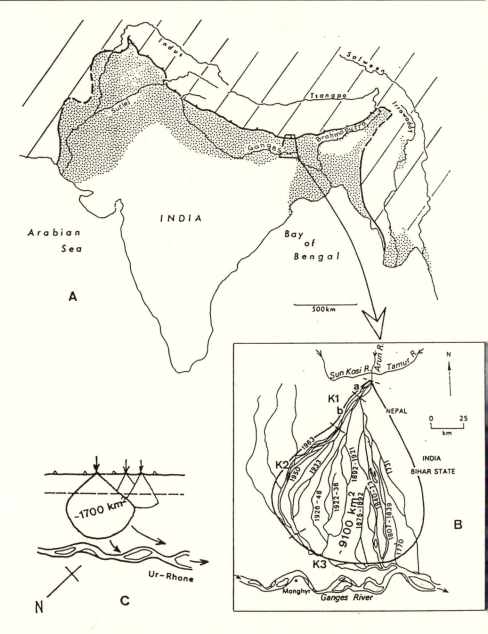

Fig. 2.6. Comparison of the Molasse of the Swiss Midland with its modern analogue south of the Himalayas (after Bürgisser 1980): (A) distribution of sediments of the molasse facies (stippled) south of the Himalayas; (B) the Kosi Fan; (C) the Miocene "Hörnli Fan," shown on the same scale as that of (B).

Alpine orogenesis did not come to a halt at the end of the Eocene Epoch. The plate-tectonic theory postulates continuous deformation of the Alps, Midland, and Jura since the beginning of the Cretaceous Period, 130 Ma ago. Foreland basins such as the Molasse were subjected to compression while the Alps were being continuously thrust forward (Hsü 1979). The thickest Molasse is found in the synclinal trough just north of the Alpine front, and these oldest Molasse sediments have been folded, overturned, and thrust under the Helvetic nappes (Figure 2.4).

The modern theory of the origin of the Molasse basin has resolved the Flysch-Molasse controversy. The oldest marine Oligocene sediments in the Swiss Midland are commonly a monotonous sequence of marl and shale, with thin intercalations of sandstone and siltstone (Figure 2.8). They are present, in Canton Bern, for example, between the Subalpine Molasse (*sensu stricto*) and the Ultrahelvetic Flysch. The controversy was centered on the question of whether those rocks are Subalpine Molasse (Gerber 1925) or Subalpine Flysch (Blau 1966). We now see that the controversy is more semantic than substantial, because the change from foredeep Flysch to foreland-

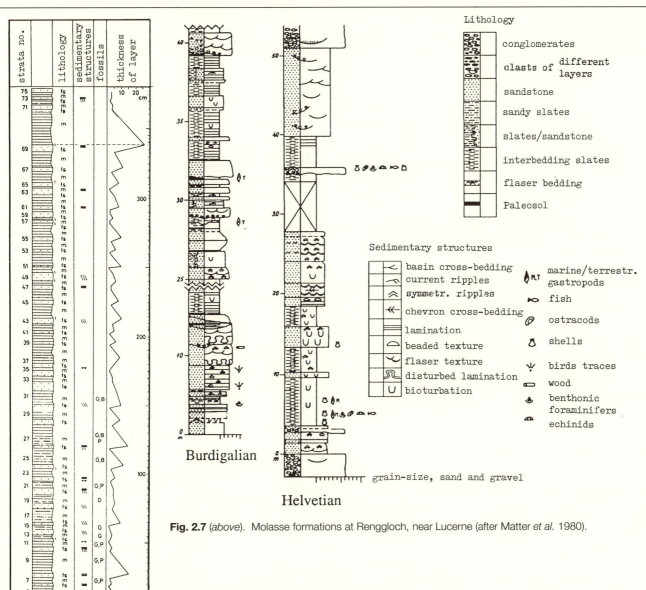

Burdigalian

Helvetian

Fig. 2.7 (*above*). Molasse formations at Renggloch, near Lucerne (after Matter *et al.* 1980).

Fig. 2.8 (*left*). Type locality of Jordisboden Marl at Jordibruch, Bern (after Blau 1966).
The interbedded fine-grained sandstone (fs) and marl (m) of the Lower Marine Molasse sequence are
sedimentologically similar to the Flysch.

basin Molasse sedimentation was a transitional phenomenon. With the gradual infill of
a deep-sea trough, the slope angle of the trench wall gradually became less steep. The
processes of subaqueous slumping and turbidity-current sedimentation, depositing
Flysch beds, occurred less and less frequently. The predominant sediments laid down
became the hemipelagic muds, forming the monotonous shale or marl sequences of
transitional character. Those cropping out south of the Alpine front are commonly
called Flysch, because they were involved in the Alpine deformation (e.g., *flysch du
val d'Illiez* of the Valais, *Dachschiefer* of Glarus, etc.). Those north of the Alpine front
were considered Subalpine Molasse since the days of Bernhard Studer (1825), until
Blau (1966) preferred to call these rocks *Subalpiner Randflysch*.

3.

The Swiss Alps

The word *Alp* is used by Swiss farmers for the mountain meadows where their cows take their summer vacation. To foreign tourists the Alps are rocky cliffs and snow-capped peaks. Swiss geologists think of their Alps as something wonderful and unique: Alpine geology, unlike Swiss cheese, is not for export, as my colleague Rudolf Trümpy used to tell us. Swiss "dialects" (in daily life and in geology) have rendered the Alps mysterious and distant, but foreigners have not been deterred from taking a reductionist approach to search for universality in the geology of the Swiss Alps. The first intruder was Marcel Bertrand from France, and his brilliant idea on the Glarus Overthrust revolutionized geology and laid the foundation stone for Alpine tectonics.

Glarus Overthrust

In the mid-nineteenth century Arnold Escher von der Linth, an excellent observer but a conservative thinker, suggested that the abnormal superposition of Jurassic limestone on Tertiary Flysch in the Glarus region could be explained by postulating a mushroomlike fold, also called the "Glarus double fold." This anomalous stratigraphical order had been known to Swiss geologists since the 1840s. The famous Martin's Loch is a hole in a Middle Jurassic limestone, and the bottom of the hole is the upper surface of an Eocene sedimentary formation called Flysch (Figure 3.1). Above the limestone is a thick sandstone formation, called Verrucano, and it is Permian, even older than Jurassic. How did the Jurassic limestone get on top of the Eocene Flysch? And how did the Permian get on top of the Jurassic?

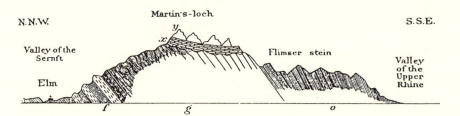

Fig. 3.1. Glarus Overthrust. This sketch of a geological profile from Sernf Valley across the Segnas Pass to the Rhine Valley, made by Murchison in 1849, shows that Arnold Escher von der Linth had discovered that the stratigraphical sequence near Martin's Loch is inverted: the metamorphosed detrital rocks (y), called Verrucano, are Permian, the mylonitized marble (x) is Middle Jurassic, while the nummulitic limestone (f) and slates (g) are Eocene.

A half-century after Hutton's theory of the earth, geologists had come far enough to be able to accept the postulate of rock deformation. Originally horizontal strata could be folded into anticlines and synclines. In the more extreme case, beds could be overturned, and the reversed stratigraphical superposition at Martin's Loch was thus thought by Escher (1866) to be the overturned limb of a syncline. Later, when the same Verrucano/Jurassic/Flysch sequence was found almost everywhere in Canton Glarus, Escher had to postulate a mushroomlike "double fold," and compared the presumably inverted succession to the underside of the crown of a mushroom (Figure 3.2).

This interpretation was *ad hoc* and tendentious. Marcel Bertrand, for one, was not convinced. Learning from Eduard Suess that the geological structures of Belgian coalfields are characterized by thrust faults that have all moved in the same direction, Bertrand introduced in 1884 the overthrusting theory to explain the Glarus geology (Figure 3.3).

Leistchamm   Foostock   Flimserstein

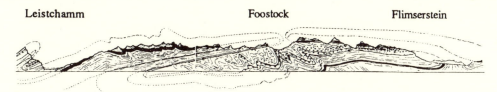

Fig. 3.2. The Glarus double fold of Sernf Valley (after Heim 1878). In Canton Glarus numerous mountain peaks (Foostock, Flimstein, etc.) are underlain by the Permian Verrucano, whereas the Eocene Flysch crop out, under a mylonitized Jurassic limestone, on valley bottoms. Escher and his student Albert Heim postulated a recumbent anticline north and another one south of the Foostock, forming a so-called double fold.

The difference between Escher's double fold and Bertrand's overthrust can easily be resolved by today's technique of using geopetal structures, such as graded bedding, which determine the top and bottom of a stratum. In an overturned syncline, all beds are inverted. In an overthrust pile, however, an older sequence is placed by thrusting on top of a younger sequence, but individual beds within each sequence should still be right-side up. We can now go to Glarus, examine the graded bedding of Flysch, and conclude that the Flysch sequence has not been overturned; it has only been under-thrust, or "tucked under," older rocks. Bertrand did not know such geopetal structures, but he made his postulate on the basis of comparative tectonics. It was easier for him to visualize one slab of rock formations being pushed on top of another slab like one blanket above another, than to imagine an extrusion of rock formations from the depths to form a mushroom fold. The theory of nappe tectonics was thus born.

Although Bertrand's revolutionary paradigm was resisted by Albert Heim, Escher's successor at ETH, for some two decades, the breakthrough was nevertheless quickly achieved. Bertrand's nappe theory was extended by Schardt (1893) to the Prealps, by Lugeon and Argand (1905) to the Penninic Alps, and by Termier (1903) to the Eastern Alps. Finally, even Heim accepted the theory and included it in his synthesis of Swiss geology (1919–1922). The edifice of Alpine geology, as we know it today, is a refine-ment of the structures erected on the foundations laid by those masters.

Good exposures in the Alps permitted accurate mapping of overthrusts. The evi-dence in the Alps led to the conclusion that Alpine thrusts have been displaced dozens or even hundreds of kilometers. Considering the very large friction at the base of the overthrust, Bertrand's postulate of a giant slab sliding on top of another seemed to be physically impossible. Physicists (e.g., Smoluchowski, 1909) made rough calcula-tions, and they found that a 100-km-long slab would be broken into many segments by the frictional resistance to overthrust movement. Modern studies of the overthrusting mechanics have indicated, however, that those calculations were wrong, because vari-ous mechanisms that could reduce the friction were not considered.

An excellent outcrop, exposing the Glarus Overthrust, is found near Schwanden, Canton Glarus (Figure 3.4). The Jurassic limestone under the Verrucano in Glarus is no longer a sedimentary rock; it is a mylonite. The Glarus Overthrust did not move as one solid block sliding over another along a fractured surface; the movement was not frictional gliding but simple shearing of the mylonite. The deformation took place under very low stress (Hsü 1969a, 1969b); thus a very large slab could be pushed forward, riding on the shearing of the mylonite, because the resistance was not friction but superplasticity, which was vanishingly small (Schmid 1975).

Although the prototype of the Alpine nappes is the Glarus Overthrust, which is a flat thrust sheet, the term nappe has also been used to describe large recumbent folds such as those found in the Helvetic Alps. There is a difference of opinion concerning the lateral continuation of such folds. Cylindricists are those who find a correlation be-tween the recumbent folds in eastern Switzerland and those in western Switzerland (or France). A broad correlation is, in many instances, justified, but few geologists are so

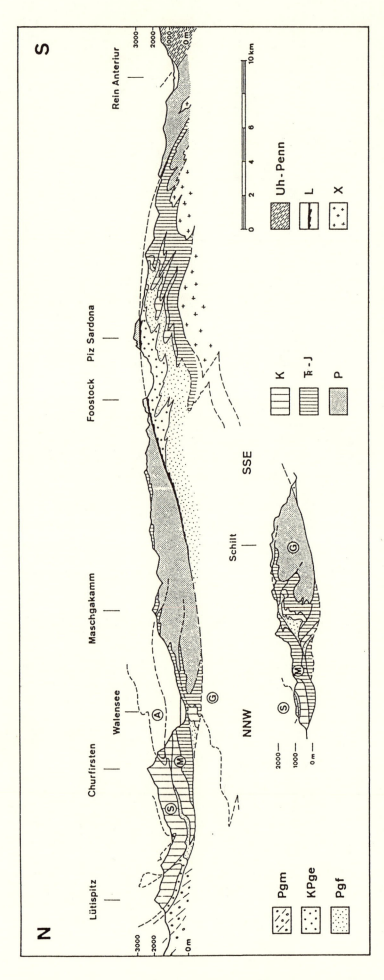

Fig. 3.3. The Glarus Overthrust (after Trümpy 1991). In 1884 Marcel Bertrand offered a simple and elegant explanation, postulating that both the Permian and Jurassic formations on the top of Glarus mountain peaks have been overthrust above the Flysch from the south. Bertrand's paper laid the foundation for the theory of Alpine nappes. Pgm: Subalpine Molasse; KPge: Cretaceous to Eocene thrust slices; Pgf: Helvetic Eocene and Flysch; K: Cretaceous formations; Tr-J: Jurassic and Triassic formations; P: Permian; Uh-Penn: Ultrahelvetic Pennsylvanian; L: Lochseitenkalk (thickness exaggerated); X: pre-Pennsylvanian basement.

Fig. 3.4. Section at Lochseite near Schwanden, Glarus (after Heim 1929): (1) Eocene Flysch; (2) Lochseit limestone-mylonite; (3) fault gouge; (4) green conglomerate of Verrucano; (5) red conglomerate of Verrucano.

naive today as to seek a one-to-one correspondence between the Säntis/Mürtschen/ Glarus complex of eastern Switzerland and the Morcles/Diablerets/Wildhorn complex of the west.

The nappe theory has given us several simple rules. Unidirectional vergence implies that the rocks in the highest nappe have traveled the farthest from their original site of deposition. We use the expressions internal and external to designate the paleogeographical positions of higher and lower nappes. The more internal is more southerly where the Alpine trend is east-west, as in the Austrian and Swiss Alps, and more easterly where the trend is north-south, as in the French Alps.

The simple relation between structure and paleogeography breaks down where the overthrusting fails to follow one single shear plane. The more southerly Ultrahelvetic is, for example, thrust above the Helvetic, along a plane that has also been called a thrust plane of envelopment. Subsequently, a Helvetic element could be overthrust along another shear surface, such as the Glarus Overthrust, and be placed above the Ultrahelvetic.

In many instances, the shearing has been pervasive, or penetrative. Originally coherent units of sedimentary formations were broken into slabs, blocks, and fragments. They were then mixed chaotically and carried along *en bloc* in a ductilely sheared matrix. Where a "nappe," or what has been called a nappe, is neither a thrust sheet nor a recumbent fold, but a mélange (Hsü 1968), one cannot use the simple rule that the highest slabs in such a mixture are necessarily more internal in paleogeography than those lower down.

The simple rule that the higher nappe is more internal also breaks down where the vergence of tectonic transport is reversed. Where rocks are overthrust southward, the higher unit should have a more northerly paleogeography. We shall address this question when we discuss the Schams paradox (Chapter 10).

Theories of Mountain-building

There are sedimentary rocks, even marine deposits, on top of mountains, but, as we have mentioned previously, a mountain is not simply a "pile of dirt." James Hutton postulated periodic upheavals of sea bottom, and his conclusion was based upon the observation of an unconformity in Scotland: beds of muddy sandstone (called graywacke) standing upright are truncated by erosion and overlain by red beds—the Old Red Sandstone of Devonian age. Elie de Beaumont (1830) saw in unconformities the record of revolutionary mountain-building convulsions of very short duration. Peaceful reigns of sedimentation were repeatedly interrupted by orogenic phases of global revolutions (Stille 1924; Bucher 1933). Hall (1859) found that the deformed sedimentary sequence in the Appalachian Mountains is thicker than the flat-lying coeval sequence on the Allegheny Plateau. He explained the difference with his theory of geosynclines, postulating that the Appalachian sediments in a geosyncline had been buried deep enough to become hot enough, and therefore weak enough, to be heaved up to form mountains. Dana (1873), on the other hand, thought that geosynclines did not come into existence because of the weight of sedimentary load; geosynclinal subsidence owed its origin rather to the deformation of the earth's crust under compression, which led ultimately to mountain-building. Nevertheless, both Hall and Dana believed there was a stage of geosynclinal subsidence prior to orogenic deformation, i.e., geosynclines are precursors to mountain-building.

Alpine geologists were never very impressed by Hall's geosynclinal theory, because the sedimentary sequence in the Alps of any epoch is not necessarily thicker than that elsewhere, as noted by Suess as early as 1875. Argand (1911), apparently influenced by the Dana school of thought, postulated continuous compressional deformation of the Alps from the Carboniferous until the Cenozoic. Others, such as Joly (1926) and, Bucher (1933), suggested alternate states of extension and compression on earth. Whereas global synchronism was difficult to verify, the concept of regional extension, combined with the theory of continental drift, led to Staub's (1928) postulate of geosynclinal subsidence under extension. The discovery of Mesozoic extensional structures in the Alps provided support for this interpretation (Trümpy 1960), so that Argand's "theory of embryonic tectonics" under continuous compression was challenged by young geologists of the postwar era. The model that emerged portrayed a geosynclinal stage of extension during the late Paleozoic, Mesozoic and early Tertiary, before the geosynclinal sequence of the Alps was deformed in the Late Eocene during an episode of regional compression. The deformation of the Molasse and the folding of the Jura Mountains are anti-climax, if not "post-orogenic." This simplistic postulate of a single episode of mountain-building is, however, contradicted by detailed analyses of the sedimentary record and by radiometric dating of metamorphic rocks in the Alps. The evidence clearly points to crustal deformations not only in the Eocene, but also during the Cretaceous and during the Oligocene and Miocene. Trümpy (1973) had to introduce the term Meso-Alpine for the "paroxysmal" late Eocene deformation, and the terms Eo-Alpine and Neo-Alpine for mountain-building "phases" before and after the Eocene.

The *Leitbild* of mountain-building was changed radically after the introduction of the plate-tectonic theory in 1968. Orogenic movements are now considered the consequence of displacements of lithospheric plates, which have been moving at linear or nearly linear rates for tens or hundreds of millions of years. There have been no global catastrophes of violence, as Hutton or Elie de Beaumont once envisioned, nor was mountain-making restricted to orogenic phases of short duration, as Stille (1924) dogmatically insisted. The plate-tectonic model of Alpine orogenesis postulates two stages of deformation since the Late Paleozoic: an earlier extensional stage of rifting (Late Permian, Triassic, and Early Jurassic) and of seafloor spreading (Middle and Late Jurassic) followed by a compressional stage of lithospheric subduction (Early Cretaceous to the present). The Tethys, an ancient Mediterranean Sea some 500 to 1000 km across, was eliminated, causing the collision of Central Europe and North Italy (Mediterranean plate). The subsidence of the Molasse foreland basin and the folding of the Jura Mountains were not "post-orogenic," but "post-collisional." Their genesis is related to the continuing compression after the continental collision in the western Alps.

Classification of the Swiss Alps

The formation of nappes involves both large horizontal displacement and internal strain of rock units. There are two major kinds of rock bodies: (i) the massive rocks constituting the basement, and (ii) the layered, sedimentary formations that form the cover of the basement.

A first-order tectonic classification is based upon two criteria: whether there has been significant horizontal displacement of a rock unit from its place of origin (allochthonous or autochthonous), and whether the basement is involved in the displacement and/or deformation (basement or cover nappes). A second-order criterion, based upon the degree of metamorphism and internal strain, distinguishes rigid-basement nappes from mobilized-basement nappes.

Still another criterion of distinction is the paleotectonic and paleogeographical framework of the sedimentary strata involved in the deformation. Active-margin sequences or accretionary prisms are sedimentary strata laid down at a plate junction where oceanic lithosphere is underthrust under continental lithosphere. Such sequences tend to form mélanges. Passive-margin sequences are sediments laid down on a continental margin within the interior of a lithospheric plate, such as the European or African Atlantic margin. They may be detached from the basement to form cover nappes, as in the case of Helvetic deformation, or still attached to the basement of rigid-basement nappes, as in the case of some Austroalpine deformations. Other sequences are the sedimentary cover of submarine swells, island arcs, the sedimentary

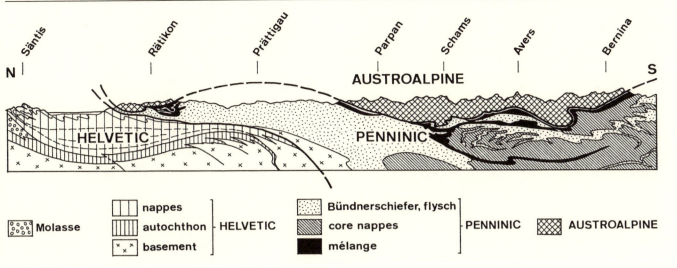

Molasse | nappes] HELVETIC | Bündnerschiefer, flysch] | AUSTROALPINE
| autochthon | | core nappes | PENNINIC
| basement | | mélange]

Fig. 3.5. Tectonic superposition of nappes across the eastern Swiss Alps (after Staub 1924).

infill of back-arc basins, etc. These are deformed to make up cover nappes, or Flysch nappes.

The Swiss Alps have been subdivided into the following tectonic units:

1. The autochthonous and parautochthonous massifs (basement of passive margin, rigidly deformed; massifs)

2. The Helvetic nappes (cover of passive margin, detached and deformed; cover nappes)

3. The Flysch nappes (accretionary wedge under active margin, detached and deformed; Flysch nappes or Wildflysch melange)

4. The Prealpine nappes (cover of a submarine swell, Brianconnais/Subbrianconnais, detached and deformed; cover nappes)

5. The Penninic core nappes (basement of passive margin, penetratively deformed and metamorphosed; mobilized-basement nappes)

6. The Bündnerschiefer (cover of passive margin and ocean floor, deformed and metamorphosed; septa between core nappes)

7. The Ophiolite Mélanges (cover and basement of ocean floor, fragmented and mixed; mélanges)

8. The Austroalpine nappes and the Southern Alps (cover and basement of overriding plate, rigidly deformed; rigid-basement and some cover nappes)

9. The Sesia-Lanzo zone (cover and basement of overriding plate, deformed and metamorphosed; mobilized-basement nappes)

10. The Bergell granite

The distribution of the major tectonic units is shown in the tectonic map in Chapter 1, and their superposition in Figure 3.5.

Autochthonous and Parautochthonous Massifs

The autochthonous and parautochthonous massifs are large basement slices rising out of the Mesozoic cover in the external zone of the Alps. The largest is the Aar massif of the Bern and Uri Alps in central Switzerland (Figure 1.7). The basement of the Aar massif consists of granites and gneisses. They have not been much affected by the Alpine metamorphism; the radiometric dates of the basement rocks are commonly Late Paleozoic. The sedimentary cover that has not been stripped off the Aar massif is very thin. A contact between the cover and the basement is well exposed at Haldenegg near Erstfeld in Canton Uri (Figure 3.6). The basal sandstone, up to 1 m thick, is coarse grained, with a few scattered pebbles. The overlying sequence consists of interbedded sandstone, dolomite, and shale beds of Triassic age.

Fig. 3.6. The Triassic sedimentary cover of the Aar massif, near Erstfeld (after Heim and Heim 1917): (1) basement (Erstfeld gneiss); (2) cross-bedded sandstone; (3)–(5) sandstone and shale; (6)–(10) dolomite with intercalated sandstone and shale beds.

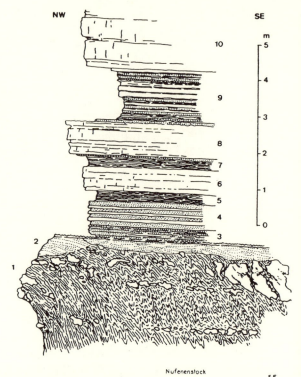

Fig. 3.7. Profile across the Gotthard massif (after Reinhard and Preiswerk 1934). Gneisses and schists constitute the basement of the massif, and the Mesozoic sedimentary cover consists of phyllites, schists, carbonates, and Bündnerschiefer.

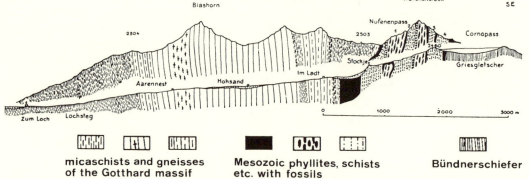

micaschists and gneisses of the Gotthard massif   Mesozoic phyllites, schists etc. with fossils   Bündnerschiefer

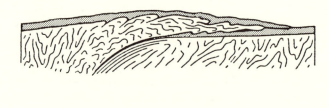

Fig. 3.8. A-subduction (after Ampferer and Hammer 1911). The concept of subduction was first proposed by Ampferer to account for the missing basement that must have been present below the sedimentary sequences now forming cover thrusts such as the Helvetic nappes. A-subduction designates the underthrusting of one continental lithosphere under another, and B-subduction designates the underthrusting of oceanic lithosphere under continent.

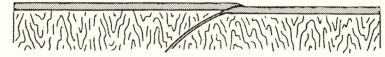

South of the Aar massif lies the small Tavetsch and, further south, the larger Gotthard massif. The basement rocks in these have been overthrust onto the Aar massif and are thus parautochthonous. They have also been more affected by the Alpine metamorphism.

The Mesozoic sedimentary cover on the south side of the Gotthard massif consists of Triassic shallow marine and Jurassic hemipelagic sediments (Figure 3.7). The Jurassic facies is comparable to that of the Ultrahelvetic nappes, and it is commonly assumed that the Ultrahelvetic Mesozoic once lay atop the Gotthard massif. The Jurassic sediments of the Helvetic nappes are, in contrast, typically shallow marine, but most of them could not have been derived from atop the Aar massif, which has its own cover. The Helvetic sediments are thus believed to have been largely the former covering beds on top of the Tavetsch, and perhaps of the northern part of the Gotthard massif (Trümpy 1960, p. 850).

The width of the Helvetic facies belt should have been at least 100 km; thus it was considerably wider than the Tavetsch basement. One cannot escape the conclusion that much of the original Helvetic basement was thrust under the Gotthard massif. This process, involving "telescoping" of adjacent lithospheric plates, when one is tucked under another, has been called *Verschluckung* in German (Ampferer and Hammer 1911) and *subduction* in French and English (Amstutz 1952; White *et al.* 1970). Subduction involving an underthrusting of continental lithosphere, such as in the case of the "swallowed" Helvetic basement, is called Ampferer-subduction or A-subduction (Figure 3.8), in contrast to the B-subduction of oceanic lithosphere down a Benioff zone.

The sedimentary cover of the Gotthard massif becomes increasingly deformed and metamorphosed in the southerly direction, and the Jurassic metasedimentary rocks of the Cornopass area are lithologically indistinguishable from the Bündnerschiefer of the lower Penninic nappes (Figure 3.7).

Helvetic Nappes

The term Helvetic nappes refers to cover thrusts and recumbent folds in the High Limestone Alps of Switzerland, which extend from Lake Geneva to the Rhine Valley (Figure 1.7). The sedimentary strata in these units are characterized by a stratigraphy that has also been designated Helvetic, being largely Mesozoic sediments deposited on a passive continental margin. Nappes that include sedimentary sequences characterized by a stratigraphy distinct from the Helvetic belong to higher tectonic units; they are called Ultrahelvetic, Penninic, or Austroalpine.

The Mesozoic and Eocene cover of the Mont Blanc massif and of the western Aar massif is stripped off and forms a large recumbent fold, commonly known as the Morcles nappe of western Switzerland (Figures 3.9 and 3.10). To the east, where no evaporite detachment horizon is present, the sedimentary cover of the Aar massif is preserved as parautochthonous slices, little displaced with respect to their substratum. Higher Helvetic nappes are derived from an outer continental margin underlain by the basement rocks of the Tavetsch and part of the Gotthard massif; these include the Diablerets and Wildhorn nappes in western Switzerland, and the Axen and Drusberg nappes in the Lake Lucerne area. In eastern Switzerland, the pile of nappes, from bottom to top, contains the Glarus, the Mürtschen, and the Säntis nappes.

There is a temptation to correlate the three western nappes with the three in the east, but such a correlation is at best hazardous. On the whole, the overthrusts have all been transported from south to north. The sediments of the highest nappes originally occupied a most southerly paleogeographical position prior to their deformation, and they are thus commonly called South Helvetic, in contrast to the North Helvetic sediments of the lowest nappes.

The Helvetic nappes are décollement structures. Involved in the deformation are only the sedimentary strata of the stripped-off sedimentary cover. The most prominent

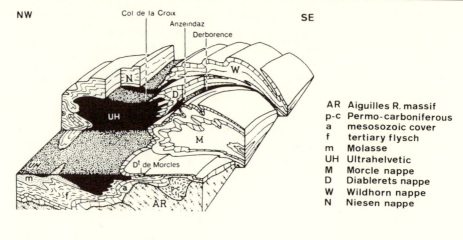

Fig. 3.9. Block diagram of the Helvetic nappes of western Switzerland (after Caron and Escher 1980). AR: Aiguilles Rouges massif; p-c: Permo-Carboniferous; a: Mesozoic cover; f: Tertiary Flysch; m: Molasse; UH: Ultrahelvetic nappes; M: Morcles nappe; D: Diablerets nappe; W: Wildhorn nappe; N: Niesen nappe.

AR Aiguilles R. massif
p-c Permo-carboniferous
a   mesosozoic cover
f   tertiary flysch
m   Molasse
UH Ultrahelvetic
M   Morcle nappe
D   Diablerets nappe
W   Wildhorn nappe
N   Niesen nappe

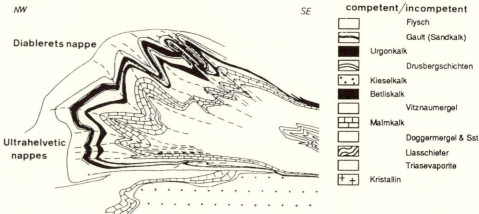

Fig. 3.10. A profile through the Morcles nappe (after Dietrich and Casey 1989).

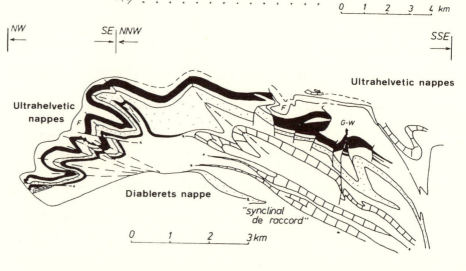

Fig. 3.11. A profile through the Diablerets nappe (after Dietrich and Casey 1989): F: Flysch; G-W: Gault to Wangschichten. Symbols as in Fig. 3.8.

feature of the Morcles nappe is the overall wedge shape of the fold nappe (Figure 3.10). The thickness of the nappe is less than 3 km at its most southern outcropping part, near the so-called root zone, but more than 6 km at the northern front. The Diablerets nappe is not a fold nappe; the inverted limb is cut off by a thrust fault at the base of the nappe (Figure 3.11). The development of a basal thrust at the base of a recumbent anticline is a step toward the genesis of a flat thrust like the Glarus Overthrust. The highest Wildhorn nappe is highly disharmonic: between the folds in the Upper Jurassic Malm limestone and the folds in the Lower Cretaceous competent beds lies a thick mass of incompetent Valanginian marls (Figure 3.12). The architecture of the nappe is further complicated by the presence of several internal thrust faults.

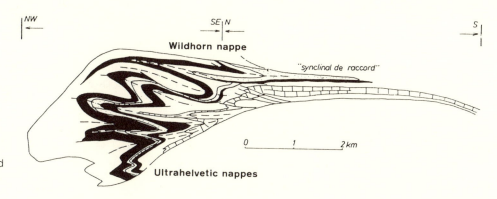

Fig. 3.12. A profile through the Wildhorn nappe (after Dietrich and Casey 1989). Symbols as in Fig. 3.8.

The internal deformation of the cover nappes is manifested by the variation in bed thickness. The Upper Jurassic Malm limestone of the Morcles and Diablerets nappes, the thickest of the competent layers that form the skeleton of the nappe, shows a variation in bed thickness considerably greater than that related to sedimentation (Figures 3.10 and 3.11). The Wildhorn nappe, on the other hand, is characterized by the variation in the thickness of the incompetent beds (Figure 3.12).

Ultrahelvetic and Flysch Nappes

The sediments on the outer continental margin, in the area south of the realm of Helvetic sedimentation, should have been laid down in deeper water, and they are characterized by sedimentary sequences with a considerable thickness of hemipelagic shale or Flysch. They are called Ultrahelvetic in order to denote a paleogeographical position "ultra" (toward the ultimate limit or beyond) that of the Helvetic realm of sedimentation, and the Ultrahelvetic strata are superposed tectonically on top of the Helvetic nappes.

The Ultrahelvetic strata hardly ever form recumbent folds; they have been deformed by flat thrusts, parallel or subparallel to the bedding planes, or by penetrative shearing. Mesozoic strata, commonly Helvetic in origin, are present locally as giant exotic blocks in the Ultrahelvetic shales and siltstones, constituting a rock unit called *Wildflysch*. Other formations, such as various types of Flysch of more internal origin, are also found as slabs in such a tectonic mixture.

The tectonic superposition of Ultrahelvetic rocks above the Helvetic took place before the décollement deformation of the latter. Emplacement of the Ultrahelvetic nappes on an unfolded and undisplaced European foreland by gravity sliding used to be a favorite mechanism invoked to explain the superposition (for example, Trümpy 1960). Ramsay (1989, p. 41) pointed out, however, that "the inclination of the basal thrust surface of the Ultrahelvetic nappes must have been subhorizontal over a transport-directed distance of at least 50 km." A consideration of the physical criteria for yielding shows that gravity sliding could not take place on such a gentle slope (Hsü 1969a, 1969b). The origin of the Wildflysch is now attributed to underthrusting at a convergent plate margin, to be discussed in Chapter 5.

Prealpine Nappes

The highest cover nappe north of the Aar massif has sedimentary sequences with a stratigraphy significantly different from that of the Helvetic margin sequences. The erosional remnant of those nappes forms the top of the Mythen in central Switzerland, and the same tectonic unit can be traced westward to western Switzerland. The original controversy on the origin of the klippes was resolved, but a new debate raged for more than half a century: German-speaking Swiss, headed by Rudolf Staub (e.g., 1958), favored an Austroalpine origin, or a paleogeographic site of deposition south of the

Piedmont Ocean, for the Median Prealp sediments. French geologists, led by Ellenberger (1952), Lemoine (1953), and Debelmas (1955), found the equivalent of the Median Prealp sediments in the Brianconnais unit of France. With the demise of Staub, a consensus was reached: The rocks of the Median Prealps were once the sedimentary cover of the Brianconnais Swell in the middle of the Tethys Ocean (Trümpy 1960). During or after the Eocene collision, this Brianconnais/Subbrianconnais sequence was stripped off of the swell and carried northward under the Austroalpine nappes to form the Prealpine nappes.

Penninic Nappes

The major core nappes of the Pennine Alps, according to Emil Argand (1911), are as follows, in descending order (see Figure 7.2):

VI. Dent Blanche nappe
 V. Monte Rosa nappe
IV. Bernard nappe
III. Monte Leone nappe
 II. Lebendun nappe
  I. Antigorio nappe

The basement of the Antigorio, Lebendun, and Monte Leone once underlay the outermost margin of Europe. Their sedimentary cover consisted mainly of hemipelagic muds, which are now phyllites and schists.

The core of the Bernard nappe was the basement of the intra-oceanic Brianconnais Swell, which stood above sea level briefly during the Middle Jurassic. The Bernard is now enveloped by a metamorphosed sedimentary cover, which can be correlated with the Brianconnais/Subbrianconnais unit in the French Alps. The Brianconnais Swell was separated from the European margin by a deep-water basin called the Valais Trough. South of the Brianconnais was an open sea, the Piedmont Ocean, which was underlain by oceanic lithosphere.

The core of the Monte Rosa crust was originally also continental basement. Whether it was situated north or south of the Piedmont Ocean is still a matter of controversy. The postulate of a southerly origin, as will be advocated in Chapter 10, explains more convincingly the presence of high-pressure metamorphic minerals in this nappe.

The Dent Blanche, the highest of Argand's six Penninic nappes, is considered today to be an Austroalpine element. The unit has experienced a tectonic evolution similar to that of the Lower Austroalpine rigid-basement nappes.

The mobilized-basement nappes of eastern Switzerland are also known collectively as Penninic, but they have been given different names. "Cylindricists" have tried to correlate the core nappes of the east with those of the west. Their noncylindrical nature is, however, manifested by the deformation of the sedimentary cover of the eastern extension of the Brianconnais Swell: the Schams nappes are thrust southward, not northward like the Median Prealpine nappes. We shall return to the puzzling geology of the Schams area in Chapter 10.

Bündnerschiefer, Ophiolites, and Mélanges

Paleogeography is reconstructed on the basis of recognizing lateral facies changes in sedimentary formations. The Triassic sediments of Europe were deposited almost everywhere near sea level, from northern Europe to the Mediterranean south. These terrestrial or shallow marine deposits formed the sedimentary cover of the continent of Pangea. The tectonic processes starting during the Early Jurassic effected the separation of Africa from Europe. The Jurassic sediments on the Helvetic/Ultrahelvetic European margin vary from shallow marine to open marine southward. The coeval deposits on the Austroalpine margin give evidence of a northward deepening. There was an open ocean between the two margins, the Tethys Ocean. The ocean was divided by the

Brianconnais Swell into a northern "Valais Trough" and a southern "Piedmont Trough." The sediments deposited in those basins are mainly fine-grained muds, with intercalated layers of thin siltstone-turbidite. The hemipelagic sediments in eastern Switzerland (for example, Canton Graubünden) are less metamorphosed, having been changed to phyllites called *Bündnerschiefer*. The coeval sediments of western Switzerland have been subjected to higher degrees of metamorphism, and the shining luster of the mica schists has given the formation the well-known name *schistes lustrés*.

The Alpine ophiolites are the basic and ultrabasic igneous rocks that once constituted the deep-sea floor of the Tethys Ocean. They include basalts, gabbros, peridotites, and serpentinites. The ocean sediments overlying the ophiolites are radiolarite and pelagic limestone. They consist of fossilized skeletons of radiolarians, tintinnids, globigerinids, nannoplankton, and/or other unicellular organisms that once lived in the surface waters of the Tethys Ocean.

The original width of the Helvetic margin has been estimated by unfolding the nappes. The practice of palinspastic reconstruction places the different fragments of a rock sequence at their original geographical positions. We can thus conclude that the Helvetic margin was originally on the order of 100 km wide (Trümpy 1980). The Bündnerschiefer and the ophiolites, however, do not form coherent rock packages: the igneous rocks, the radiolarian chert, and the sandstone, being more brittle during deformation, were broken up into fragments or slabs, and they are embedded in a phyllite or schist matrix that was sheared by ductile deformation. This type of tectonic structure has been called mélange, or ophiolite mélange. A palinspastic reconstruction of the mélange paleogeography is all but impossible. The width of the Tethys Ocean was thus once a problem that had no solution.

The theory of seafloor spreading has brought to us a new criterion with which to answer this puzzle. The width and orientation of seafloor magnetic lineation can be used to calculate the distance and direction of continental displacement, and the positions of continents such as Africa and Europe can be determined on the basis of computer calculations (e.g., Le Pichon 1968). Using this technique, we now believe that the width of the Tethys should have been about 1000 km (Hsü 1989).

What happened to the wide expanse of the Tethyan ocean floor?

It was swallowed, or—to use modern geological terminology—subducted, down the Benioff zone.

The subduction of the Tethys was not an episodic event, but a long, continuous process, which started at the beginning of the Cretaceous and continued until the complete swallowing of the Tethys caused the continents on both sides to collide. This is what we propose to call Eo-Alpine deformation.

As oceanic lithosphere plunged down beneath continental lithosphere, the ocean sediments (Bündnerschiefer) and ocean crust (ophiolites) were broken up and/or pervasively sheared in the Benioff zone to make mélanges. The evidence of fragmentation, shearing, and mixing, during the Eo-Alpine subduction and Neo-Alpine post-collision deformation, will be discussed in Chapter 9. Radiometric dating of metamorphic minerals in the mélange has largely verified our postulate of Cretaceous subduction.

Austroalpine Nappes

After the elimination of the Tethys Ocean, the southern plate rode on top of the northern plate. The leading edge of the former was sliced into numerous slabs, and the Austroalpine nappes of Switzerland are some of those slabs.

Because they were thrust up and thus on top, the rocks of the Austroalpine nappes, as a rule, are not much metamorphosed. The metamorphic minerals and structures in the basement rocks were formed during the Paleozoic, and they were not recrystallized or changed into new minerals during the Alpine deformation. Displaced slabs of such basement rocks are also known as rigid-basement nappes.

The sedimentary cover of the Austroalpine nappes either is still attached to the basement, or has been stripped off to form cover nappes between rigid-basement nappes. The sedimentary sequences have retained their coherence, and palinspastic reconstruction of the Austroalpine paleogeography is possible.

The contact between the Austroalpine and Penninic nappes is not easily determined, because the lowermost Austroalpine nappes, being more deeply buried, were more or less metamorphosed during the Alpine deformation, and they may display geological structures not dissimilar to those of the Penninic nappes. We have mentioned that the Dent Blanche and Monte Rosa nappes are tectonic slices of the southern Tethyan margin and should thus be correlative to the Austroalpine, yet the great master Emil Argand grouped them together with the other Penninic nappes of the Valais. The same controversy about the tectonic classification of the Margna nappe raged for years. Margna was once considered Penninic, because it occupied a tectonic position beneath a Penninic ophiolite mélange, but it is now considered Austroalpine, because its sedimentary cover indicates an original paleogeographic position on the southern shore of the Tethys.

The degree of metamorphism of the Monte Rosa indicates that the rocks of this nappe were once taken down to a depth of 100 km. The same is true of the rocks in the Sesia-Lanzo zone. Continental basement rocks in both units were carried down the Benioff zone in the Early Cretaceous. Later, when the ocean lithosphere began to be thrust under, the deeply subducted continental crust could rise again and be exhumed. There was some tectonic mixing during the process of ascent, so that the Monte Rosa is now enveloped in ophiolite mélanges. On the whole, however, those slabs of southern basement have been thrust, like the Austroalpine nappes of the east, over and above the Penninic core nappes.

Subduction of continental or oceanic lithosphere may bring silicate rocks down to a depth where partial melting is possible. Volcanism and granitic intrusions are thus common phenomena in the geological evolution of the overriding plate in plate interaction. The history of the Alps gives, however, little evidence of such activity. The Bergell batholith is one such intrusion. The batholith seems to have a "floor," having been thrust over lower Penninic nappes, and it has a roof where magma intrudes into superjacent rocks.

South of major Tertiary fault zones, such as the Tonales Line, are the Southern Alps. The stratigraphy of their sedimentary cover shows close affinity to that of the Austroalpine nappes; both belong to the southern continent. The Southern Alps lie mainly outside of Switzerland. Their geology constitutes a chapter of the geology of Italy, and will not be discussed in detail in this book.

4.

Two Helvetic Unconformities

Regression and transgression are words used in geology to denote the retreat or advance of shoreline. Where new land is made by sedimentation, the shoreline moves toward the sea; it can be said to have retreated. This process is called regression. Conversely, where marine waters move gradually inland, the shoreline advances, and the result is called transgression. The deposition of shoreline sand and shallow marine carbonate sediments on an erosional surface is typical evidence of transgression. As a result of this transgression, a shallow marine deposit overlies older rocks of various ages. This contact relation involving a hiatus or missing record in geological history is called unconformity, and the unit above the unconformity overlaps successively older formations beveled by erosion.

This chapter tells the stories of two unconformities. The pre-Tertiary unconformity is a record of a marine transgression. The unconformity below the uppermost Cretaceous Wang Formation is, however, a very different feature and has recorded a very different process of submarine erosion.

Einsiedeln Schuppen Zone

The Sihl River runs parallel to Lake Zürich and joins the Limmat near the main Zürich railroad station. The northwesterly course of this superposed stream can be traced back to the time of deglaciation. The Limmat glacier once covered not only the city of Zürich but also the 125-m-deep Lake Zürich basin, and the Sihl was a melt-water stream between the glacier and its side moraine. Driving up the Sihl Valley, one crosses the Molasse plateau of the Swiss Midland, the zone of Subalpine Molasse and Subalpine Flysch, and reaches Lake Sihl near the ancient monastery of Einsiedeln. On the right side of the Sihl, under the Sattelberg, are four prominent ledges of nummulite limestone. (The term nummulite denotes a group of relatively large single-celled organisms, foraminifera, that evolved during the Tertiary. Their calcareous tests are disk-shaped, and the largest such "disks" are several centimeters across.) This feature caught the attention of Arnold Escher, the first professor of geology at our institution. Escher's geological sketch (Figure 4.1) was later published in a monograph by Franz Joseph Kaufmann (1877), a pioneer in Swiss geology.

In 1848 Murchison visited the Sihl Valley in the company of Escher. The partially covered intervals between the nummulitic limestone banks are dark gray marl, and Murchison thought that the marl was interbedded with the limestone and was, therefore, also Tertiary. Escher, who worked in the area, thought that there may have been

Fig. 4.1. Einsiedeln schuppen zone, exposed on the slope between Steinbach and Euthal (after Kaufmann 1877). f: Foraminifer formations; n: nummulite limestone; R: gully cut into marls. The four layers of nummulite limestones belong to four different slices of the schuppen zone. The shaly intervals between the limestone banks include not only Eocene Globigerina Marl, but also Upper Cretaceous Amden Marl.

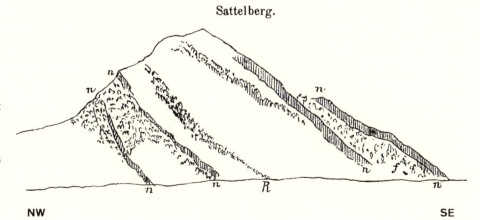

Sattelberg.

NW   SE

tectonic complications (see Kaufmann 1877, p. 113), but the opinion of the foreign expert prevailed.

Nummulite fossils are common in the limestone. Mayer-Eymar (1898) correlated the Einsiedeln nummulite fauna with the "Parisian" (Lutetian) of the Paris basin, which had been dated Middle Eocene. Therefore, geologists of the last century concluded that the "Einsiedeln Beds," including both the limestone and the marl, were Middle Eocene (Kaufmann 1877; Arn. Heim 1908).

When a Cretaceous fauna was discovered in the marl between nummulitic limestone beds on the Sattelberg, a contradiction in stratigraphical paleontology was apparent. If the marl and limestone were interbedded and constituted a formation of continuous deposition, we had to assume either a Cretaceous or a Tertiary age for those strata. The Cretaceous age of the invertebrate fauna seemed unquestionable, because it includes an ammonite specimen from eastern Switzerland; ammonites could not be Tertiary, because they died out at the end of the Mesozoic. Louis Rollier, a former professor of stratigraphy at the ETH, therefore claimed during the Zürich Meeting of the Swiss Society for Natural Science in 1918 that the Einsiedeln nummulites must have evolved during the Late Cretaceous. Prominent geologists in the audience, such as Hans Schardt and Maurice Lugeon, violently objected to this simplistic explanation; they pointed out the possibility that Cretaceous marl beds might have been overthrust on top of Tertiary limestone banks.

Arnold Heim had preferred a tectonic solution at one time, but he was not able to delineate fault planes between the marl and the limestone beds. Therefore, he and Rollier coauthored a monograph on "Cretaceous nummulites" in 1923, and the mistaken notion that the limestone and marl belong to one and the same formation was upheld for another decade or two. Nevertheless, nowhere except in the tectonically complicated Alpine region are nummulite fossils ever found in presumably Cretaceous formations.

The situation became even more confusing when a globigerina fauna in the marly interval was identified as Late Eocene. The marl is at least in part Eocene, only slightly younger than the limestone (see Leupold 1937). Did we have Tertiary ammonite then? No. The paradox was finally resolved when Jeannet, Leupold, and Beck (1935) followed a suggestion by Escher, Schardt, Lugeon, and Albert Heim and explored the possibility of tectonic complications on the Sattelberg.

The paleontologists made a detailed study of the nummulite faunas in the four limestone banks in the Sihl Valley, and found them identical. Some evolutionary development would be expected if the four were different in age. The presence of an identical fauna in the apparently successive layers is a convincing argument that the same limestone unit has been repeated by faulting. Murchison was wrong; the "Einsiedeln Beds" do not constitute a single formation.

Leupold introduced the term Einsiedeln Flysch in 1937, because the term Flysch had been used to designate tectonic units in the Swiss Alps such as the Wildflysch. Leupold recognized that the Einsiedeln Flysch is a tectonic mixture. The three formations involved in the repetition by overthrusting are, in descending order,

Upper Eocene Globigerina Marl
Middle and Lower Eocene Nummulitic Limestone
—unconformity—
Upper Cretaceous (Senonian) Amden Marl

The ammonite fossil was found in the Amden Marl of Late Cretaceous age, and the Eocene formations on the shore of Lake Sihl unconformably overlie this marl. The presence of four limestone beds between marl intervals indicates that the Cretaceous marl has been thrust above the Eocene formations four times. Such closely spaced overthrust slices constitute what has been called a schuppen zone.

When the idea that Flysch is a lithostratigraphic unit began to prevail in the 1950s

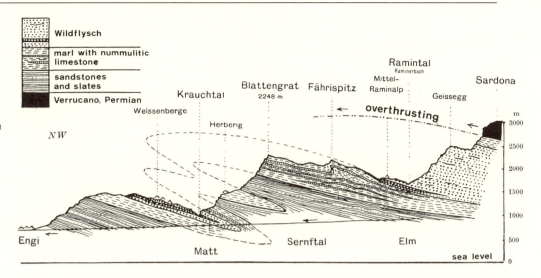

Fig. 4.2. Blattengrat schuppen zone, exposed on the right side of the Sernf Valley (after Oberholzer 1933). The four nummulite limestone layers of the Blattengrat, like the four near Einsiedeln, also belong to four slices of a schuppen zone. The Wildflysch under Piz Sardona is now considered a tectonic mixture of largely Cretaceous formations; this Ultrahelvetic unit was overthrust above the Helvetic before the folding of the latter. The highest tectonic unit is the Permian Verrucano formation of the Glarus nappe.

the term "Einsiedeln Flysch" had to be changed again. The faulted sequence on the right side of the Sihl is now referred to as the Einsiedeln schuppen zone. A schuppen or imbricated structure is very common in a foreland thrust belt. This Prealpine schuppen zone can be traced eastward to the Sernf Valley in Canton Glarus, well exposed on the Blattengrat, where exactly four nummulitic limestone bands are also exposed (Figure 4.2). The "Blattengrat Beds," once called "Blattengrat Flysch" (Leupold 1937), are now designated the Blattengrat schuppen zone.

Early Tertiary Transgression

Intercalated in the limestone formation is a green sand (Figure 4.3). Studer cautioned us as early as in 1851 that this green sand is not to be confused with the Cretaceous green sand in the Helvetic Alps, which had been called "Gault." Both are green because of the presence of a green mineral, glauconite. The "Gault" has yielded Cretaceous fauna, and it lies between ammonite-bearing formations. The green sand of the Einsiedeln schuppen zone, however, has large foraminifers belonging to the genus *Assilina*, a relative of the genus *Nummulite*, and it is a transgressive sand, early Tertiary in age. This Tertiary nummulitic limestone/green-sand formation is easily recognizable in the High Calcareous Alps, but it is not everywhere exactly the same age, nor is the age of the Cretaceous marl below the unconformity everywhere the same.

Arnold Heim (1908) summarized the observations on the ages of the strata above and below the Cretaceous/Tertiary unconformity, and he noticed that the Tertiary limestone/sand formation is invariably older and the Cretaceous marl below the unconformity invariably younger in a higher Helvetic nappe than those in a lower nappe. In the Wageten Kette and Mürtschen nappes, two of the lowest Helvetic nappes, the limestone/sand formation is Middle Eocene and the strata below the unconformity are lower Upper Cretaceous (Cenomanian, Turonian, Coniacian) Seewerkalk. In the Axen nappe, the middle of the three major Helvetic nappes, Middle Eocene limestone/sand overlies the upper Upper Cretaceous (Santonian/Campanian) Amden Marl. In the Einsiedeln schuppen zone, the pre-Tertiary formation is still Amden Marl, but the overlapping bed is a Lower Eocene limestone. In the Drusberg nappe, the highest of all Helvetic nappes, Paleocene green sand overlies unconformably either the Amden Marl or the uppermost Cretaceous (Maastrichtian) Wang Formation (Leupold 1967a). Since the rocks of a higher nappe must have come from a more southerly site, the stratigraphy indicates that the gap in the unconformity between the Tertiary and Cretaceous increases from south to north. The paleogeographical reconstruction is shown in Figure 4.4. Farther north, in the Jura Mountains, where Eocene quartz sand and bauxitic clay overlie Jurassic formations, the gap is still larger.

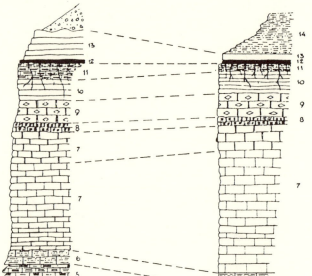

Fig. 4.3. Stratigraphy of the Einsiedler schuppen in the Lake Sihl Valley (after Leupold 1967a). 1: Amden beds; 2: Paleocene lithothamnian limestone; 3: middle green sand; 4: nummulite limestone; 5: *Assilina* limestone; 6–9: nummulite limestone beds; 10: Lutetian limestone; 11: hematitic nummulite limestone; 12: Steinbach bed; 13: upper green sand; 14: marl, glauconitic near base.

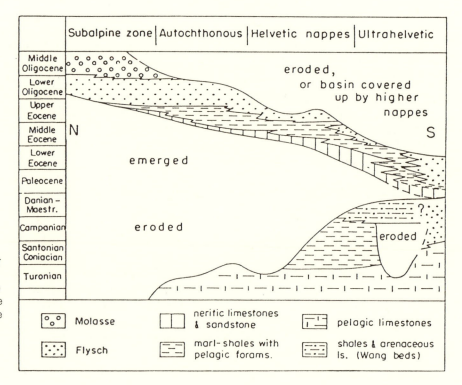

Fig. 4.4. Transgressive unconformity at the base of the Tertiary (after Trümpy 1960). The basal nummulitic limestone and/or green sand beds above the unconformity are Paleocene in the Ultrahelvetic, but as young as Late Eocene in the autochthonous. The Cretaceous strata under the unconformity are, on the other hand, progressively younger in higher nappes.

Above the Helvetic limestone/sand formation is Globigerina Marl. It is a hemi-pelagic sediment, so called because at least half of the sedimentary particles "swam their way out" there. The Helvetic site of deposition sank to a greater depth and the shoreline became more distant. Nummulites and other fossil organisms of the shallow sea could no longer survive and flourish, and sand grains could no longer be carried so far out from the shore. Deposition of carbonate and quartz sand was replaced by the sedimentation of silt- and clay-sized particles, and of fossil skeletons of small, one-celled organisms called *Globigerina*. This is also a genus in the order *Foraminifera*, but is very different from nummulites. Globigerina tests are small, rarely more than a fraction of a millimeter across. More important, they are swimmers, or planktonic organisms, whereas nummulites are benthic, or bottom dwellers. The Globigerina Marl also has different ages in Helvetic nappes; again there is a systematic trend, the marl on the highest Helvetic nappe invariably being the oldest (Figure 4.4).

The Cretaceous/Tertiary unconformity on this northern margin of the early Tertiary Tethys Ocean indicates that the margin had been uplifted and subjected to erosion prior to the deposition of limestone and sand. This interruption of a subsidence trend has been called "Paleocene restoration" by Rudolf Trümpy (1973). The "restoration" did not happen at the same time everywhere. The most northerly part of the margin remained above sea level and became inundated by a transgressing sea only at a much later date, some 15 million years after the erosion and inundation of the south Helvetic area. Later the Helvetic sea bottom subsided, permitting the sedimentation of the Globigerina Marl.

My colleague Daniel Bernoulli has proposed a mechanism to explain the "Paleocene restoration" in terms of plate tectonics. The erosional unconformity indicates that the Helvetic margin was elevated at the beginning of the Tertiary. He considers the uplift a manifestation of the elastic bending of the European plate as it was thrust under the Mediterranean plate (Figure 4.5).

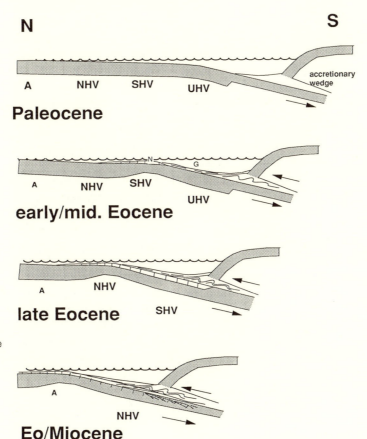

Fig. 4.5. Northward shift of the bending of the axis of uplift and the pre-Eocene unconformity of the Helvetic margin (after a postulate by Daniel Bernoulli, personal communication, 1988).
A: autochthonous; NHV: North Helvetic; SHV: South Helvetic; UHV: Ultrahelvetic; N: nummulite limestone; G: globigerina marl.

It is commonly postulated that the Tethys Ocean was subducted under a southern continent since the beginning of the Early Cretaceous (see Chapter 12). Toward the end of the Cretaceous, the Tethys ocean-lithosphere was largely consumed, so that the European margin came close to the southern Tethyan margin. Lithospheric plates have a certain rigidity, and the subduction of the rigid European plate caused bending, and consequently an uplift of the Helvetic margin. During the Paleocene the last of the ocean lithosphere was thrust under into an accretionary wedge, and the European coastline lay within the Ultrahelvetic realm of sedimentation, where neritic limestone and shallow marine sands were deposited (Figure 4.5). With the advancing of the southern continent during the Early Eocene, the Ultrahelvetic realm was pulled down, and deep marine Flysch sediments were deposited at the plate margin. The hinge of the uplift and the European coastline had shifted northward in the meantime, and now lay within the South Helvetic realm, where Eocene Nummulitic limestone and green sand were laid down (Figure 4.5). With the further progression of the subduction during the Late Eocene, the hinge migrated farther northward; the site of coastal sedimentation shifted now to the autochthonous realm, while deep-water Globigerina Marl and Flysch were deposited on top of older coastal sediments of the South and North Helvetic realms (Figure 4.5). The autochthonous region was the last to be dragged under and was thus exposed the longest to erosion, so that more older sediments were removed by erosion before the Eocene transgression. The observation that the gap of pre-Tertiary unconformity increases toward the north, a conclusion made by Alpine geologists of the classic era, was to find a satisfactory explanation half a century later in the modern plate-tectonic theory.

"Wang Transgression"

An unconformity exists not only at the top of the Upper Cretaceous, but also within the Upper Cretaceous, as noted by Kaufmann in 1886 and Arnold Heim in 1908. The Maastrichtian Wang Formation overlies older strata ranging from Late Cretaceous to Middle Jurassic in age (Figure 4.6). Unlike the Tertiary transgressive formation, the Wang Beds are not shallow marine or shoreline deposits, but consist mainly of hemipelagic marls. The microfossils in the marls are mainly planktonic foraminifers, and they belong mainly to genus *Globotruncana*. From studying the relative abundance of swimmers and bottom dwellers, namely the ratio of planktonic to benthic foraminifera, paleontologists have concluded that the hemipelagic sediments rich in planktonic forms were deposits on open marine waters hundreds, if not thousands, of meters deep (see Oberhänsli-Langenegger 1978). This conclusion is supported by the finding that the benthic foraminifers in the marls are deep-sea bottom dwellers. Intercalated in the marls locally are thin turbidite layers or thick breccia beds. Another noteworthy feature of the Late Cretaceous sedimentation on the Helvetic passive margin is the southward thinning of the Upper Cretaceous formations, indicative of a reduced sedimentation rate on far-offshore bottom (Figure 4.6).

The normal pattern of passive-margin sedimentation is characterized by a more complete, and thus thicker, shallow marine sequence on the far side of the margin, where the subsidence rate is greatest. The thickness distribution of the Jurassic and Lower Cretaceous shallow marine deposits of the Helvetic margin obeys this rule (Figure 4.7). Toward the end of the Early Cretaceous (late Aptian/Cenomanian) and the beginning of the Late Cretaceous (Cenomanian), however, the continental margin subsided more rapidly and the rate of sedimentation could not keep pace with subsidence. The carbonate shelf was drowned. First glauconite sands (nodular beds and Turrilites beds in Figure 4.6) were laid down on the shelf, followed by the deposition of pelagic limestones and hemipelagic marls (namely Seewerkalk, Seewerschiefer, Amden Marl, and Wang Formation) on the bottom of the former shelf, which had subsided to become continental slope or continental rise. All of these deep marine deposits are thinner on the outer margin, where the sedimentation rate was slower because of re-

Fig. 4.6. Pre-Wang unconformity (after Bolli 1944). E: Eocene; W: Wang beds; Wb: Wang breccia; A: Amden beds; SS, US, RS, OS: Seewen formations; T: Turrilites beds; KS: upper nodular beds; Fi: Fidersberg beds; K: Lower nodular beds; L: Lochwald bed; Tw: Twirren beds; F: Fluhbrig beds; B: Brisi breccia. Note that Upper Cretaceous formations are thinner in more southerly realms of sedimentation on the outer Helvetic margin, and they are absent altogether in the Ultrahelvetic. The pre-Wang erosion is related to submarine erosion during the Late Cretaceous on continental margin.

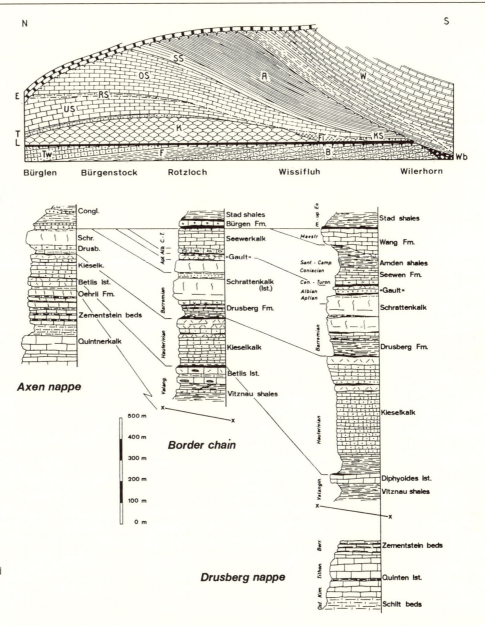

Fig. 4.7. Stratigraphical columnar sections of the Upper Jurassic and Cretaceous formations of the Helvetic nappes around Lake Lucerne (after Trümpy 1980).

duced sediment supply to areas far from shore. This consideration explains why the Upper Cretaceous strata are thinner in higher Helvetic nappes (Figure 4.6). In the highest, the Drusberg nappe of the Wilerhorn region, the Upper Cretaceous is largely missing and the Wang Formation overlies various Lower Cretaceous formations, ranging from Albian to Hauterivian in age (Arn. Heim 1908; Bolli 1944). The hiatus is still more remarkable in the Ultrahelveltic nappes of the *Zone des Cols* region. The term "Ultrahelvetic" denotes that the sediments in those nappes were deposited at the most southerly parts of the Helvetic continental margin, where the Maastrichtian Wang Formation is in direct contact with Upper Jurassic (Malm) or Middle Jurassic (Dogger) formations.

Although the expression "Wang Transgression" has appeared in Swiss geological literature (e.g., Staeger 1944; Badoux 1945), the overlapping deposition of the Wang Beds is not comparable to that of a shoreline transgression. The older sediments were not subjected to subaerial erosion prior to the Wang deposition. Sedimentary breccias, found at the base or within the Wang Formation, are not the coarse clastics piled up on a transgressive beach; they are not "transgressive breccias." My student Heidi Diefen-bach (*Diplomarbeit*, 1988) has presented incontrovertible evidence that the breccia beds are submarine slide deposits, laid down at the foot of a steep continental slope.

The idea that erosional unconformity always represents the work of uplift and sub-aerial erosion is a simplistic explanation. Ocean drilling of the Atlantic margins has shown that an unconformity can also be the work of submarine erosion. On the West African continental margin, for example, Miocene hemipelagic sediments, similar in lithology to those of the Wang Formation, unconformably overlie deep-marine de-posits, which range in age from early Tertiary to Jurassic (Figure 4.8). The outer mar-

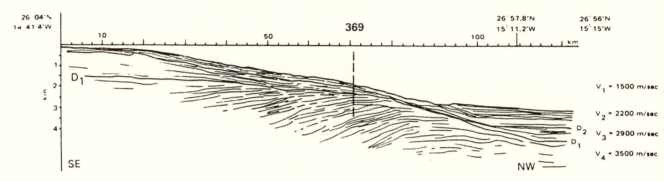

Fig. 4.8. Submarine erosion on the African continental margin (after Lancelot and Seibold 1977). The seismic profiling record shows a hiatus in sedimentation on the passive margin west of the Spanish Sahara. The gap in the record between Lower Cretaceous and Lower Miocene on the continental rise was verified by drilling. The unconformity is now taken as evidence of erosion by bottom currents.

gin has never been subjected to uplift and subaerial erosion since the Jurassic. The pre-Miocene unconformity is submarine and has resulted from submarine erosion caused by the vigorous circulation of bottom currents at the foot of a steep continental slope (Hsü 1982). Those currents, also called contour currents, because they flow par-allel to the contour line, are strong enough to erode previously deposited deep-sea muds, marls, or even limestones. When the current off the West African margin was no longer sufficiently vigorous to prevent erosion, Miocene hemipelagic mud was de-posited on the eroded bottom, overlapping sediments of different ages. Considerations of comparative sedimentology suggest that the Wang Beds were laid down under sim-ilar conditions of submarine erosion.

The over-steepening of the continental slope, a consequence of the submarine ero-sion of continental rise, has led to frequent submarine slumping. Fallen masses of lithified sediments would break as in landslides and produce angular fragments, whereas soft sediments would slide down like snow avalanches. The mixture of the fragments and the mud produces breccia deposits, which have been sampled by deep-sea drilling at sites below submarine escarpments. The breccias of the Wang Forma-tion are thought to have a similar origin.

Palinspastic Reconstruction of the Helvetic Paleogeography

Palinspastic reconstruction is the attempt to determine the relative positions of the rocks in various allochthonous tectonic units prior to their deformation. When Marcel Bertrand established the predominantly northward vergence of the Swiss Alps, the simple rule for palinspastic reconstruction was to place the depositional site of sedi-ments in higher nappes in a more southern position. However, in places, where south-ward thrusting has taken place, the higher tectonic unit has a more northerly origin, and proper consideration has to be given to such exceptions. Yet even if we could recog-nize all the intricacies of the geometry of the internal structure of a mountain belt and could determine the proper displacement direction of each tectonic slice, we would still be stopped if the geological record is woefully incomplete, as it is in the case of tectonic mélanges. This difficulty is well known to anyone who has tried to fit the pieces of a jigsaw puzzle together. If the pieces are complete, or if only a few pieces are missing, a geometrical fitting, difficult as it may be, is ultimately possible by matching the complicated outlines of neighboring pieces. If, however, more than half of the pieces are lost, complete reconstruction is impossible unless a picture has been painted on the pieces before they were cut apart. Assuming that the picture depicts a certain natural object, the fragmentary image on each piece helps us position it approximately. We could in fact reconstruct an unknown picture even if less than half or a quarter of the puzzle pieces were available. This is done in archeology: wall paintings of ancient Crete, for example, are restored on the basis of a few fragments because archeologists are familiar with the theme and patterns of the Cretan arts. Palinspastic reconstructions are worked out similarly, based upon knowledge of the pattern of the sedimentary-facies variation.

As mentioned previously, the geological concept of facies was introduced by Amand Gressly; he wrote the following in 1837:

> I noticed in the rock formations of the Jura regions very variable modifications, both lithological and paleontological, which serve to interrupt the lateral uniformity that has been maintained for different stratigraphic units. There are two types of modifications that I shall call facies (i.e., aspects) of stratigraphic units: one is the lateral modification of lithology and the other that of fauna. (free translation by the author)

These two aspects of lithological and paleontological modifications are today called lithofacies and biofacies, respectively. Lithofacies are rock records of sedimentary environments (Moore 1949). Since different sedimentary environments exist at any given time, the sedimentary deposits of a given time are manifested by different facies. The Middle Eocene of the autochthon and of the lower Helvetic nappes is, for example, represented mainly by a facies of sands that were deposited on beaches or shallow marine sand shoals. The rocks of this age in the higher Helvetic nappes are nummulitic limestone, the sediments on shallow marine carbonate banks. Coeval sediments of still higher nappes are mainly hemipelagic muds and turbidites. Applying our knowledge of the actualistic change of sedimentary environments from sand beaches to shallow marine shoals and carbonate banks, and that of mud deposition on continental slopes and continental rises, we can conclude that the sediments of the lower Helvetic nappes were originally laid down in the more northerly parts of the Helvetic realm of sedimentation, and we therefore use the term North Helvetic to designate this paleogeographical province. Likewise, the conclusion was reached that the sediments of the upper Helvetic nappes were laid down in the South Helvetic paleogeographical province, and the Ultrahelvetic nappes in the Ultrahelvetic province to the south of the Helvetic. This palinspastic reconstruction on the basis of lateral facies variation is not only in agreement with tectonic reconstructions showing that higher nappes came from more southerly sites, but also verified by the story of two unconformities: The sea inundating the Helvetic margin lay to the south of Eocene Europe. The Jura region and the Swiss Midland were dry land during most of the Middle Eocene age. The coast lay somewhere within the autochthonous or the North Helvetic realm. The South Helvetic carbonate banks of the Eocene were situated offshore, and the Ultrahelvetic was then the realm of sedimentation on continental slopes and continental rises (Figure 4.5).

Alpine geologists have been making palinspastic reconstructions since the beginning of the century. Arnold Heim (1908), for example, was not deceived by the apparently similar positions of the Wageten and Aubrig thrust slices in the Wägital. Both were displaced to the Alpine front, but their internal stratigraphies are different. Heim was able to recognize, on the basis of the Mesozoic/Tertiary sedimentary facies, that the Wageten is a slice sheared off from the parautochthon, and that the Aubrig, an element in the Border Chain position (see Figure 4.9), represents one of the highest Helvetic nappes.

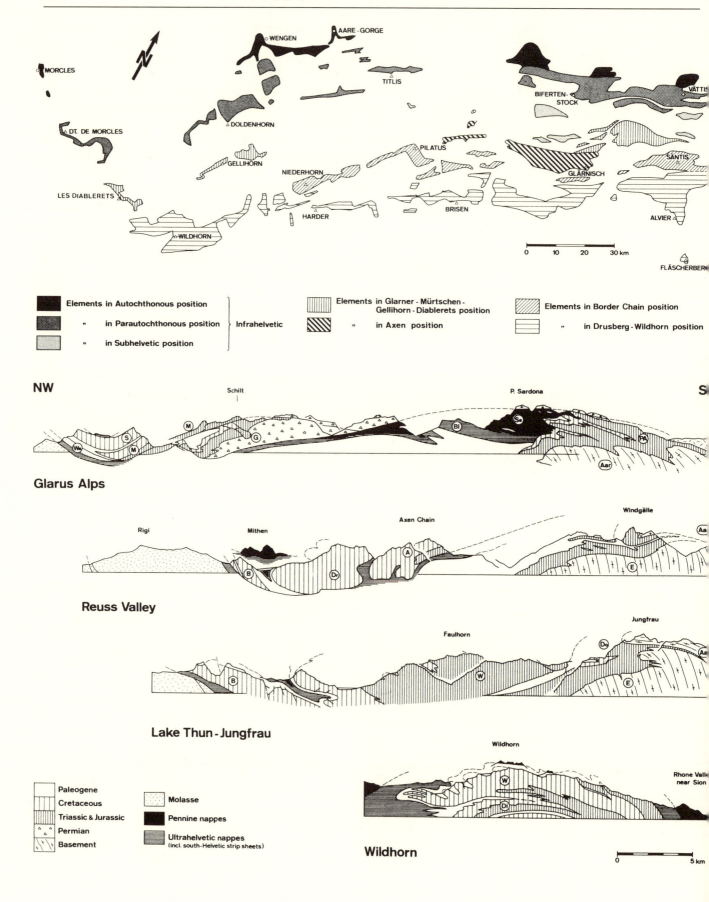

A palinspastic synthesis of the Helvetic paleogeography of eastern and central Switzerland has been presented by Trümpy (1969). The Aare Gorge was cut into, and Wengen is a village built on, the autochthon. The sediments of the South Helvetic realm, according to this reconstruction, originally lay 60 km south of the Aar massif (Figure 4.9). This sedimentary cover was sheared off and thrust forward over and across the autochthon, and is now found as the deformed strata of the Drusberg and Wildhorn nappes and of the Border Chain (Figure 4.10). Pilatus lies now north of the Aare Gorge, and Säntis north of Vättis. Their positions on the map show their paleogeographical positions prior to the 50–100 km displacements during the Alpine deformation. Trümpy's 1969 synthesis symbolizes the crowning achievement of the Alpine geologists of the last century.

ON OPPOSITE PAGE:

Fig. 4.9 (*above*). Palinspastic reconstruction of the Helvetic paleogeography (after Trümpy 1969).

Fig. 4.10 (*below*). Schematic structural cross-sectional profiles across the Helvetic Alps (after Trümpy 1980). A: Axen; B: border chains; Bl: Blattengrat; Di: Diablerets; Do: Doldenhorn/ Morcles; Dr: Drusberg; E: Erstfeld; G: Glarus nappe; M: Mürtschen nappe; PA: parautochthonous; S: Säntis nappe; Sa: Sardona; W: Wildhorn nappe; Wa: Wageten.

5.

Flysch and Wildflysch

Some words, such as truth, democracy, etc., are used so often that they begin to lose their exact meaning. Truth exists because we can recognize falsehood, and democracy is appreciated by people who suffer under tyranny. The word *Flysch* is such a well-used word; in fact, it was misused so much that Berhard Studer, who invented the word in 1827 and recognized that it was poorly defined, proposed as early as 1848 that it should eventually disappear from the geological literature. Nevertheless, the term Flysch is too useful to be dropped, and the concept of Flysch has been packaged for global export, as Flysch formations are found in all orogenic belts of the world. Flysch does exist.

The Meaning of the Word Flysch

The term Flysch was introduced by Bernhard Studer to describe an informal stratigraphic unit of dark gray shale and intercalated sandstone that overlies an Upper Jurassic formation in the Simmen Valley, Canton Bern. Later he wrote that Flysch in local dialect means a shaly rock (Studer 1872). Although the term was originally intended as a "petrographic" or lithological term, we have to remember that such terms were often used during the nineteenth century to designate stratigraphic units, and some, such as chalk, green sand, coal measure, etc., are still with us. In fact, Flysch was considered a formation by Studer in 1827, when he wrote:

> A sequence of interbedded gray sandstone and dark gray shale is present, parallel to the Alpine trend, in an outcrop belt that extends from the Simmen Valley between Erlenbach and Zweisimmen, to Hundsrücken and the Valley of Ablantschen, the Saanenmöser, and back to the Simmen Valley between Rougement and Château-d'Oex, and thence to the Valley of Mosses up to the village of Sépey. It is convenient to use the name Flysch to designate the rocks of this formation. (free translation by the author)

The name Flysch has been adopted to designate thick sequences of interbedded sandstone and shale elsewhere, and this widespread adoption can be traced to two reasons: (1) Flysch-like formations are widespread in the Alps, the Apennines, and the Carpathians, and as they are very similar in certain lithological, biological, and tectonic aspects, even today they could still be lumped together as a super group of formations of somewhat different ages. (2) The absence of megafossils, which provided the only criterion for dating during the nineteenth century, prevented the distinction of one Flysch formation from another, so all shared the same designation.

As long as Flysch was not assigned a very specific stratigraphical significance, there was no confusion. The first crisis came when Studer was being persuaded by Arnold Escher that the Alpine Flysch was a not a "petrographic" (lithostratigraphic) unit, but a formation (chronostratigraphic) of a certain age. Studer (1848) wrote that he went with Escher during the autumn of 1833 to the mountains of Entlebuch, where they found rocks, apparently identical to the Flysch of the Simmental, resting on top of the Nummulitic Limestone. Escher assumed a normal superposition, and this opinion was reinforced by Murchison (1850); the Flysch was, therefore, dated Late Eocene. Studer used the term Flysch to designate Upper Eocene formations in the Alps in his monograph *Geologie der Schweiz*, published in 1853.

The notion of an Upper Eocene Flysch formation in the Swiss Alps persisted until 1908. The hypothesis involved the assumption of normal superposition of the Flysch above the Helvetic Nummulitic Limestone. With the introduction of the theory of nappe tectonics by Marcel Bertrand in 1884, the assumption was no longer self-evident. In 1898 Schardt noted the presence of exotic elements of Mesozoic age in the

Wildflysch of central Switzerland. Shortly thereafter, Cretaceous *Inoceramus* was identified in a calcareous shale (*Leimerschichten*) of the Wildflysch (cited in Beck 1911). This development led Buxtorf in 1908 to postulate that the Habkern Wildflysch and the Schlieren Flysch of central Switzerland had been emplaced tectonically above the Helvetic Nummulitic Limestone.

The breakthrough in Flysch tectonics was achieved by a specialist in Eocene microfossils. Jean Boussac's work with foraminifers in the tectonically little disturbed Paris Basin established the basis for micropaleontological dating of various Flysch formations. His friend, Arnold Heim, quickly realized that a large part of the Alpine Flysch above the Helvetic Nummulitic Limestone is older, and must therefore belong to a higher tectonic element. The overthrust contact is not clearly visible, because the Upper Eocene Globigerina Marl is lithologically indistinguishable from the marly shale of the Flysch, but the tectonic superposition can be determined on the basis of stratigraphical paleontology. Heim developed his theory of nappe-enveloping (*Deckeneinwicklung*) in 1911. The theory recognized the fact that the recumbent folds of Helvetic strata are enveloped by the Flysch, which seemed to be a stratigraphically younger unit. In fact, the Flysch is older, and it must have been emplaced as a nearly flat thrust sheet above the Helvetic sediments. Referring to the model of northward vergence, Heim postulated that the Flysch was deposited in an open marine realm, called Ultrahelvetic because the site lay farther south, beyond the carbonate platforms of the Helvetic continental margin. The Flysch nappes were likewise called Ultrahelvetic. The Helvetic strata and the tectonically superposed Ultrahelvetic Flysch were later folded together and overthrust to form the nappes of the High Limestone Alps.

The complexity of Flysch geology was revealed as a result of advances in Swiss geology during the first half of this century, when micropaleontological investigations led to the recognition of Flysch formations with ages ranging from Early Cretaceous to Oligocene and with vastly different tectonic and paleogeographic settings of their depositional environment. In keeping with Studer's original intention, Flysch became the lithological designation in a binominal stratigraphical nomenclature. In most cases it is preceded by a geographical name, as in Niesen Flysch, Gurnigel Flysch, Schlieren Flysch, Sardona Flysch, Wägital Flysch, Prättigau Flysch, Vorarlberg Flysch, etc.; in others it is modified by the tectonic position, as in Subalpine Flysch, North Helvetic Flysch, Klippen Flysch, Simmen Flysch, etc. A few are characterized by some distinct feature, such as Wildflysch, Helminthoid Flysch, etc. Some Flysch formations are designated sandstone (Taveyannaz Sandstone, Altdorf Sandstone, etc.) because shaly interbeds are rare or absent. Yet all of these formations are thick, mainly detrital, marine sequences in the Alps—the one feature common to all Flysch formations.

Flysch as a Recurrent Facies

Flysch is a very distinct sedimentary facies. Although Studer originally intended to use the term as a synonym for gray shales (*Graue Schiefer*), thick shale formations in the Helvetic Alps, such as those of Middle Jurassic (Aalenian), Early Cretaceous (Valanginian Marl), or Eocene (Globigerina Marl), are not considered Flysch, nor are the thick shale formations (e.g., Bündnerschiefer, *schistes lustrés*) of the Pennine Alps. As Sujkowski (1957) pointed out to geologists outside of Switzerland, "the name Flysch is a facies denomination for a marine deposit composed of innumerable alternations of sharply divided pelitic and psammitic layers." The interbeds can be thin or thick, but they are characterized by the sharpness of the lower boundary, and the constancy of the bed thickness as far as one can see at an outcrop. This feature of the Flysch is so distinctive that one has no difficulty in identifying a Flysch formation, even when driving past a roadcut at a speed of 100 km per hour. The only exception to this rule is Wildflysch. A former student, Andreas Bayer (1982), studied its genesis and found that Wildflysch is commonly not a sedimentary formation, but a tectonic mixture of various exotic blocks in a Flysch or a shaly matrix.

When one takes a closer look at the Flysch in an outcrop, one finds numerous distinctive features in addition to the regular bedding. Graded bedding is one of the most characteristic, and bottom markings indicative of current erosion are very common on the base of sandstone or siltstone interbeds. The Flysch shales are as a rule hemipelagic, but the coarser-grained interbeds are turbidity-current deposits, or turbidites. Turbidite beds are so common in Flysch formations that some students consider the two synonymous. In fact, the term turbidite describes the mode of origin of the bed, namely a bed deposited by a turbidity current, whereas Flysch designates a formation, or a tectono-stratigraphic facies.

A turbidity current is an underwater current; it flows because the current is a suspension mixture of detritus and water, having a density greater than that of the ambient fluid. A dust cloud is a suspension current; the suspension of dust and air has a density greater than that of air. Turbidity currents are not avalanches, but suspension currents generated by avalanches. The mass of snow in an avalanche moves under the weight of solid snow, but the suspension cloud of mixed snow and air moves under the weight of the greater density of the suspension. Submarine avalanches deposit coarse breccias at the base of continental slope; the Wang breccia discussed in the last chapter is one such example. Fine detritus from an avalanche could, however, mix with ambient water to form a submarine turbidity current, which could transport sand and silt detritus to abyssal plains dozens or hundreds of kilometers distant.

The conclusion, based mainly upon sedimentological interpretations, that the coarser interbeds in Flysch formations are deep-sea turbidites has been verified by other lines of evidence, such as paleontology and tectonics. Deep-sea deposits are present, of course, not only in the Alps, but also elsewhere. Formations displaying features characteristic of the Alpine Flysch are common in other mountain belts. There is, therefore, a need to use the term flysch to describe a recurrent facies, namely, a tectono-stratigraphic facies that appears again and again in the geologic record. I suggest, therefore, that the term flysch be used (with lower-case f) to designate such recurrent facies (Hsü 1970, p. 9); the flysch facies is a sequence of thick marine shales with alternating sandstone and/or some impure limestone layers, which constitute a well-bedded sequence in an Alpine-type mountain chain with a tectonic setting, and sedimentological features similar to those of the Alpine Flysch in their more typical development.

North Helvetic Flysch

The Flysch masses of the northern Helvetic belt attain a thickness of some 2000 m. They range from Late Eocene to Early Oligocene in age, and in places the Flysch has a lithology transitional to that of the overlying Molasse. They crop out mostly in parautochthonous or autochthonous areas.

The oldest North Helvetic Flysch consists mainly of sandstone—Taveyannaz Sandstone. The formation is named after its type locality at Alp Taveyannaz in the Diablerets of western Switzerland, where this Upper Eocene (Priabonian) formation is folded in the Diablerets nappe. In neighboring areas of the Lower Valais, the Taveyannaz Sandstone is identified in the parautochthonous Morcles nappe. The site of Taveyannaz deposition extends into the parautochthon and autochthon eastward. This unit is very interesting because of the very high content of volcanic detritus (some of the clasts measure up to 10 cm in diameter), and is notable for its green color in all the outcropping areas between the Rhône and the Rhine.

We mentioned in the last chapter that the Tertiary sedimentation in this region began with Middle Eocene shallow marine sedimentation (Nummulitic Limestone and glauconite sand). The shelf subsided rapidly during the Late Eocene, when the Globigerina Marl was deposited. The European plate was underthrust, and the North Helvetic belt became a foredeep during the latest Eocene and Early Oligocene (Figure 5.1). Detritus

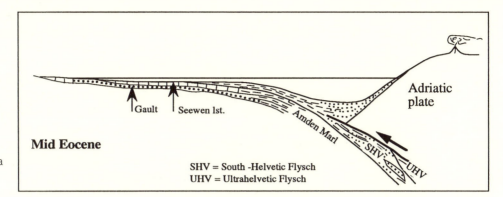

Mid Eocene

Gault   Seewen lst.

Amden Marl

SHV = South -Helvetic Flysch
UHV = Ultrahelvetic Flysch

Adriatic plate

SHV

UHV

Fig. 5.1. Flysch sedimentation in a foredeep on active continental margin.

from southern highland accumulated first on mountainous coast, before being resedimented as a sand avalanche or as turbidite on the deep-sea bottom.

In the classical theory of "embryonic tectonics" (Argand 1916), Flysch was considered a foredeep deposit. Schardt (1898) and Arbenz (1919) referred to Flysch as the orogenic sediments deposited in front of rising and advancing nappes. Sedimentological evidence indicates a southerly source supplying detritus to the North Helvetic foredeep (e.g., Radomski 1961). While the existence of a coast range south of the foredeep seems unquestionable, we have not been able to find in the Alps the terrane supplying the andesitic detritus. None of the tectonic units above the lower Helvetic nappes include thick volcanic formations. The very complex tectonics and paleogeography of the Eocene Flysch cannot be interpreted out of context, without reference to the regional framework. We shall, therefore, leave the discussion now and return to this question of Flysch paleogeography in the last section of this chapter.

The Taveyannaz Sandstone is overlain by Oligocene turbidite formations, called the Val d'Illiez Formation, Altdorf Sandstone, and the Elm Formation in western, central, and eastern Switzerland, respectively (Cadisch 1953). They are overlain by shale formations. In eastern Switzerland the Roofing Slate (*Dachschiefer*) of Glarus is characterized by its fish fauna of Oligocene age (Oberholzer 1933). In western Switzerland, the shale of the Val d'Illiez grades upward into the Red Molasse (*molasse rouge*). The trend of decreasing size of the detritus in these formations suggests that the submarine slope became less steep as the basin was shoaled by sedimentary accumulation during the Early Oligocene. The shallow marine sediments above the sandstone and shale in the Val d'Illiez are no longer considered Flysch, but are grouped with the Lower Marine Molasse.

South Helvetic Flysch

The term South Helvetic Flysch was first applied by Herb in 1962 to describe an interbedded turbidite sandstone and shale formation stratigraphically above the Globigerina Marl in the Säntis nappe of the Amden district. The marl is middle to late Middle Eocene in age, and the Flysch is thus Late Eocene.

The so-called Einsiedeln Flysch, Blattengrat Flysch, and Ragaz Flysch are not flysch formations, but schuppen zones. The Globigerina Marl, Nummulitic Limestone, and Amden Marl of these zones are not Flysch, but the Blattengrat Sandstone, or the turbidite sandstone and shale interbeds above the Globigerina Marl, in the Blattengrat and Einsiedeln schuppen zones is South Helvetic Flysch (Leupold 1966; Bayer 1982). Other occurrences in central Switzerland have been described by Bayer; they include the South Helvetic Flysch of the Lake Lucerne region (Lielibach type) and that between Engelberg and the Lake of Thun (Südelbach type).

In addition to turbidite sandstone, mud breccia and polygenic breccia are common in the South Helvetic Flysch. Mud breccia is formed by submarine slumping. The mud

clast consists of somewhat consolidated Globigerina Marl, indicating its derivation
from a Helvetic realm of sedimentation. The polygenic breccia consists of lithothamnian (a red alga) debris, granite pebbles, arkosic (quartz plus feldspar) sand, as well as
displaced foraminiferal tests and glauconite grains. The provenance of the materials is
Helvetic. The facies distribution of the Upper Eocene Helvetic deposits suggests, however, that the northern slope of the Flysch basin was covered by a thick sedimentary
sequence. The granite debris could have been derived only from a steep southern shore.
Bayer (1982) concluded therefore that a "swell" or an island chain must have existed
south of the Helvetic realm of sedimentation. The idea of foredeep sedimentation is
thus equally applicable to explain the genesis of the South and the North Helvetic
Flysch formations, even though the highland beyond the South Helvetic Flysch basin
was underlain by granite basement, and that south of the North Helvetic Flysch was a
terrane of active andesitic volcanism.

In many places between the Aare and the Rhine, the South Helvetic Flysch is the
youngest formation in the highest Helvetic nappes. This Flysch formation was commonly sheared off under the Penninic nappes during the Alpine deformation. Flysch
sandstone and shale occur as exotic slabs in a shaly or marly matrix containing Middle
and Upper Eocene globigerina faunas. The South Helvetic Flysch was thus mapped as
an Upper Eocene Wildflysch formation. Bayer, in his 1982 dissertation, did much to
clarify the geology of Wildflysch. He pointed out that this unit between the Aare and
the Rhine is commonly not a sedimentary formation but a tectonic unit. The Wildflysch zone may be up to 1000 m thick, but each unit consists of imbricate slices
derived mainly from rocks of the Eocene South Helvetic zone, i.e., Globigerina Marl
and South Helvetic Flysch. Also introduced into the Wildflysch by tectonic mixing are
rare slabs of Nummulitic Limestone, Amden Marl, or Wang Formation of South
Helvetic origin, and blocks of Habkern granite and slices of Leimern marl and limestone derived from the Penninic realm. Since Wildflysch is not a valid lithostratigraphic unit, Bayer proposed the term Habkern Mélange to designate this type of tectonic mixture in central Switzerland.

Bayer's interpretation shows that the tectonics of the Habkern Mélange and the
Einsiedeln schuppen zone are comparable. Both consist of imbricated slices. The former is dominated by exotic blocks of South Helvetic Flysch in a Globigerina Marl
matrix; the latter consists mainly of Nummulitic Limestone and Amden Marl. The
tectonic mixing took place in shear zones under major nappes.

"Ultrahelvetic Flysch"

The name Ultrahelvetic Flysch has been used to designate many Flysch formations
that have been overthrust above the highest Helvetic nappes. Some are tectonic mixtures or Wildflysch, such as the Habkern Mélange; others are Flysch formations of
uncertain paleogeographic origin. All such Flysch that have been called Ultrahelvetic
in the geological literature will be referred to as "Ultrahelvetic Flysch" in this text for
the sake of convenience. However, the term Ultrahelvetic Flysch, *sensu stricto*, will be
used to designate only those Flysch formations whose original site of deposition has
been reliably traced to an area between the South Helvetic and North Penninic realms
(Homewood 1977). "Ultrahelvetic Flysch" of uncertain origin will be referred to by its
local designation—for example, Sardona Flysch, Schlieren Flysch, etc.

Ultrahelvetic Flysch, sensu stricto

"Ultrahelvetic Flysch" formations commonly occur as tectonic slices in the shear zone
between the main Helvetic and Penninic nappes. The Mesozoic rock slabs found in
adjacent tectonic slices could be identified as South Helvetic or Ultrahelvetic (Tercier
1925; Badoux 1945; Guillaume 1957). Some "Ultrahelvetic Flysch" formations include sediments whose facies and age are such that, according to our present knowl-

edge, they may not, or could not, have been deposited in the Ultrahelvetic realm. Those now excluded from the Ultrahelvetic (*sensu stricto*) are, for example, Gurnigel Flysch, Schlieren Flysch, Wägital Flysch, and "Sardona Flysch" (Caron 1976; Homewood 1977).

The Ultrahelvetic Flysch (*sensu stricto*) includes the Meilleret Flysch of the Zone des Cols, Adelboden Flysch, and Leissigen Flysch of south of the Lake of Thun; they are all involved in the deformation of the nappes of the Internal Prealps. They have been designated "Ultrahelvetic Flysch of the Schlieren-Gurnigel type" because of their similarity to those sandstone flysch formations of the External Prealps. Their age is distinctly different: the Schlieren and Gurnigel are older, ranging from Maastrichtian to early Middle Eocene, while the Ultrahelvetic Flysch, *sensu stricto*, is Middle Eocene (Homewood 1977; Bayer 1982), intermediate in age between the Schlieren and the South Helvetic.

The sandstone interbeds of the Ultrahelvetic Flysch (*sensu stricto*), consisting of detritus derived from submarine and subaerial source areas, were deposited in a deep-sea basin by turbidity currents (Hsü 1960; Homewood 1977). The more distal turbidite beds are intercalated in the Globigerina Marl of the South Helvetic realm. The facies distribution of Middle Eocene sediments indicates that the Ultrahelvetic Flysch, (*sensu stricto*) was the deposit of a marginal sea south of Europe. On the north side was the Helvetic passive continental margin, with a broad continental shelf and a gentle continental slope; on the south side was a mountainous island that supplied detritus for Flysch turbidite sedimentation. This southern elevation has been called "Marginal Basement High" by Homewood (1977) and Bayer (1982). The pattern of Ultrahelvetic, (*sensu stricto*) sedimentation during the Middle Eocene is thus similar to that taking place during the Late Eocene in the South Helvetic realm, and during the Oligocene in the North Helvetic realm (see Figure 4.5).

"Sardona Flysch" or Sardona Mélange

In the early days of Swiss geology, when a normal superposition of the Flysch above the Helvetic formations was assumed, all the rocks of the flysch facies in Canton Glarus were called Glarus Flysch. Field work by J. Oberholzer led Arnold Heim in 1908 to propose a threefold subdivision of this unit (Figure 5.2). Since the work of Boussac in 1909, it has been known that only the lower part of the Glarus Flysch is normally superposed: The Matt-Engi Sandstone/Dachschiefer complex of Leupold (1966) is North Helvetic Flysch, which overlies conformably the older Tertiary strata of the autochthonous region. The middle part of the Glarus Flysch, or the "Blattengrat Beds" of Leupold (1966), is the Blattengrat schuppen zone. The upper part, the "Wildflysch" of Heim (1911) or the "Sardona Flysch" of Leupold (1966), is a shear zone of tectonically mixed rocks, including a sandstone formation that could be considered Ultrahelvetic (*sensu stricto*). Following the practice suggested by Bayer (1982) that such tectonic mixtures be called mélange, the "Sardona Flysch" will be referred to as Sardona Mélange in this text (see Table 5.1).

The "Sardona Flysch" was investigated by Leupold (1937, 1942) and several of his

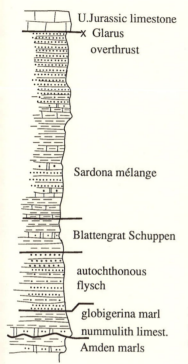

U.Jurassic limestone
Glarus overthrust

Sardona mélange

Blattengrat Schuppen

autochthonous flysch

globigerina marl
nummulith limest.
Amden marls

Fig. 5.2. The threefold division of the "Glarus Flysch." The "Glarus Flysch" is not a formation, but a pile of thrust slices and mélanges. The three tectonic units are, in descending order, Sardona Mélange, Blattengrat schuppen, and autochthonous Flysch.

Table 5.1

Evolution of Flysch Terminology in Glarus

Heim 1908	Leupold 1966	Bayer 1982 and This Text
Upper Glarus Flysch	"Sardona Flysch"	Sardona Mélange incl. Ultrahelvetic Flysch (*sensu stricto*)
Middle Glarus Flysch	"Blattengrat Beds"	Blattengrat Schuppen incl. South Helvetic Flysch
Lower Glarus Flysch	Sandstein/Dachschiefer	North Helvetic Flysch Matt-Engi Sandstone

students. Even though the stratal superposition was completely disrupted by shearing during overthrusting, the components can be dated by fossils. The tectonic slices could be identified as belonging to four major groups of stratigraphic units; they are as follows:

(1) Flysch formations (Mörderhorn Beds, Plattenkalk Sandstone, Upper Sandstone Flysch). These Upper Paleocene to lower Middle Eocene Flysch have been dated by displaced nummulite faunas correlative to those found in the Schlieren Flysch of central Switzerland (Wegmann 1962). The Flysch formations at Sardona are apparently the facies equivalent of, but are older than, the Ultrahelvetic Flysch (*sensu stricto*) and the South Helvetic Flysch. This difference in age is a manifestation that the outer, or more southern, continental margin of Europe subsided earlier (see Figure 4.5). The age of Flysch sedimentation varies systematically from the oldest in the more internal, or southerly, position to the youngest in the autochthonous region. The older Flysch in the Sardona Mélange suggests that this unit occupies a higher tectonic position than the Ultrahelvetic Flysch (*sensu stricto*) of central and western Switzerland.

(2) Quartz sandstones (*Oelquarzit, Sardona Quarzit*). These sediments are Paleocene and are apparently the facies equivalent of, and are older than, the Lower Eocene Assilina Sand of the higher Helvetic nappes. This fact reinforces the postulate of an earlier subsidence date of the outer European margin, resulting in a progressively northward transgression of the Helvetic realm during the Early Tertiary.

(3) Black shales and "*Sideroliteskomplex.*" The uppermost Cretaceous (Maastrichtian) elements at Sardona include black shales and thin-bedded calcareous sandstone, as well as thick, coarse sandstone and breccia containing clasts of sedimentary and granitic rocks. These Maastrichtian sediments are the lateral equivalent of the Wang Formation in the highest Helvetic nappes. The presence of turbidite beds of this age at Sardona suggests a depositional site farther down the continental slope toward the abyssal plain. The fact that the time span represented by the unconformity between the Tertiary and the Cretaceous is smaller here than in the Helvetic realm reinforces the postulate of a more internal position (than the Helvetic) for the deposition of Sardona sediments.

(4) Hemipelagic marls and pelagic limestones. These Upper Cretaceous sediments are dated by planktonic foraminifers as ranging from Campanian to Cenomanian in age. The Sardona deposits are on the whole more calcareous than the coeval Amden Marl, Seewerschiefer, and Seewerkalk in the highest Helvetic nappes. This decrease in terrigenous content could be explained by a position more distant from shore, namely, a position more internal than the Helvetic.

In discussing the pre-Wang unconformity, we mentioned that the time gap is largest in the Ultrahelvetic in western Switzerland. Assuming uplift and erosion prior to Wang deposition, Leupold (1937, 1942) suggested that there was a "swell" ("South Helvetic Swell") between the Helvetic and Ultrahelvetic. We have adopted the modern concept of submarine erosion to explain the discordance below the Wang Formation, and conclude that Leupold's "South Helvetic Swell" did not exist. Contour currents for submarine erosion are most active on the continental rise at the foot of a steep continental slope, where the greatest hiatus in sedimentary record can be expected. A deep-sea sedimentary sequence farther away from the continental margin (in the direction of the abyssal plain) should thus be again more complete. The absence of pre-Wang erosion suggests that the depositional site of the Sardona hemipelagic sediments was located south of the Ultrahelvetic (*sensu stricto*) realm, in general agreement with the interpretations based on other lines of evidence.

The heterogeneous nature of the rocks in the Sardona Mélange has long been recognized. Arnold Heim (1911) and Oberholzer (1933) used the term "Wildflysch" because of the presence of numerous Cretaceous limestone beds, apparently intercalated in the Flysch, and they recognized their similarity to the Seewerkalk Formation, and also to the Leimern limestone and marl in the "Wildflysch" (Habkern Mélange) of central Switzerland. The results of investigations by Wegmann (1962) and Leupold (1966) indicate, however, that the various components of the Sardona Mélange could all have been derived from an Ultrahelvetic realm of sedimentation.

Arnold Heim (1911) postulated a "Wildflysch nappe" as one giant nappe, and assumed that the nappe was displaced in the early Tertiary, northward from its original Ultrahelvetic realm of deposition, and overthrust above the Helvetic strata, before both were folded and thrust to form the Helvetic nappe complex of the High Calcareous Alps. This is his theory of nappe envelopment (*Deckeneinwicklung*).

The Wildflysch of Central Switzerland (Habkern Mélange)

The presence of lenticular masses of pelagic limestone and marl in the Wildflysch was first described by Kaufmann (1886), who named the rocks after their type locality at Leimern. The pelagic sediments are intimately associated with Upper and/or Middle Eocene Flysch (Leupold 1966; Bayer 1982). Beck (1918) postulated, after the Cretaceous age of the limestone and marl was recognized, that the Wildflysch constitutes a tectonic mixture (Figure 5.3). This opinion was reinforced by Buxtorf (1918), who wrote, "Light gray, lenticular Leimern beds of Late Cretaceous age are embedded in dark gray shale to constitute the Wildflysch. We have to conclude that the Wildflysch is a tectonic mixture of Cretaceous exotics in an Eocene shale matrix, unless we could prove that the shale is also Cretaceous." The Eocene age of the Flysch is now certain, and Bayer confirmed, in his 1982 dissertation, the foresight of Beck and Buxtorf: the Wildflysch of central Switzerland, like that of the east, is largely a tectonic mélange (Figure 5.4). Bayer proposed the term Habkern Mélange in place of Wildflysch, to emphasize that it is a tectonic and not a lithostratigraphic unit.

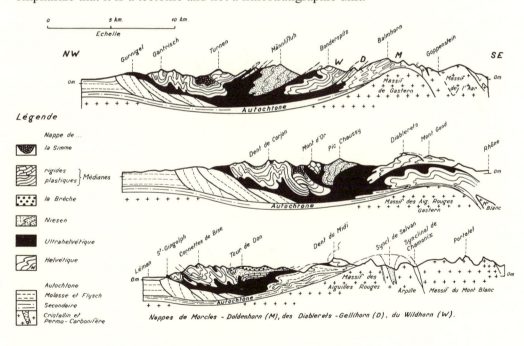

Fig. 5.3. Structural sections across the Prealps of western Switzerland (modified after Badoux 1967). Note the envelopment of Helvetic nappes by Ultrahelvetic mélanges and schuppen zones.

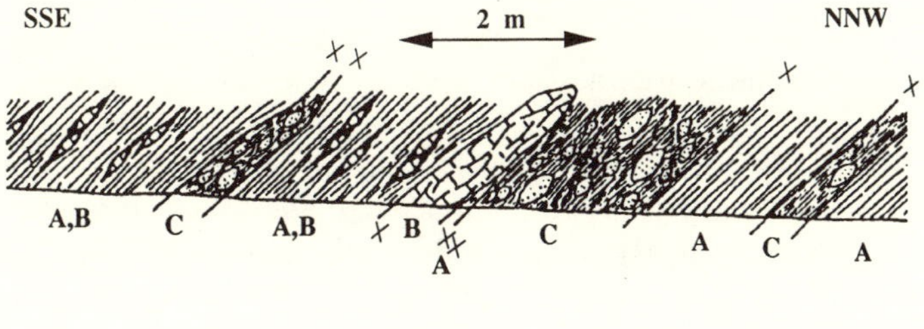

Fig. 5.4. Sörenberg Mélange (after Bayer 1982). This mélange is a tectonic mixture of (A) globigerina marl, (B) Leimern limestone and marl, and (C) Schlieren Flysch.

The different origins of the exotic blocks in the Habkern permit its subdivision. Bayer described three such mélange units in central Switzerland: (1) Leimern Mélange, (2) Sörenberg Mélange, and (3) Iberg Mélange.

The Leimern beds consist mainly of pelagic sediments in a sheared shaly matrix, with some exotic slabs of Flysch sandstone. The original sequence, prior to its tectonic fragmentation and mixing, ranges in age from Cenomanian to Paleocene. Such a sequence, as we shall discuss later, is known from the Penninic nappe of the Median Prealps, and this fact led Bayer to postulate a South Penninic origin for the Leimern beds and for the Schlieren Flysch, which has been thrust above the Leimern Mélange. It has been recognized, however, that the Upper Cretaceous sediments of the Ultrahelvetic (*sensu stricto*) are also largely pelagic. Lenticular layers of limestone and marl are common in the Sardona Mélange, and those pelagic sediments, as we have just discussed, are believed to have been derived from an Ultrahelvetic site of deposition. A pelagic sequence ranging from Cenomanian to Paleocene has also been described from the Ultrahelvetic of western Austria (Richter 1957). I prefer, therefore, to adhere to conventional wisdom and place the Leimern in the Ultrahelvetic or North Penninic, because I find no suitable position in the South Penninic realm that could accommodate the sedimentation sites for the Leimern beds and for the Schlieren. The Leimern Mélange, in my opinion, thus has an origin similar to that of the Sardona Mélange, except that the slabs of Flysch sandstone in the Leimern are derived from the Helvetic and not from the Ultrahelvetic realm of deposition. In 1968 I proposed that only those tectonically sheared mixtures that include exotic elements are mélanges, *sensu stricto*, and those that contain no exotics could be called broken formations. Strictly speaking, therefore, the Leimern is thus a true mélange, whereas the Sardona is a broken formation.

In addition to the Flysch and Leimern limestone fragments, the Wildflysch of central Switzerland may have other exotic elements. The Middle Eocene Globigerina Marl is commonly present as the matrix, whereas granite and Paleocene Schlieren Flysch are present locally as exotic slabs or blocks. One such mélange, lying directly under the Schlieren Flysch nappe, is well exposed in the vicinity of Sörenberg, and is called by Bayer the Sörenberg Mélange.

The origin of the granite, also known as Habkern granite, is a most puzzling aspect of Swiss geology. The pink granite occurs as exotic blocks; the largest found near Habkern has a volume of 13,000 m^3. It has a radiometric age of 270 Ma, and is thus commonly referred to as Hercynian, signifying its origin during that late Paleozoic mountain-building. The Habkern is, however, sufficiently different from other Hercynian granites of the Swiss Alps that the home of the exotics cannot be definitely identified. Debris of Habkern granite is common in the Schlieren Flysch, suggesting that exotic granite blocks may have been deposited by submarine slides before they were tectonically mixed in the mélange.

The Iberg Mélange, cropping out in Canton Schwyz, underlies mainly the nappe of the Median Prealps (Klippen Decke), and consists of a tectonic mixture of the Klippen blocks and Helvetic Flysch.

Schlieren Flysch

The Flysch formations exposed on the drainage divide between Kleinemme and Sarner Aa were called Schlieren Flysch by Kaufmann in 1886, after its type locality in the valley of the Schlierenbach west of Alpnach in Obwalden. Schaub's studies of the displaced foraminifers in the turbidite sandstone beds have established the age of this 1500-m-plus thick sequence as Maastrichtian to Early Eocene (late Yprésian).

The Schlieren Flysch displays features typical of the sediments of the flysch facies. The sequence is an alternation of even-bedded sandstones and shale. Graded bedding is very clearly manifested by the sandstones, especially the coarse-grained variety in the Lower Schlieren. The arkosic detritus and the displayed large foraminifers suggest a source area of granite terrane fringed by shallow marine sand beaches and shoals.

The undersurface of the sandstone beds is well exposed in the Schlierenbach section, and bottom markings indicating the direction of current transport are well exposed. The preferred orientation of the paleocurrents ranges from SW-NE to W-E, and is thus more or less parallel to the Alpine trend of central Switzerland. Based upon the assumption that the Schlieren sediments were deposited in a WSW-ENE trending trough, the currents must have been flowing along the axis of the trough, or with longitudinal transport when the Schlieren turbidite layers were laid down (Trümpy 1960).

Hsü (1960) postulated a southern source for the Schlieren, and his paleogeographical reconstruction indicates that the Schlieren was deposited in the same setting as the Helvetic and Ultrahelvetic Flysch (*sensu stricto*), namely in a deep-sea marginal basin beyond the Helvetic shelf. Consequently, we could consider the Schlieren Flysch Ultrahelvetic, at least *sensu lato*, if not *sensu stricto*. Such an interpretation correlates the Schlieren with the Paleocene/Lower Eocene Flysch of the Sardona Mélange. The source area of the Habkern granite, also called Habkern Island (Hsü 1960), is the Paleocene equivalent of the Eocene Marginal Basement High of Homewood (1977).

Winkler (1981) recognized the existence of two detrital populations within the Schlieren Flysch. He postulated a westerly source for the pink granite debris, and a southerly source for the andesite-tonalite detritus. In his reconstruction, the southerly highland was a part of a volcanic island arc.

Bayer (1982), following Caron (1972) and Homewood (1977), assumed a South Penninic origin of the Leimern sediments in the Habkern Wildflysch, and thus an even more southerly position for the Schlieren Flysch. This assumption is not necessary if we accept the postulate of an Ultrahelvetic Leimern. The Flysch sediments of South Penninic origin, such as the Simmen Flysch or the South Penninic Flysch of eastern Switzerland, contain ophiolite debris. Neither the Schlieren sandstone nor the Leimern marl contains chrome spinel or other detritus from ophiolite; there is no observational evidence for a South Penninic origin of the Leimern and Schlieren.

Gurnigel Flysch

The Gurnigel Flysch of Bern and western Switzerland is dominated by its sandstone interbeds, so it was called Gurnigel Sandstone by Studer (1827). Later Gerber (1925) grouped the turbidite sandstone formation together with a mélange (Wildflysch) and called the unit Gurnigel Flysch. This Flysch formation is so similar in age and lithology to the Schlieren that the two have commonly been considered to belong to the same formation, and to have been deposited in the same basin (Trümpy 1960; Leupold 1966). More recent work has indicated, however, that the youngest Gurnigel is Middle Eocene (Lutetian), somewhat younger than the youngest Schlieren Flysch (uppermost Lower Eocene).

The Gurnigel Flysch crops out in the External Prealps only, and its tectonic position cannot be determined precisely by an ordering on the basis of nappe superposition. The Flysch was considered Ultrahelvetic because the Mesozoic rocks in the Wildflysch of Gurnigel have a sedimentary facies similar to that of the Mesozoic in the highest Helvetic nappes (Guillaume 1957). Assuming that the Flysch and the Mesozoic were deposited at the same place, an Ultrahelvetic origin for Gurnigel seemed well established.

Studies of the sedimentary structures by Crowell (1955) and by Hsü (1960) revealed that the feldspathic and granitic sand debris of the Gurnigel were transported by southerly flowing turbidity currents, a transport direction opposite to that of all Helvetic and Ultrahelvetic Flysch turbidite. The granite debris in the Gurnigel bears no resemblance to the Aar granite, and as the northern Helvetic shelf was not sufficiently steep for submarine slumping to generate turbidity currents, the debris could not have come from the northern continent. Hsü (1960) suggested, therefore, that the Habkern Island (Homewood's Marginal Basement High) south of the Schlieren Flysch basin was the source terrane. Yet blocks of the Mesozoic sedimentary rocks in the Wildflysch of the Montsalvens district are definitely Ultrahelvetic (Guillaume 1957). This paradox can be resolved if one assumes that the Ultrahelvetic Mesozoic blocks in the Wildflysch and the Gurnigel Flysch have not been derived from the same source terrane.

Impressed by the similarity in age and lithology between the Gurnigel and the Flysch of the Sarine nappe, the lowest unit of the Simmen nappes, Caron (1972, 1976) suggested a South Penninic origin for the Gurnigel. He also cited the opinion of earlier workers (Tercier 1925) that the Habkern granite could be the basement of the Austro-alpine realm. In making this comparison, he overlooked the fact that although flysch sediments of similar age may have similar sedimentology, they could have been deposited in separated basins, such as the Miocene/Pliocene turbidite formations of the flysch facies in various Southern California basins (Hsü 1958). Moreover, Caron completely ignored the evidence of a northerly source, presented by Crowell and by Hsü, for the Gurnigel detritus when he postulated an Austroalpine source. Finally, the absence of ophiolite debris, which is common in the coarse clastics of the Simmen nappes, also speaks against a correlation with a South Penninic Flysch.

In 1971 Hsü and Schlanger postulated a deep-sea trench south of an island arc as the site of Gurnigel Flysch deposition. The trench should have originated when the northern Penninic ocean floor was subducted under the European continent during the Late Cretaceous. This north-dipping subduction should also have caused the rifting of a segment of the European margin to form the Habkern Island Arc and the Schlieren basin behind the arc. The hypothesis places the Gurnigel south of Homewood's Marginal Basement High, which, as we have indicated, is equivalent to our Habkern Island. This position has been reserved by Homewood for the North Penninic Niesen Flysch. The more calcareous Niesen includes more hemipelagic sediments and more limestone turbidite than the Gurnigel. One could speculate that the upper beds of the Niesen and the Gurnigel are coeval sediments of different facies in the same basin. Another possibility is that the Niesen is a South Penninic Flysch, an idea suggested by Rudolf Staub (1937).

Wägital Flysch

The Flysch strata above the Einsiedeln schuppen zone and below the Klippen Decke in Canton Schwyz are known by the name Wägital Flysch (Leupold 1966). They have been compared with the Schlieren and Gurnigel of central Switzerland. Like those latter, the Wägital consists of two tectonic units: a turbidite-sandstone formation (Wägital Flysch, *sensu stricto*) is overthrust above a Wägital Wildflysch. The Wägital Flysch ranges from Campanian to Maastrichtian in age, in part coeval with and in part older than the Schlieren and Gurnigel Flysch (Kuhn 1972). The Wildflysch is a mélange, including sediments of Middle Eocene age.

A study of the current markings at the base of the sandstone beds indicates that the currents depositing the turbidite layers flowed almost due south (Hsü 1960). This fact reinforces the consensus that the Wägital is a North Penninic Flysch. Hsü and Schlanger (1971) suggested that both Gurnigel and Wägital were deposited in a deep-sea basin south of an island arc on the southern border of the Ultrahelvetic.

Niesen Flysch

The Niesen Flysch is known in the Prealps from the Rhône to the Lake of Thun over a distance of 60 km along the strike. It lies between the Ultrahelvetic and the Médianes nappes. This 2000-m-thick formation has been studied by de Raaf (1934), McConnell (1951), Lombard (1971), and Ackerman (1986). The Niesen Flysch, *sensu stricto*, includes turbidite sandstone, lime-mud turbidite, boulder conglomerate, and hemipelagic marl and shale. Microfossils ranging from Albian to Paleocene have been reported, but the main part of the Flysch is probably only Campanian-Maastrichtian (Homewood 1977), and thus coeval to the Wägital Flysch.

The sediment transport was commonly believed to be directed from northwest to southeast, indicating the derivation of the sediment detritus from a highland more external than the basin (Lombard 1971; Homewood 1977). If this highland is the Marginal Basement High of Homewood, the Niesen should be North Penninic. A South Penninic origin has, however, been suggested by Staub (1937) and by Cadisch (1953), because green-schist detritus of presumably southerly origin is common in the Niesen Flysch. If, as Niggli (1953) reported, blue-schist detritus is also present in the Niesen, the latter alternative seems even more attractive (Hsü 1989). A Wildflysch or mélange is present at the base of the Niesen Flysch nappe, but its geology gives little additional clues to the origin of the Niesen. The Mesozoic exotic blocks in the Wildflysch seem to have been derived from the Ultrahelvetic (Lugeon 1938).

Other Flysch Formations

Thick sediments of the flysch facies were deposited not only on the northern margin of the Tethys Ocean during the Late Cretaceous and Early Tertiary, but also in other parts of the ocean at the same time or earlier. They are known by local names, e.g., Prättigau Flysch, Simmen Flysch, South Penninic Flysch, etc. They will be discussed in later sections.

Flysch Tectonics

The Ultrahelvetic Flysch, *sensu lato*, wraps around the recumbent folds of the Helvetic nappes, as if the Flysch had been superposed on undeformed Helvetic strata as a horizontal sheet before both units were folded and overthrust to form the complex nappe-complexes of the Helvetic/Ultrahelvetic Alps (Heim 1911). Since thin slices of flysch rocks could hardly be pushed onto the Helvetic from behind, the tectonic emplacement of the "Ultrahelvetic Flysch" has been considered evidence of gravity-sliding deformation (Figure 5.5). Lugeon (1943) and Badoux (1967) spoke of diverticles. "Diverticulation" is portrayed as a process in which successive layers, each detached from its underground along a weak layer, slide one after another under their own body force down an incline. Aside from physical considerations, which rule out the feasibility of

Fig. 5.5. A structural profile across an Ultrahelvetic "nappe" (after Badoux 1945). This section across the Pommerngrat shows clearly that the Mesozoic rocks do not form a recumbent fold or a nappe in the Simmen Valley. They are slabs in a Wildflysch, which is not a sediment deposited by gravity-sliding, but a tectonic mélange formed by underthrusting. The slabs are detached mainly from Helvetic units; they are as follows. 1: Wang beds; 2: intercalated lenses of Aptychus limestone (Valanginian?); 3: black shale (Aalenian); 4: limestone and marl (Bajocian); 5: micalcesous sandstone (Flysch), with intercalated lenses of *Discocyclina*-bearing limestone; 6: marl with ammonite fossil (Oxfordian); 7: Capionella limestone (Malm); 8: Flysch; 9: black shale (Aalenian, like 3); 10: limestone and marl (Bajocian, like 4); 11: Wildflysch; 12: black shale (Aalenian); 13: Wildflysch; 14: limestone lense (Malm); 15: marl (Oxfordian); 16: limestone lense (Malm); 17: nodular limestone ("Sequanian"), with ammonite fossil; 18: Wang beds; 19: limestone lense (Malm); 20: limestones (Maastrichtian, Barremian); 21: black shale (Aalenian); 22: Flysch; 23: dolomite, rauwacke (Triassic).

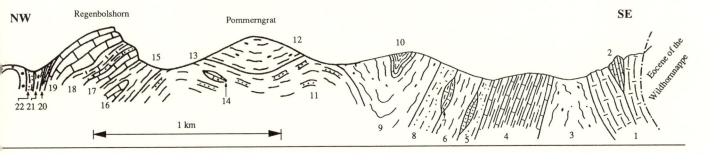

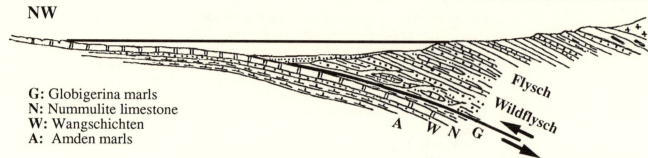

G: Globigerina marls
N: Nummulite limestone
W: Wangschichten
A: Amden marls

Fig. 5.6. The thrusting of a Helvetic sequence under an Ultrahelvetic accretionary wedge, before the envelopment (*Deckeneinwicklung*) during recumbent folding.

gravity-sliding as a significant tectonic process (see Hsü 1968, 1969a), I reject this ad hoc explanation, because "diverticulation" has no Recent analogue; its proponents are speculating without reference to any place where such a process is actually going on. With the innovation of the plate-tectonic theory, we can again appeal to uniformitarianism, or actualism, which is the foundation of geology: tectonic mélanges are the product of shearing where one lithospheric plate or crustal wedge is tucked under another by underthrusting (Figure 5.6). This kind of process is taking place in subduction zones (B-subduction) or zones of crustal underthrusting (A-subduction). One need not appeal to the imaginary and physically impossible process of gravity-sliding to explain the structure of the Ultrahelvetic thrust slices.

The subduction processes west of North America illustrate this principle of Wildflysch genesis by underthrusting. Offshore, near the plate boundary between the Pacific and North American plates, deep-sea sediments, including thick turbidite layers, are underthrust to form accretionary prisms, signifying that ocean sediments have been accreted to the base of continent. Underthrusting of younger accretional prisms under older ones is a tectonic process called underplating. Continuing underplating during the Quaternary caused the uplift of older accretionary prisms, namely, the Tertiary flysch formations and mélange units that are now exposed in the mountains of the Olympic Peninsula of the state of Washington.

An even more appropriate example can be seen in the geology of Taiwan. The mountains of the island mark the suturing of the Philippine and Eurasian plates, and the northern tip of the back-arc South China Basin has been eliminated by the arc-continent collision (see Chapter 14). Thrust under a volcanic island arc of the Philippine plate is an accretionary wedge of Wildflysch-like rocks in the Central Range. This tectonic mixture, of Miocene hemipelagic shales and Paleogene sediments that had been deposited on the continental shelf of the eastern Asian margin, is an analogue of the Alpine Ultrahelvetic. The leading edge of the Tertiary passive-margin sequence is thrust under the accretionary wedge of the Central Range, and the bulk is pushed westward to form the thrust belt of the western foothills, analogous to the Alpine Helvetic. At the boundary between the accretionary wedge and the cover-thrust belt, in the area of Chukung in central Taiwan for example, the younger passive-margin sequence (cf. Helvetic) has been thrust under a mélange unit of deep-water sediments (cf. Ultrahelvetic), and the two are folded together to form recumbent anticlines in a nappe envelopment like that postulated by Heim.

Using the plate-tectonic theory to interpret the Flysch tectonics, we suggest that the Ultrahelvetic Flysch did not travel on its own by downslope movement to the Helvetic realm. Instead, the Ultrahelvetic Flysch formed accretionary prisms (e.g., Sardona and Schlieren). This process took place during the early Tertiary before the Late Eocene arc-continent collision, and the result was the mélange called Wildflysch. With the elimination of the back-arc basin north of the Habkern Arc, the Helvetic passive-margin sequence was thrust under the Ultrahelvetic accretionary wedge, mainly during the Oligocene. With the relentless northward march of the Austroalpine and Penninic nappes, the Ultrahelvetic and Helvetic strata were in part tucked under the overriding plate, and in part pushed out, over and beyond the top of autochthonous massifs, to form the cover thrusts in the High Limestone Alps.

Flysch Paleogeography

The Aegean Sea is an actualistic analogue of the Paleocene paleogeography of the Helvetic/Ultrahelvetic realm (Hsü and Schlanger 1973). On the north side of the Aegean is the European passive continental margin, which was formed when a crustal slice was rifted apart from Europe, in late Miocene time, to form the Hellenic Island Arc, which extends from the Peloponnesus to Crete and Rhodes (Hsü and Ryan 1973). To the north of the arc is a back-arc basin, the South Cretan Basin (Figure 5.7). The back-arc region has since been subjected to extensional stress related to the back-arc seafloor spreading. There is active back-arc volcanism, building up the islands of Santorini. The detritus for the sediments of the back-arc South Cretan Basin is derived mainly from the south (Crete), with minor contributions of volcanic debris from Santorini. To the south of the arc is an active plate-margin, with the Hellenic Trench marking the zone of subduction. The inner slope of the Hellenic Arc is also very steep, close to 40 degrees, and the presence of talus at the foot of the slope indicates that submarine slides occur frequently. Flysch turbidites are accumulating on both sides of the Hellenic Arc. The deep basins of the back-arc basins contain volcanic detritus from

Fig. 5.7. The eastern Mediterranean as an actualistic model for the Paleocene Flysch paleogeography (after Hsü and Schlanger 1971). The arrows in the figure indicate the sediment-transport direction as determined by flute casts on the bottom side of Flysch turbidite beds. The placement of Niesen Flysch is uncertain (see text for explanation).

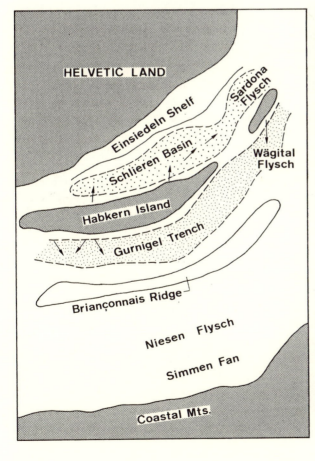
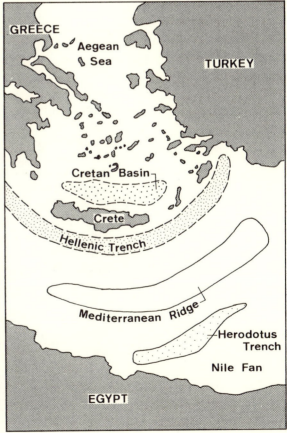

the back-arc volcanoes. The source of the fore-arc sediments is the nonvolcanic island chain, and these sediments contain little or no volcanic debris.

Using the Aegean model, the Schlieren Flysch is assumed to have been deposited in a back-arc basin like the South Cretan Basin: the granite debris is derived from the Habkern Island to the south, and volcanic debris from back-arc volcanism (Figure 5.7). The Gurnigel is a fore-arc basin, or a trench like the Hellenic Trench; the Gurnigel Flysch has no volcanic detritus because the source terrane to the north is the non-volcanic island arc.

Extension prevailed in the Schlieren back-arc region during the Paleocene. Compressional tectonics became active, however, in the Eocene, when the deep-sea floor of the back-arc basin was thrust under its own frontal arc. The Eocene Helvetic/Ultrahelvetic realm under compression finds an actualistic analogue in the region of the South China Sea: the floor of this back-arc basin has been thrust eastward under the Philippine Arc. The gradual elimination of a back-arc basin by such a subduction process, leading eventually to arc-continental collision, has been called back-arc basin collapse (Hsü *et al.* 1989). That the Schlieren back-arc basin collapsed during the Eocene is evidenced by the deformation of the Ultrahelvetic Flysch and mélanges, as discussed in the preceding section.

Numerous Alpine geologists considered the Wildflysch a sedimentary deposit, and they assumed the presence of a granite mountain on the southern shore of the Late Eocene Helvetic Sea, because giant blocks of Habkern granite, in a matrix of Globigerina Marl, are found in the Wildflysch. Schardt (1898) painted a vivid picture of granite blocks peeling off the steep front of advancing nappes and being dumped into the trough of Globigerina Marl deposition. However, according to the scenario presented in this book, the Eocene coastal mountain was not underlain by the Habkern granite. The granite blocks must, therefore, have been emplaced earlier, in the late Maastrichtian and/or Paleocene, when high-angle faulting induced by the extensional stress prevailing in the back-arc region could have produced a granitic coast range on an island arc, and when the coarse turbidites (containing granite debris) of the Schlieren and Gurnigel were deposited. After the basin-collapsing process was initiated in the Eocene, the accretionary wedge formed by the underthrusting of Ultrahelvetic and South Helvetic Flysch caused the rise of a coastal range south of the site of Globigerina Marl sedimentation, and this mountain should have been underlain, during the Oligocene, by Flysch and mélanges, while the older granite terrane receded (relatively southward) farther and farther inland. Once again we can cite the Cascade Range of Washington as an actualistic analogue. The coastal Olympic Mountains are underlain by Eocene, Oligocene, and Miocene rocks that contain abundant granitic debris. The Recent flysch sediments do contain finer granite debris that has been transported by rivers to the coast and from there by resedimentation to the deep sea, but no granite blocks could be supplied to form Recent submarine breccias because no granites crop out on the coast; the granite basement of the Cascade Range now lies more than 50 km from the shore. It is thus not surprising that the Helvetic/Ultrahelvetic Eocene sedimentary breccia contains only sedimentary rocks derived from advancing Penninic nappes, but no granite blocks (e.g., Bayer 1982). Later, in the Late Eocene and Oligocene, the autochthonous/North Helvetic Flysch trenches were bordered on the south by volcanic terranes, as evidenced by the abundant andesitic detritus in the Taveyannaz Sandstone.

6.

The Prealps

Klippe is an English geological term that denotes an outlier, or the erosional remnant of an overthrust mass, that has been isolated from the main overthrust mass by erosion. A fenster is, by contrast, the outcrop of an underthrust mass, exposed by erosion, and surrounded on all sides by overthrust. Both klippe and fenster are words derived from German, testifying to the advanced state of Alpine tectonic studies during the development of geology.

The Klippes

Klippe is the German word for cliff. The relevant cliffs are the ones under the peaks of the Mythen, Big and Little. They rise abruptly from grassy slopes that overlook a bend of Lake Lucerne, and are a landmark well known to tourists. The grassy slopes are underlain mainly by Flysch. On the cliff face are, in ascending order, gypsiferous shale and dolomite, limestones, and red marly limestone.

Escher noted the similarity of the red limestone at the top of the Big Mythen to the Seewerkalk of the Helvetic nappes, and thus assigned a Cretaceous age to this formation (see Studer 1853, p. 182). Yet at the foot of the cliff is Nummulitic Limestone and Flysch of Eocene age (Figure 6.1). Escher and Studer were both confounded by this apparently abnormal stratal superposition.

Kaufmann was fascinated by this twin-peaked mountain. He opened his 1877 monograph with this declaration:

> The peaks of the Mythen, so majestic and wonderful in appearance, have been able to hide their inviolable inner secrets, despite many futile attempts by geologists.

Kaufmann noted the similarity of the gypsiferous shale and dolomite at the base of the cliff to the Keuper Formation of the German Triassic; he also reported finding

Fig. 6.1. Mythen (after Heim 1922). Panoramic view from Lake Lucerne in central Switzerland. Mesozoic rocks are exposed on the cliff faces of the Big and Little Mythen, but Tertiary nummulitic limestone and Flysch are found at the foot of the cliffs and in the meadows surrounding the peaks. Hans Schardt proposed the theory of Prealpine thrusting to resolve the Mythen dilemma. This sketch by Heim was selected for illustration because it is one of the most beautiful in the Alpine literature. Investigations since 1922 have revealed, however, that Heim erred slightly: the Rotenfluh schuppe, which is now correlated to the Upper Penninic "Arosa schuppen zone" (see Trümpy 1980), is thrust over, not under, the Mythen schuppe.

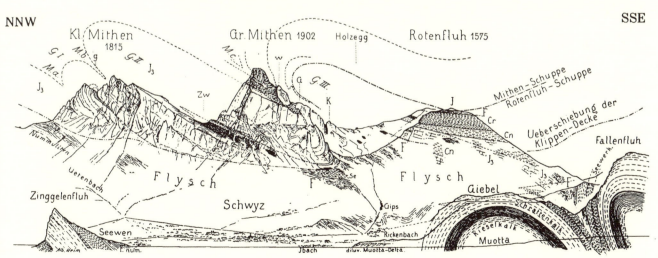

Triassic plant fossils in the shale. The overlying sandy limestone formation contains abundant fossil belemnites: these are cephalopods and, like the modern nautilus, they were swimmers or planktonic organisms. This limestone formation could thus be dated Jurassic and was considered by Kaufmann to be a correlative of the Brown Jura of southern Germany. Above that is a white massive limestone with belemnite and ammonite fossils. This "White Jura" was compared with the Malm or Upper Jurassic of western Switzerland. The red limestone at the top is now called *couches rouges*. In addition to the light and dark red limestone and marly limestone, there are some white and green limestone intercalations. *Inoceramus* fossils, similar to those in the Seewerkalk of the Helvetic, have been found in this mainly Upper Cretaceous formation.

A naive person might assume that the Mesozoic rocks of the Mythen rose as ancient islands above an ocean of Eocene Flysch. Murchison at one time entertained such an idea when he tried, in vain, to rationalize the inverse superposition of Permian rocks above the Tertiary Flysch in Glarus; he thought that the younger sediments were plastered on the side of a steep cliff of older rocks. Beach sedimentation at the foot of cliffs is a common feature on mountainous coast, and this relation of younger sediment unconformably overlying an older topographic relief is called buttress unconformity. In such a situation, however, the younger sediment should consist largely of coarse detritus coming down from the steep slope. The Nummulitic Limestone at the foot is not a sediment above a buttress unconformity, so this idea was never seriously entertained by geologists to explain the origin of the Mythen.

A second way out of the dilemma was to assume that the Mythen rocks rose like a plug along steeply dipping fault planes. Favre did suggest such a mechanism in 1865 to explain the klippes of the Savoy Prealps (see Bailey 1935). Structural geologists today consider such a mechanism physically impossible. Furthermore, the sedimentary sequences on the two sides of a fault should not be very different, but the Mesozoic formations of the Mythen are very different from those of the Helvetic nappes, which crop out on all sides of the klippe.

A third alternative was to assume a normal stratigraphical superposition of the rocks exposed on the cliff face above the Tertiary Nummulitic Limestone and Flysch. Until 1893 Hans Schardt tried to bully his way through by ignoring other people's stratigraphical evidence; he thought that the Mythen carbonate sequence was Tertiary. His postulate created serious new problems: The Mythen rocks are not correlative, lithologically or paleontologically, to the Tertiary Molasse of the Swiss Midland. On the contrary, the Mythen rocks resemble some of the Mesozoic formations of western Switzerland, and the resemblance is too real to be ignored. Schardt had to abandon his *ad hoc* explanation, which had led to nowhere; it was contradicted by more and more new observations.

Less than a decade after Bertrand proposed the Glarus Overthrust, Schardt (1893) finally accepted the interpretations of paleontologists and recognized the possibility of an abnormal stratigraphic superposition of the Mythen rocks on the Helvetic. He correlated the Mythen Mesozoic with a similar sequence in the Median Prealps of western Switzerland, and postulated a southern origin for those allochthonous rocks. He thus had to propose that the Big and Little Mythen have been thrust above the Helvetic nappes as an overthrust complex—the nappes of the Median Prealps (*Klippen Decke*).

A century has gone by since the publication of Schardt's revolutionary manifesto. We are beating a dead horse in Switzerland when we discuss "buttress unconformity," "faulting," or "assumption of normal superposition" in connection with the klippes of the Mythens. Yet tectonics is in such a sorry state in certain countries that these ridiculous ideas are still being presented as "theories" to explain the geology of klippes (see Hsü 1989).

The Prealpine Nappes

Hans Schardt, working in the Prealps of western Switzerland toward the end of the last century, enunciated a "law of the Prealps" (see Bailey 1935, p. 84), namely:

Everywhere within the Prealps the substratum, upon which Triassic, Permian, or Carboniferous rests, is formed of a more recent formation, generally Tertiary Flysch.

Once we recognize that this surface separating the Prealpine rocks from their substratum is an overthrust, the logical conclusion is that the Prealps consist of a complex of nappes. The following order of tectonic superposition has been tentatively suggested (in descending order):

Simmen nappe
Breccia nappe
Nappes of the Median Prealps
    Nappe of the Rigid Median Prealps
    Nappe of the Plastic Median Prealps
Niesen nappe
Gurnigel nappe (nappe of the External Prealps)
Col nappes (nappes of the Internal Prealps)

Applying the rule of unidirectional thrusting—that the higher, or more internal, nappe has come from a more southerly source—the sediments of the Simmen nappe should have been deposited at the most southerly site and those of the Col nappes at the most northerly site. All of these nappes were piled above the Helvetic nappes, which in turn were thrust over the autochthonous massifs.

Like the Helvetic nappes, the nappes of the Prealps are all strip sheets, or décollement nappes, involving the deformation of sedimentary covers stripped off from their basement. Unlike the Helvetic nappes, however, they rarely form recumbent folds; they commonly occur as thrust sheets. The Prealpine nappes are folded, but we hardly ever see the overturned limb of a large recumber on the top of the Prealps, and the contrast in the style of deformation is shown in Figure 6.2.

The sedimentary rocks of the Prealpine nappes were deposited in the Mesozoic/Early Tertiary ocean between the European (Helvetic) and the Italian (Austroalpine) margins. Except for the Col nappes, which are considered Ultrahelvetic, the Prealpine

a

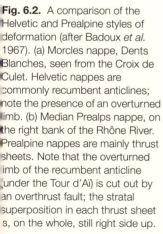

Fig. 6.2. A comparison of the Helvetic and Prealpine styles of deformation (after Badoux *et al.* 1967). (a) Morcles nappe, Dents Blanches, seen from the Croix de Culet. Helvetic nappes are commonly recumbent anticlines; note the presence of an overturned limb. (b) Median Prealps nappe, on the right bank of the Rhône River. Prealpine nappes are mainly thrust sheets. Note that the overturned limb of the recumbent anticline (under the Tour d'Aï) is cut out by an overthrust fault; the stratal superposition in each thrust sheet is, on the whole, still right side up.

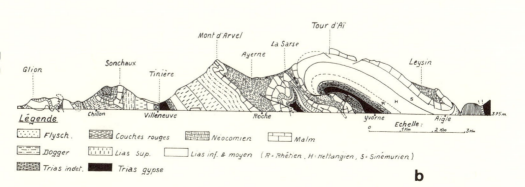

b

nappes are now interpreted as Penninic cover nappes; they were the original sedimentary cover above the basement of Pennine realms—basement now found in the gneissic core of Penninic mobilized-basement nappes. But this conclusion was arrived at only after much painstaking field work by many Swiss and French geologists.

The Nappes of the Median Prealps

The Klippen nappes are also called the nappes of the Median Prealps, because they occur between the internal and external zones of the Prealps. The upper of the two is also called the *Médians rigides*, or Rigid Median Prealps, while the lower is called the *Médians plastiques*, or Plastic Median Prealps, referring to the more rigid and plastic modes of deformation of the two units.

We can discern from Kaufmann's description of the rocks of the Mythen that their stratigraphy is very different from that of the Helvetic strata.

The Paleocene/Eocene Nummulitic Limestone is a typical formation of the Helvetic Tertiary. No such limestone is found in the Klippen nappes.

The Cretaceous of the Helvetic nappes is very thick, reaching, in the words of Bailey (1935), "mountainous proportions." Particularly impressive are the thick and massive limestone beds, the Schrattenkalk Formation (also called the limestone of the Urgonian facies); they form many of the cliffs or ledges in the High Calcareous Alps. The Cretaceous of the Klippen nappes is, in contrast, very thin. The absence of Schrattenkalk, the massive cliff-former, is one of the most arresting scenic peculiarities of the Prealps; the Lower Cretaceous is, in fact, entirely missing in some places.

The Jurassic formations of the autochthonous, Helvetic, and Ultrahelvetic constitute a typical passive-margin sequence on the northern margin of the Tethys. Whereas the Jurassic sediments of the autochthonous and Helvetic realms of sedimentation are mainly marine limestones, the Ultrahelvetic Jurassic (of the Col nappes, for example) includes thick hemipelagic shales. The Jurassic formations of the nappes of the Median Prealps are, however, distinctly different from those of the European margin; their lithology and facies change are shown in Figure 6.3.

Fig. 6.3. Palinspastic cross section through the Mesozoic formations of the Median Prealpine nappes in western Switzerland; vertical scale exaggerated 4× (after Trümpy 1960). The sediments of the Brianconnais and Subbrianconnais facies of the nappe of the Median Prealps are characterized by two unconformities—an upper of submarine erosion origin in the Brianconnais between the Upper Cretaceous and Upper Jurassic, and one of transgressive origin between Middle Jurassic and Middle Triassic. The stratigraphic hiatus is smaller in the Subbrianconnais domain of deeper water sedimentation. The Triassic formations shown below the basal thrust plane of the nappes have not been stripped from their Paleozoic basement and are now found on the front of the Great St. Bernard nappe in the Valais.

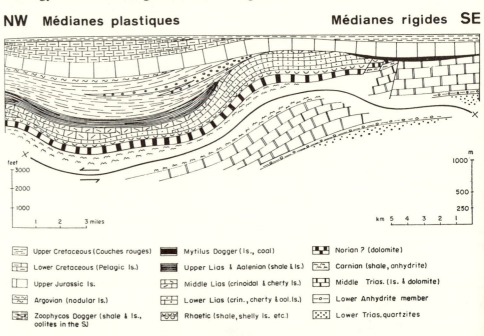

The Upper and Middle Jurassic formations of the upper nappe, or the Rigid Median Prealps consist of shallow marine "Malmkalk" of Tithonian to Oxfordian age, and Mytilus Beds, also known as Mytilus Dogger, of Callovian to late Bajocian age. The brackish marine Middle Jurassic sediments are interbedded with coal seams. At the base of Mytilus Dogger is a layer of lateritic soil, indicative of subaerial weathering in Middle Jurassic time (Badoux and Weisse 1959).

The Upper and Middle Jurassic formations of the lower nappe, or the Plastic Median Prealps, consist of massive limestone and nodular limestone ("Argovian"), ranging from Tithonian to Oxfordian in age, and of *Zoophycos* Dogger, of early Oxfordian to early Toarcian age (Felber 1984). *Zoophycos* is a trace fossil, or the fossilized tracks and trails left behind by mud-feeding organisms, and *Zoophycos* Dogger is a thick sequence of shale, marl, and silty limestone, whose undersurface is marked by *Zoophycos* trails. *Zoophycos* Dogger contains ammonites and other marine fossils, having been deposited in a water environment deeper than that of *Mytilus* Dogger. The deformation of the thinner-bedded *Zoophycos* sequence is more plastic than the deformation of *Mytilus* Dogger, and hence the two modifiers "plastic" and "rigid" have been used to denote the lower and higher nappes of the Median Prealps, respectively.

The Lower Jurassic, or Lias, is locally present in the Rigid Median Prealps, and it consists of argillaceous limestone and calcareous shale, a sequence more than 100 m thick on the Stanserhorn, where crinoid faunas are found. The Upper Triassic of the nappes of the Median Prealps consists of Triassic anhydrite, shale, and honeycombed dolomite called *Rauwacke* (rough stone). The Middle Triassic consists of algal limestones and dolomites, and the Lower Triassic is a quartzite formation. The Permo-Carboniferous of the Prealpine nappes is absent in the Prealps, because the detachment surface of the Plastic Median Prealps is commonly an Upper Triassic gypsum and that of the Rigid Median Prealps is a Lower Triassic gypsum (Trümpy 1960, p. 885).

The search for the Prealpine root, as Bailey (1935) noted, "has proved a long and exciting chase." The sediments of the nappes of the Median Prealps were deposited in a terrane more southerly than the Ultrahelvetic and more distant from the European shore. The Ultrahelvetic Middle Jurassic formations are hemipelagic, but the Jurassic in the nappes of the Median Prealps is shallow marine or continental. The stratigraphy of the Prealpine Jurassic sediments indicates a reversal of the facies trend. Instead of a thicker, more complete, and deeper-water sequence, the sequence is thinner, less complete, shallow-marine, and partly subaerial; the lateral facies change indicates shoaling instead of deepening southward. This facies change implies that the Prealpine sediments were not laid down on the European margin; they were either the sediments of the southern margin (Austroalpine) or those of an island or island chain in the Tethys Ocean (see Figures 6.4 and 6.5).

Schardt offered two hints. On the one hand, he thought that the nappes of the Median Prealps were the prolongation of the nappes "which form a part of the Rhätikon massif"—the Falknis and Sulzfluh nappes east of the Rhine. On the other hand, he compared the Prealpine nappes with the rocks of the Penninic units in the Brianconnais district south of Mt. Blanc. The correlation with the Falknis/Sulzfluh was confirmed during the first half of this century (Cadisch 1953; Staub 1958), and the achievements of the postwar generation of French geologists have established the correlation of the Rigid and Plastic Median Prealps with the Brianconnais and Subbrianconnais units of the French Alps (Ellenberger, 1952; Lemoine 1953; Debelmas 1955). Schardt was vindicated, yet a new controversy was unleashed: Staub (1958) continued to consider the Falknis/Sulzfluh sequences the sedimentary cover of the Lower Austroalpine nappes, while geologists of the younger generation correlated the Brianconnais/ Subbrianconnais sequence with the sedimentary cover of the Penninic Great St. Bernard nappe.

The detached cover nappes of the Brianconnais zone are not far removed from their underlying Permo-Carboniferous, and the Brianconnais Carboniferous zone extends northward into Switzerland, where it is called *zone houillère* (Barbier 1948). These sediments have always been considered the sedimentary cover of the Middle Penninic Bernard nappe (Argand 1916). Such a reconstruction indicates the Penninic origin of the Brianconnais.

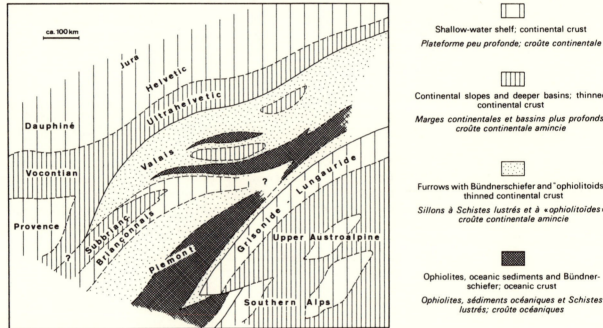

Fig. 6.4. Hypothetical cross section through some Prealpine nappes at the end of the middle Jurassic; vertical scale exaggerated 2.5× (after Trümpy 1960). The Brianconnais was an intraoceanic swell, dividing the Mesozoic Tethys into a northern (Valais) and a southern (Piedmont) ocean.

Fig. 6.5. Reconstruction of the Mesozoic paleogeography by R. Trümpy (1980), showing the relative positions of the various facies belts prior to their deformation. This reconstruction is on the whole still valid, except for some details and unresolved questions. There is no horizontal scale, because the amount of crustal shortening is unknown. The Jurassic Piedmont trough south of the "Brianconnais platform" could be several hundreds of kilometers wide.

The correlation of the Klippe nappe with the Falknis and Sulzfluh nappes of the Rhätikon has also not been disputed. Staub (1958, p. 23) stated:

> Falknis und Sulzfluh represent the prolongation, to the east of the Rhine River, of the Klippen nappe. This interpretation, established by Schardt, Lugeon, Steinmann, Seidlitz, and D. Trümpy, has hardly ever been challenged. Both the Sulzfluh and Falknis sequences consist mainly of the richly fossiliferous Upper Jurassic limestone and the Cretaceous *couches rouges*. Whereas the Middle and Lower Jurassic rocks are rarely present in the Sulzfluh nappe, the Falknis sequence includes considerable Middle Jurassic and Triassic strata, as well as a thicker and more complete Cretaceous sequence. The nearly continuous sedimentation from Late Jurassic to Late Cretaceous in the Falknis, and the disconformity between the Upper Jurassic and *couches rouges* in the Sulzfluh, found its correspondence in the inner and outer zones of the Median Prealps. Except for the fact that the thick Lower Jurassic of western Switzerland is thin or missing in the Klippen nappe of the Mythen, and missing altogether in the Falknis/Sulzfluh, the correlations of the Falknis/Sulzfluh with the Klippen nappe and with the nappes of the Median Prealps are excellent. (free translation by the author)

We cite this paragraph at length because Staub was expressing the consensus of the Alpine geologists when he made the correlation of the Falknis/Sulzfluh with the nappes of the Median Prealps. There has been no debate on that point. Unfortunately, Staub (1958), while accepting the correlation of the Median Prealps with the Falknis/Sulzfluh, rejected the possibility of their correlation with the Brianconnais, because he believed that the Falknis/Sulzfluh and the Median Prealpine sediments were all deposited on the Austroalpine margin. He fought a last-ditch battle to defend the prejudice of his generation, and the facts that led to his conclusions are as follows:

(1) The granite clasts in the breccias and conglomerates of the Falknis/Sulzfluh bear some resemblance to the basement of the Lower Austroalpine nappes.

(2) The sedimentary facies of the Falknis/Sulzfluh, especially that of the Cretaceous *couches rouges*, bears some resemblance to the Lower Austroalpine sequence of the Engadine. If so, they should have been deposited in adjacent areas.

Neither of these lines of evidence are, however, sufficiently compelling to place the Falknis/Sulzfluh site of sedimentation immediately adjacent to the Austroalpine. The dispute was settled after the correlation of the Brianconnais/Subbrianconnais sequences with the sedimentary cover (the Barrhorn series) of the Bernard nappe was firmly established (Iten 1948; Ellenberger, 1952): the Median Prealps and the Falknis/Sulzfluh are both correlative to Brianconnais/Subbrianconnais, and they are all Middle Penninic (see the review by Trümpy [1960]). The dogma of a Klippen nappe of Austroalpine origin passed into oblivion with the passing of its defenders.

Breccia Nappe

The klippes of the Breccia nappe are preserved in the Chablais south of Lake Geneva and in a small area in the Bernese Oberland south of the Lake of Thun.

The Breccia nappe in the Swiss Prealps was detached along a Triassic gypsiferous horizon. The Permo-Carboniferous strata originally underlying the sediments of the klippes are known only west of the Rhône (Cadisch 1953, p. 212). The Triassic rocks of the Breccia nappe klippes commonly lie on Flysch, according to Schardt's "law of the Prealps." The quartzite, *Rauwacke*, and dolomite are correlative to the Triassic rocks in the Breccia nappe in the Chablais. During the last epoch of the Triassic, called Rhät, breccia beds first appeared in the sequence, interbedded with limestone and dark gray shale.

The Jurassic of the Breccia nappes is characterized by the two breccia units, interbedded with sediments deposited under normal marine conditions. K. Arbenz (1947) recognized four units in the Hornfluh district; they are, in descending order,

Upper Breccia
Upper Shale
Lower Breccia
Lower Shale

The Lower Shale is a Lias sequence of calcareous shale, marly limestone, and thin breccia beds, and it records the history of the subsidence during the Early Jurassic, which led to the birth of the Tethys Ocean. The breccias were deposited on the steep southern slope of the Brianconnais Swell (Figures 6.4 and 6.5).

The Lower Breccia includes the thickest breccia beds of the Bernese Oberland, and the unit is up to 500 m thick. Two types of breccia have been recognized. The coarser breccias, with clasts some 10 cm across, are "framework supported": the angular fragments are in contact and supporting one another, whereas the space between the fragments is filled with mud or cement. The finer breccias are commonly mud-supported, with larger clasts scattered in a mud matrix. The finest breccias grade into very coarse sands; they commonly show graded bedding and they are poorly sorted, grain-supported sediments. The clasts are derived mainly from Triassic dolomite; quartzite and limestone components are subordinate. The age of this unit, as dated by belemnite fossils, ranges from late Early Jurassic to Middle Jurassic.

The Upper Shale consists of shale, siliceous shale, calcareous shale, and breccias, up to 200 m thick. Few index fossils have been found, and its age is estimated to range from Middle to Late Jurassic.

The Upper Breccia unit consists of mainly fine breccias, with intercalations of sandy limestone, pelagic limestone, shale, and calcareous shale. Graded bedding is very common. The age of the unit is Late Jurassic, ranging up to earliest Cretaceous (Infra-Valanginian), as determined by microfossils belonging to genus *Calpionella* in pelagic limestone.

The Lower Cretaceous is thin in the Breccia nappe of the Chablais, and is largely missing in the Hornfluh district, where the Upper Cretaceous *couches rouges* directly overlie the Upper Breccia. These pelagic sediments are dated by planktonic microfossils as Campanian and early Maastrichtian.

The *couches rouges* are overlain by a Flysch turbidite formation. The age of the Flysch of the Breccia nappe is Maastrichtian in the Hornfluh district, but Flysch sediments as young as Early or Middle Eocene in the Prealps have been considered to be the youngest sediments of the Breccia or of the nappes of the Median Prealps (Caron 1972). Numerous ophiolite slabs are found in the Flysch. They consist mainly of pillow basalt or of diabase, apparently detached from an ocean crust. These slabs were probably mixed tectonically with the Flysch during the underthrusting of the Breccia under the Simmen nappe and/or the Penninic Ophiolite Mélange.

That the Jurassic breccias of the Breccia nappe were formed by downslope movements under gravity was commonly recognized by the pioneers of Alpine geology. Two types of motion can be distinguished. In sediment-driven flow the sediment particles are propelled to move under their own weight; the fluid is carried along. A typical such deposit is the sturzstrom deposit generated by rockfall (Heim 1932; Hsü 1975). While moving downslope as an avalanche of broken blocks, momentum is transferred from one accelerating block to propel another in front. The coarse breccias in the Breccia nappe owe their origin to this mechanism. The coarse components, mainly Triassic, were already consolidated rocks before they plunged down as rockfall and were transported to the deep sea by sturzstroms. In fluid-driven flow, by contrast, the gravity of the fluid is the driving force; the solid component in such a solid–fluid mixture is carried along by the fluid force of the mixture. Mud flows, depositing mud-supported breccias, are a kind of fluid-driven flow. Turbidity currents are another kind; they are generated when debris from a solid flow or a mud flow is mixed with ambient water to form a flowing suspension.

The Breccia nappe overrides the nappe of the Median Prealps, indicating a more southerly origin. The Jurassic facies suggest that the breccias were deposited at the foot of a steep slope on the southeastern side of the Brianconnais Swell, a remnant island-arc. The missing Lower Cretaceous was once considered evidence of uplift and erosion (Arbenz 1947). We believe, however, that the absence of the Lower Cretaceous is evidence of nondeposition or erosion at the foot of a continental slope where contour current was active; both the Upper Breccia below the unconformity, and the *couches rouges* above, are deep marine sediments. The hiatus below the *couches rouges* could thus be compared with the pre-Wang unconformity of the Helvetic sequence.

The realm of breccia sedimentation remained deep during the Campanian/Maastrichtian and the Paleocene, when the Flysch sediments were accumulated, probably as submarine fans, before they were subducted to become the accretionary prism under the Simmen nappe and the Ophiolite Mélange. Ophiolite slabs were mixed with the Flysch sediments in the subduction zone during the Paleocene or the Eocene.

Simmen Nappe

The Simmen nappe, *sensu lato*, is the highest tectonic unit of the Prealps, and it is therefore also called *La Nappe supérieure des Préalps*. The nappe was thrust from the south onto and beyond the Breccia nappe, so that its frontal portion came to rest di-

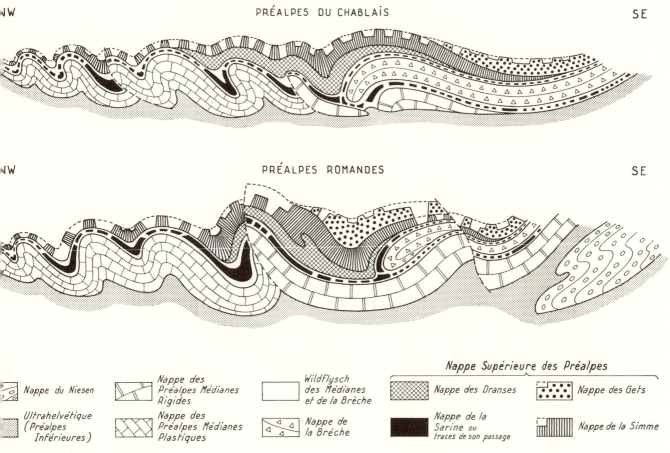

PRÉALPES DU CHABLAÏS

PRÉALPES ROMANDES

Nappe Supérieure des Préalpes

Nappe du Niesen	Nappe des Préalpes Médianes Rigides	Wildflysch des Médianes et de la Brèche	Nappe des Dranses	Nappe des Gets
Ultrahelvétique (Préalpes Inférieures)	Nappe des Préalpes Médianes Plastiques	Nappe de la Brèche	Nappe de la Sarine ou traces de son passage	Nappe de la Simme

Fig. 6.6. Schematic cross section illustrating the envelopment of the Breccia and Median Prealpine nappes by the Simmen nappe (after Caron 1972). This envelopment relation is similar to the envelopment of the Helvetic by the Ultrahelvetic. The rocks of the Simmen nappe are mainly Flysch and Wildflysch, similar to those of the Ultrahelvetic, *sensu lato*.

rectly on the Flysch of the nappes of the Median Prealps (Figure 6.6). Later, the Breccia nappe and its Simmen tectonic cover moved together once more and buried the advance guard of the Simmen nappe, pinching it into the compressed synclines of the nappes of the Median Prealps (Figure 6.7).

Caron (1972) proposed a subdivision of this nappe complex (Figure 6.8); the tectonic units are (in descending order)

Gets nappe
Simmen nappe, *sensu stricto*
Dranses nappe
Sarine nappe

The sediments in the Gets nappe are Cretaceous, and the nappe has a twofold subdivision. The lower unit, designated the Perrières Series, is an Upper Cretaceous sandstone flysch sandwiched between two mélanges (Figure 6.8). The lower mélange has blocks of granite and of ophiolite and slabs of *argilles a palombini* enclosed in a shaly or marly matrix; the Palombini marls are Lower Cretaceous. The upper mélange has slabs of granitic basement and fine-grained limestone. The higher unit, the Hundsrück Series, is a 200-meter-plus thick Upper Cretaceous sequence of coarse conglomerate and turbidite sandstone.

The Simmen nappe, *sensu stricto*, has likewise been subdivided into two units. The lower Manche Series consists of an Upper Cretaceous (Turonian) flysch-sandstone formation in normal superposition, also sandwiched between two mélange units. The lower mélange contains slabs of Albian limestone. The upper mélange is characterized by the presence of exotic slabs of radiolarite, aptychus limestone (*Biancone*), and pe-

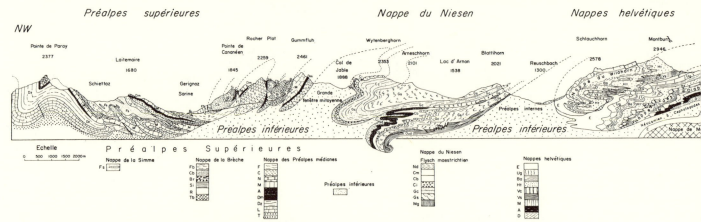

Fig. 6.7. Cross section through the Prealps (after Gagnebin 1942). In this cross section across the Prealps, the Simmen nappe, originally the most southerly, has been overthrust far to the north to lie directly on top of the Klippen nappe in the Laitemaire area, before the subsequent overthrusting of the Breccia nappe. Displaced northward later, the Breccia nappe of the Sarine Valley now seems to lie on top of both the Simmen nappe and the nappe of the Median Prealps. Such structural complications led to arguments over whether the Simmen or the Breccia was the higher nappe of more southerly origin. We now believe that this argument is unnecessary, because the simple rule that the higher nappe came from the more southerly realm is not valid where more than one generation of overthrusting has taken place.

		Series	Age
Gets nappe		Hundsrück Series	Upper Cretaceous
		= turbidites	
		= mélanges	— — —
		Perrières Series	Upper Cretaceous
Simmen nappe		Mocausa Series	Upper Cretaceous
		Manche Series	Turonian
Dranses nappe		Biot Series (Helminthoid Flysch)	Senonian
Saanen nappe		Reidigen Series	Paleocene Maastricht
Klippen and Breccia nappes		Wildflysch	— — —
		Flysch	Eocene Paleocene
		Upper limit Couches Rouges	

Fig. 6.8. Tectonic subdivision of the Simmen nappe, *sensu lato*, of the Prealps (after Caron 1972). Note that the lower nappes have younger rocks in this accretionary prism, formed by underthrusting at an active plate margin.

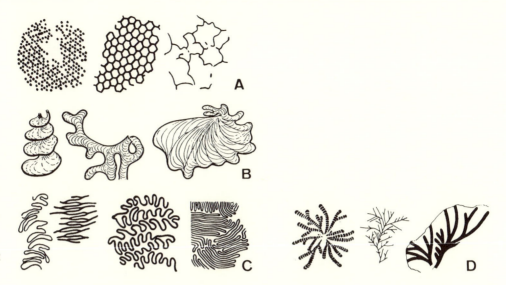

Fig. 6.9. Trace fossils (after Seilacher 1958). Some of the common trace fossils are shown here. A: *Paleodictyon*, common in Flysch; B: *Zoophycos*, typical of Zoophycos Dogger in the Median Prealps; C: *Helminthoidea*, common in Helminthoid or Simme Flysch; D: trace fossils with a dendritic pattern, common in Flysch; formerly called fucoids, but now identified as chondrites.

lagic marls, which are typical Upper Jurassic and Cretaceous sediments of the Upper Penninic or Lower Austroalpine sequence.

The Dranses nappe consists mainly of Helminthoid Flysch. This formation is a typically thin-bedded limestone flysch, characterized by trace fossils called *Helminthoid* on the bottom of turbidite beds (Figure 6.9). The age of the Flysch is Maastrichtian and older. The "basal complex" of this nappe is again a mélange with slabs of Helminthoid Flysch in a shaly matrix.

The lowest Sarine nappe also consists of Flysch. The trace fossils in this Flysch are mainly *fucoids*. Fucoids are bottom marks of uncertain origin in sedimentary rocks; the term has been used to designate doubtful trace fossils that show a dendritic pattern (Figure 6.9). The Flysch of the Sarine nappe ranges in age from Maastrichtian to Paleocene. It has been correlated with the Gurnigel and Schlieren because of their similarity in age. Our paleogeographical reconstructions suggest, however, that these coeval Flysch sediments were deposited in different basins. This lowest nappe of the Simmen nappe complex is superposed tectonically above the mélanges of the Breccia and Klippen nappes. The Simmen Flysch are South Penninic, being deposits of the "Piedmont Ocean" (Figure 6.10).

The Simmen nappe, *sensu lato*, is distinguished by the mélange blocks of Austroalpine sediments, Austroalpine basement, and oceanic basement. The Simmen Flysch sandstones and conglomerates also contain distinctive detrital components from those sources (Wildi 1985). Hsü (1989) postulated that the subduction of the Tethys Ocean began in the Early Cretaceous. The underplating of the southern continent caused the rise of a coast range on the northern border of the Austroalpine realm. The coastal mountains are underlain by granite, by Austroalpine formations, and by accretional prisms of ophiolite mélanges. The coarse detritus derived from such source terranes was transported to a foredeep on the southern border of the Penninic realm, where the turbidite deposits built submarine fans (Figure 6.11).

The growth of accretional prisms continued during the Late Cretaceous, when younger Simmen Flysch sediments were underthrust beneath older ones. The ocean floor between the Simmen and Breccia realms of sedimentation was finally consumed during the Early Eocene, when the Breccia Flysch began to be thrust under the Simmen, before the Klippen Flysch was thrust under the Breccia. After the collision of the Austroalpine and the Brianconnais, the deformed sedimentary cover of the Simmen, Breccia, and Klippen nappes was stripped off from the basement along detachment horizons of Triassic gypsum. These Prealpine nappes moved under the forward motion

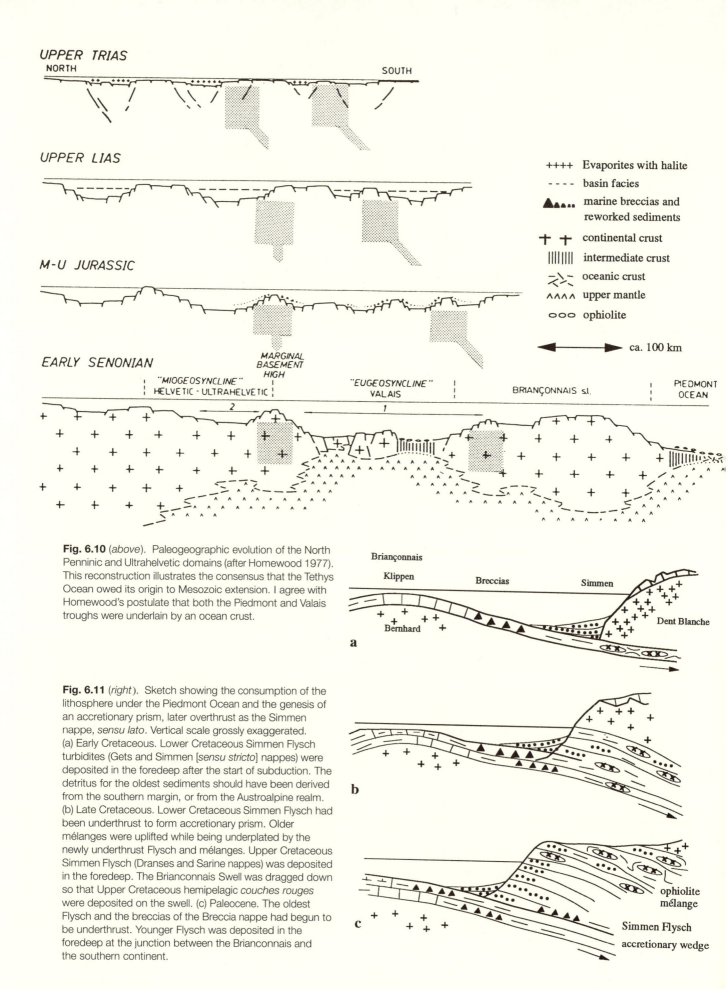

UPPER TRIAS

NORTH   SOUTH

UPPER LIAS

M-U JURASSIC

EARLY SENONIAN

MARGINAL
BASEMENT
HIGH

"MIOGEOSYNCLINE"
HELVETIC - ULTRAHELVETIC   "EUGEOSYNCLINE"
VALAIS   BRIANÇONNAIS s.l.   PIEDMONT
OCEAN

2   1

++++ Evaporites with halite

- - - - basin facies

▲▲... marine breccias and
reworked sediments

+ + continental crust

||||| intermediate crust

⌁⌁⌁ oceanic crust

∧∧∧ upper mantle

ooo ophiolite

◄─────► ca. 100 km

Fig. 6.10 (*above*). Paleogeographic evolution of the North Penninic and Ultrahelvetic domains (after Homewood 1977). This reconstruction illustrates the consensus that the Tethys Ocean owed its origin to Mesozoic extension. I agree with Homewood's postulate that both the Piedmont and Valais troughs were underlain by an ocean crust.

Briançonnais

Klippen   Breccias   Simmen

Bernhard

Dent Blanche

a

b

c   ophiolite mélange

Simmen Flysch
accretionary wedge

Fig. 6.11 (*right*). Sketch showing the consumption of the lithosphere under the Piedmont Ocean and the genesis of an accretionary prism, later overthrust as the Simmen nappe, *sensu lato*. Vertical scale grossly exaggerated. (a) Early Cretaceous. Lower Cretaceous Simmen Flysch turbidites (Gets and Simmen [*sensu stricto*] nappes) were deposited in the foredeep after the start of subduction. The detritus for the oldest sediments should have been derived from the southern margin, or from the Austroalpine realm. (b) Late Cretaceous. Lower Cretaceous Simmen Flysch had been underthrust to form accretionary prism. Older mélanges were uplifted while being underplated by the newly underthrust Flysch and mélanges. Upper Cretaceous Simmen Flysch (Dranses and Sarine nappes) was deposited in the foredeep. The Brianconnais Swell was dragged down so that Upper Cretaceous hemipelagic *couches rouges* were deposited on the swell. (c) Paleocene. The oldest Flysch and the breccias of the Breccia nappe had begun to be underthrust. Younger Flysch was deposited in the foredeep at the junction between the Brianconnais and the southern continent.

of the overriding Austroalpine nappes to their present position above the Ultrahelvetic mélanges and Flysch.

Paleogeography of the Prealps

The nappes of the Median Prealps are overthrust onto the Helvetic and the Ultrahelvetic. The simple rule suggests that they came from a more internal, or a more southerly (or easterly), position. The facies change from shallow marine limestone to hemipelagic shale indicates increasing water depth from the European continental margin outward: the Middle Jurassic of the higher Helvetic and Ultrahelvetic nappes is hemipelagic shales laid down in water hundreds if not thousands of meters deep. Yet the coeval Mytilus Dogger of the Prealpine nappes shows evidence of subaerial exposure. The southward deepening trend of the Jurassic Tethys Ocean was reversed. Were the Prealpine sediments deposited on the south shore of the Tethys or on an intra-oceanic swell?

In the French Alps, the Subbrianconnais, the Brianconnais, (*sensu stricto*), and the Mesozoic "Vanoise Series" are tectonic units sandwiched between the Ultradauphinois (=Ultrahelvetic) and the Helminthoid Flysch (cf. Simmen nappe) plus the *schistes lustrés* of the Piedmont Trough. The Middle Penninic position of this Brianconnais zone is thus determined by the tectonic superposition. The rocks of the Subbrianconnais facies are thrust under the Brianconnais, in a fashion similar to the underthrusting of the Plastic Median Prealps under the Rigid. Through a comparison of their tectonics and stratigraphy, the nappes of the Median Prealps, as well as the Falknis/Sulzfluh nappes, are correlated with the nappes of the Briancon area (Haug 1925; Gignoux and Moret 1933; Ellenberger 1952; Trümpy 1955); the sedimentary cover on a chain of islands (and/or submarine banks) has been stripped off and carried northward to form these rootless thrust slices. This Middle Penninic topographic high, the Brianconnais Swell, served to divide the Tethys Ocean into two marginal seas: north of the swell were the Subbrianconnais and the Valais Trough, and south was the "Piedmont Ocean" (Figure 6.10).

The Brianconnais Swell first came into existence as a remnant island-arc during the Middle Jurassic. The discovery of laterite intercalated in the sediments of the swell implies the existence of low-lying islands rising above shallow marine banks. Coarse siliciclastic debris, i.e., quartz, feldspar, granite, or other minerals and rock fragments rich in silica or silicates, is not common in the strata of the Prealpine Median and Breccia nappes. Yet Felber (1984) reported the presence of conglomerates and coarse-grained sandstones, in part late Bathonian in age, on the Little Mythen and on the Sulzfluh. These sediments contain clasts of gneiss and rhyolite, and their presence suggests that there were mountains of considerable relief on the Brianconnais islands.

What was the paleogeography prior to the Jurassic extension?

The Alpine realm was a shallow marine carbonate bank or an arid coastal flat during the Late Triassic. The "Piedmont Ocean" was born when the Brianconnais became separated from the northern margin of the southern continent, namely the Lower Austroalpine realm. The continental crust between the two areas became progressively thinner during the Early Jurassic, before the "Piedmont Ocean," floored by submarine basalt, came into existence (Figure 6.10). The Middle Jurassic Trough started as a narrow gulf, and has a modern analogue in the Gulf of California (Kelts 1981). Subsidence took place along normal faults on passive margins, or along vertically dipping faults (with strike-slip displacement) on transform margins. The breccias of the Breccia nappe apparently originated in rockfall down such steep fault scarps. With the start of subduction in the Early Cretaceous, the "Piedmont Trough" was steadily consumed when the ocean lithosphere of the back-arc basin was subducted along a south-dipping Benioff zone under the Austroalpine realm, forming the accretionary prism which is the Simmen nappe, *sensu lato* (Figure 6.11). At the same time, the Brianconnais remnant arc subsided. While the Simmen Flysch was being deposited in a foredeep of the "Piedmont Ocean," Upper Cretaceous hemipelagic *couches rouges* were laid down in the Breccia and Median Prealpine realms. With the further advance of the southern continent, the "Piedmont Ocean" and the Brianconnais Swell became a fore-deep for Eocene Flysch sedimentation, when the ocean floor was thrust under the advancing Austroalpine nappes. After the collision of the Brianconnais Swell (or island arc) and the southern continent, the Mesozoic sediments of the Prealpine Median and Breccia nappes were tucked under an accretionary wedge, which was made of Simmen Flysch.

The Valais Trough, a back-arc basin north of the Brianconnais remnant-arc, also subsided to a considerable depth during the Cretaceous. Occurrences of North Penninic ophiolites suggest the presence of an ocean crust under the trough, although the question is still being debated. The deformation of this Valais basin during the Late Cretaceous and/or Early Tertiary has an actualistic analogue in the collapsing of the back-arc basin between Celebes and Halmahera (Indonesia) in that the back-arc basin was bounded on both sides by Benioff zones. In the fore-arc trench where the Valais seafloor was thrust northward under the Habkern Island Arc, Gurnigel Flysch was desposited. To the south, the Valais seafloor was subducted under the Brianconnais Swell, while the sedimentary infill was peeled off to form the nappe of the Plastic Median Prealps and the North Penninic thrust slices (Ferret Schist and Niesen Flysch).

The elimination of the Valais Trough caused the collision of the Austroalpine/Brianconnais and the European margin in the Eocene. Under the overriding Austroalpine nappes, the sedimentary cover on, and north of, the Brianconnais remnant-arc was detached along Triassic evaporite horizons, and thrust on top of the Ultrahelvetic and Helvetic units to form the cover thrusts of the Median Prealps, Breccia and Simmen.

7.

The Penninic Core Nappes

The expression Pennine Alps originally referred to the western Swiss Alps, or the Valais Alps. The Alps of Canton Tessin are the Lepontine, and those of Graubünden are the Rhätic. The Pennine Alps face the Dent de Morcles, Les Diablerets, and the Wildhorn in the Bernese Oberland across the Rhône Valley. In contrast to the High Limestone Alps, the Pennine Alps are underlain mainly by metamorphic rocks. That these rocks in the region south of the Rhône constitute a pile of core nappes was recognized even before the Glarus Overthrust was postulated.

The Pennine Alps

Argand proposed calling the nappes of the Pennine Alps the Penninic nappes. With time, the nappe structure of the Valais was able to be extended to Tessin and Graubünden, so that the term has also been applied to mobilized basement and metamorphosed sediments of the Lepontine and Rhätic Alps. The lower nappes of Canton Tessin, however, are still referred to as Lepontine or Tessin nappes, but they are also Penninic, being the lowest of the Penninic complex. On the other hand, the Dent Blanche nappe, which was the highest nappe of Argand's Pennine Alps, is no longer considered Penninic; it is an Austroalpine unit, equivalent to the Lower Austroalpine nappe of Graubünden (Staub 1958). The highest tectonic unit of the Penninic nappes, in the present use of the term, is the ophiolite mélange of the Saas-Zermatt zone and of the Arosa schuppen zone, which lies directly under a Lower Austroalpine nappe.

A traveler across the Simplon Pass from the Valais to Tessin will gain the general impression that the first third of the country is underlain by steeply dipping schists. The middle third is steeply dipping gneiss, with intercalations of thin tectonic slices of schists and of dolomite marble. The last third is underlain by a granite gneiss.

The first breakthrough was made by Gerlach, a pioneer of Pennine geology; in the 1860s he recognized a recumbent fold in the Valley of Antigorio in Tessin—the core nappe of Antigorio (Figure 7.1a). His work was published posthumously in 1883. Lugeon (1902), Schardt (1904), Schmidt and Preiswerk (1908), Argand (1911), and others successfully exploited the concept of recumbent core nappes. Thanks to the opening of the Simplon Tunnel, the postulate that the gneiss in the middle third belongs to a core nappe higher than the Antigorio was verified (Figure 7.1b). This higher nappe is now called the Berisal nappe. The Berisal is considered by many a lobe of the Bernard nappe, but by others another core nappe because the Berisal in the Simplon region is separated from the Bernard by a large NW-SE trending fault (Milnes 1974a; Trümpy 1980).

Much additional work has been done in the area since the early part of the century, and the Simplon profile, as published in the 1967 edition of the *Geological Guide of Switzerland*, is shown in Figure 7.1c to illustrate the evolution of our interpretation of the nappe geometry. We can hardly claim that we have departed very much from Lugeon's or Schardt's interpretations. The same geometry is, however, susceptible to different interpretative ideas, and our theoretical understanding of the Pennine Alps has not stood still since the time of Argand. Bearth remained true to Argand's concept of core nappes. Younger workers such as Milnes (1974a, 1974b) introduced the mélange concept to interpret the geology of the Pennine Alps. The Antigorio and Bernard/Berisal have been recognized as true core nappes. Tectonic interpretation of the Monte Leone and Lebendun "nappes," however, presents some difficulties. Their basement "cores" seemed to have been separated into two or more "digitations," in the Eisten and Alpjen areas, respectively (Figure 7.1c). This type of structure had to be explained in terms of the very ductile deformation of the cores. Unlike the true core

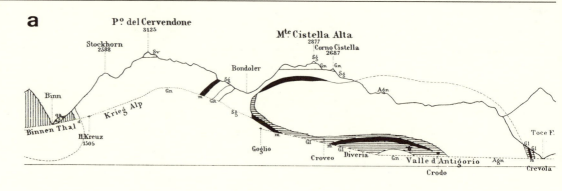

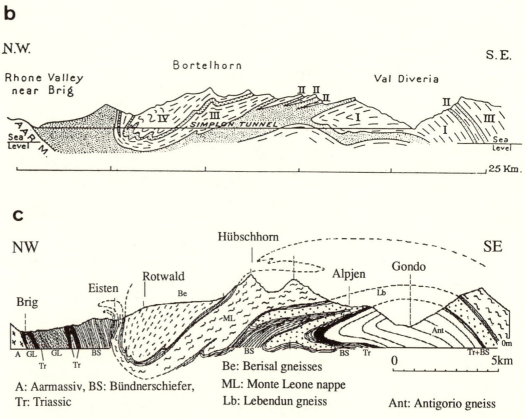

Figure 7.1. Structural cross sections across the Simplon district. The three profiles show the evolution of the interpretation of the nappe geometry in the century after 1870. The first profile (a), was based upon the work of Heinrich Gerlach (1883). This NW-SE section from the Rhône Valley near Fiesch across the Binn Valley to the Toce Valley in Italy, lying some 15 km to the northeast of the Simplon Tunnel profiles, shows that he had recognized the Antigorio nappe. The dark band wrapping around the recumbent fold is a Triassic dolomite marble. The middle profile (b), reproduced from Bailey (1935), was based upon the work of Schardt (1904) and Schmidt and Preiswerk (1908). The numbering of the nappes (I: Antigorio; II: Lebendun; III: Monte Leone; IV: Bernard) was introduced by Argand (1911). Gerlach's Antigorio nappe was verified by tunneling. Also shown is the folding of nappes III and IV under the Bortelhorn. The two basement slices are separated by metamorphosed sedimentary cover. The core-nappe concept failed to explain the fact that the one Lebendun basement slice SE of Val Diveria is represented by three Lebendun slices on the NW side. The lower profile (c), prepared by Bearth and others in 1967 for the *Geological Guide of Switzerland*, was a modification of the work of Schmidt and Preiswerk (1908). The core nappe in the synform was called the Berisal nappe. What had been mapped as the basement under the Antigorio in the tunnel was interpreted by Bearth as the crest of a digitation of the Lebendun nappe.

nappes, however, the gneisses of Monte Leone and Lebendun are not enveloped by Mesozoic sedimentary strata. Invoking the mélange concept, Milnes suggested that they were not cores of recumbent folds, but slabs of gneiss basement intercalated in a schistose matrix of a tectonic mélange. The Monte Leone and Lebendun units are neither nappes nor recumbent anticlines squeezed out of a root zone, but tectonic slices sheared off from other basement highs and mixed in a mélange before the whole pile was folded.

The Core Nappes of the Valais

The rocks of the Penninic Alps were grouped by Argand into two types: the massive quartz-feldspar gneisses are the cores of the mobilized-basement nappes, whereas layered schists, quartzite, marble, etc. are the metamorphosed sedimentary cover wrapping around the recumbent folds.

The major core nappes of the Pennine Alps, according to Argand (1911), are as follows, in descending order (Figure 7.2):

- VI.   Dent Blanche nappe
- V.   Monte Rosa nappe
- IV.   Bernard nappe
- III. Monte Leone nappe
- II.   Lebendun nappe
- I.   Antigorio nappe

Argand's concept of a root zone for all Penninic nappes, as shown in this diagram, was that each basement core of a nappe should have been the "root zone" for a particu-

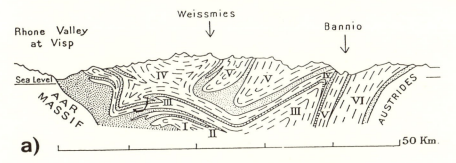

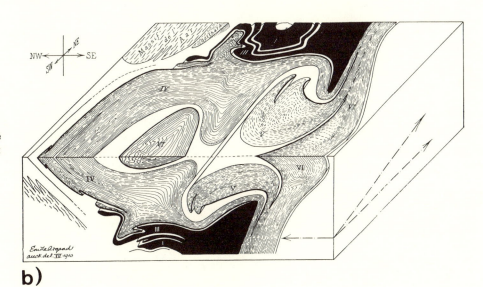

Figure 7.2. Argand's Penninic nappes. (a) Argand's structural cross section of the Pennine Alps south of Visp (after Bailey 1935). I: Antigorio; II: Lebendun; III: Monte Leone; IV: Bernard; V: Monte Rosa; VI: Dente Blanche. Although the geometric interpretation of this section has stood the test of time, new ideas have deviated somewhat from Argand's classic scheme (see text for discussion). (b) Argand's tectonic stereogram (after Argand 1911). This diagram shows the clear separation of Bernard and Monte Rosa.

lar sequence of sedimentary cover. The cores of the Antigorio and Bernard nappes, for example, may have been the Habkern Swell or the "marginal basement high" and Brianconnais Swell, respectively. Now that the Monte Leone and Lebendun are no longer considered basement cores, there is no compelling reason to postulate additional paleogeographical elevations as the homes of those thrust slices; they may have been derived from the Antigorio or the Bernard root zone, such as the Habkern or Brianconnais Swell. The Dent Blanche nappe, as mentioned previously, is Austroalpine. The only debatable question today in Penninic geology concerns the home of the Monte Rosa nappe.

The basement cores of the Antigorio and the Bernard were deformed and metamorphosed during the Alpine orogenesis. The micaceous minerals of the gneisses have a preferred orientation, or gneissosity, marking planes of penetrative shearing during the basement deformation. The gneissosity commonly conforms to the outline of the core-nappe geometry; it is thus mostly flat, as a casual observer might note while traveling across the Canton Tessin. Yet one comes across places where the gneissosity is found to steepen locally to make a turn, to form the crest or hinge of a recumbent fold (Figure 7.3).

The outline of the fold is defined not only by the core geometry, but also by the sedimentary layers that wrap around the recumbent fold. Most of these are schists subjected to various degrees of metamorphism, and are called *schistes lustrés*, or *Glanzschiefer*, because of the lustrous appearance given to the rocks by their metamorphic micas. Their original sediments have a monotonous lithology and a great thick-

Figure 7.3. Structure at the hinges of the Tessin nappes. (a) Structural cross section showing the Tessin nappes south of the Bedretto Valley in Canton Tessin. This section, some 40 km NE of the Simplon Tunnel, was published by Heim (1922). The Maggia nappe in this profile seems to occupy a position similar to that of the Bernard nappe in Schardt's Simplon profile, but it is considered a digitation of the Tessin nappes that underlies structurally the Adula nappe of southeastern Switzerland; the latter is the equivalent of the Bernard. (b) This sketch profile by Heim (1919–1922) shows that the layering of the sedimentary cover seems to wrap around the several "digitations" of the Maggia nappe. A more elegant solution is to assume that the gneiss is not the core of basement nappes, but thrust slices of granite (Lebendun), interlayered with slabs of marble (Triassic), embedded in the schistose matrix (Mesozoic Bündnerschiefer) of a mélange. The layers are separated by shear planes, resulting from penetrative shearing, and not by bedding planes.

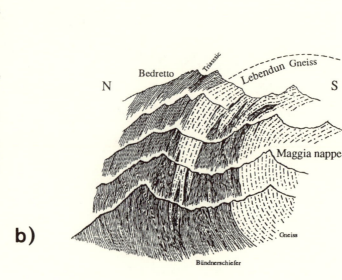

ness, the metamorphism has practically eliminated all fossils, and the bedding planes have been largely destroyed by pervasive shearing. Consequently, it is all but impossible to work out the geometry of structures by studying the schists. Fortunately, thin layers of metasedimentary rocks with a distinct lithology, such as quartzite and marble, are present here and there. Layered rocks consisting of quartz or of carbonate minerals are as a rule sedimentary rocks. Even Gerlach was able to correlate the highly metamorphosed marble with the Triassic dolomite outside the Pennine Alps. He was able to identify the Antigorio nappe, because he noted the presence of the marble both above and below the Antigorio gneiss.

When the strata are mapped in detail, one perceives that the layering of sedimentary strata was disrupted prior to the folding (Figure 7.3). The recumbent folds of the Penninic nappes, as mentioned previously, are not all recumbent anticlines. Many are recumbent *antiforms*, because the folded layers are not a stratified sedimentary sequence, but a pile of penetratively sheared metamorphic rocks that are not present in their normal stratigraphic superposition. The structure of the antiforms indicates that there was an earlier stage of simple-shear deformation by underthrusting, prior to the folding of the mélange units around the core nappes.

After the Antigorio nappe, the Bernard nappe is the most prominent core nappe of the Pennine Alps. The recumbent nature of the nappe was verified through a correlation of the normal and overturned limbs at the outcrops and in the Simplon Tunnel (Figure 7.1). A remarkable aspect of the Bernhard nappe is that the vergence, or direction of tectonic transport, seems to be southward, or in a direction opposite to the general trend of Alpine deformation. This apparently reverse vergence is manifested by the north-dipping layered rocks between the Bernard (IV) and Monte Rosa (V) nappes (Figure 7.4). The usual explanation, ever since the days of Argand, is backthrusting: After the nappes were piled on top of one another, the whole nappe pile was folded. The mobilized basement of the Bernard nappe was deformed ductilely. Like toothpaste spilling out of the confines of the tube, the ductile basement core was thrust southward to rest on top of the Monte Rosa nappe, while the latter was being pushed northward and squeezed under the Bernard (Figure 7.2b). In his interpretation Argand considered the Monte Rosa nappe a distinct core nappe, separated by its own sedimentary cover from the Bernard nappe (see Figures 7.2 and 7.4). Staub (1937), however, citing the field work by Huang (1935), came to the conclusion that the basement cores of the two are not separated by sedimentary covers. He considered Bernard and Monte Rosa two parts of one giant basement nappe, the so-called Mischabel nappe, corresponding to the Tambo/Surreta of Graubünden. In the schemes of both Argand and Staub, the "Piedmont Ocean," remnants of which are represented by the ophiolites of the Saas-Zermatt zone, was located south of the Bernard and Monte Rosa realms of sedimentation.

Amstutz (1971) ventured an unorthodox interpretation and postulated an ocean between Bernard and Monte Rosa. He thought that this ocean was consumed by subduction along a north-dipping Benioff zone, leading ultimately to a Cretaceous continental collision when the Bernard nappe was thrust southward on top of the Antrona ophiolite and both of these were thrust onto the Monte Rosa nappe. Bearth's (1974) field mapping indicated to him, however, that the Antrona ophiolite occupies a tectonic position below the Monte Rosa nappe, not above, as postulated by Amstutz. Correlating the Antrona ophiolite with that of the Saas-Zermatt zone, which is tectonically superposed above the Monte Rosa, Bearth dismissed the possibility of a southward thrusting of the Bernard over the Monte Rosa. Milnes *et al.* (1981) accepted Bearth's correlation of the two ophiolites, but they concluded nevertheless that Bernard and Monte Rosa were separated by an ocean, because they discovered a thin septum of mélange between these two nappes. Their Monte Rosa, situated south of the ocean, is thus Austroalpine. I agree with this interpretation and shall discuss the geology of the Monte Rosa unit in Chapter 9.

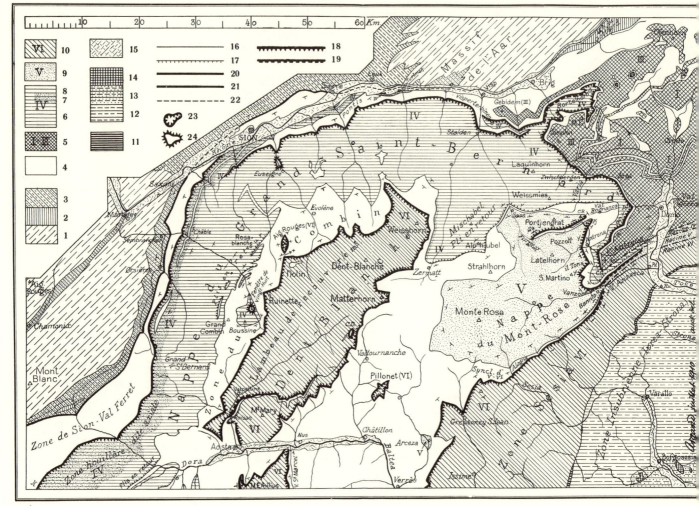

a)

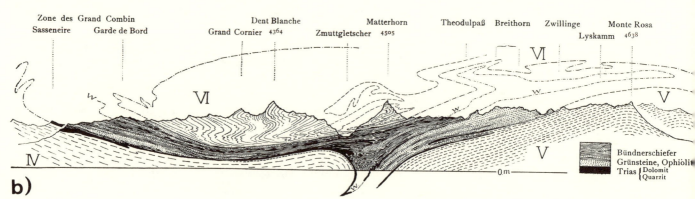

b)

Figure 7.4. Geology of the Zermatt region. (a) Sketch map of the Pennides of the Valais, showing the distribution of tectonic units (after Argand 1911 and Bailey 1935). Argand made his tectonic stereogram on the basis of this map. The relation between the Bernard/Monte Rosa (IV and V) and Simplon (I–III) nappes in the area northeast of Weismies is somewhat uncertain because of faulting. 1: Autochthonous massifs; 2: parautochthonous massifs; 3: Helvetic nappes; 4: Upper Penninic schists and mélanges; 5: Antigorio (I), Lebendun (II), Monte Leone (III); 6–8: Bernard (IV); 9: Monte Rosa (V); 10: Dent Blanche (VI); 11: Canavese zone; 12: Ivrea zone; 13–15: Southern Alps. (b) A structural cross-section from the Dent Blanche to the Matterhorn and Monte Rosa, showing the north-dipping schists between the Bernard and Monte Rosa nappes (after Heim 1919–1922). This structure is commonly considered evidence of back-thrusting during Neo-Alpine deformation. Southward vergence during Eo-Alpine deformation is an alternative explanation.

Except for the Antigorio and the Bernard/Monte Rosa, the Penninic nappes in western and central Switzerland do not have a massive basement core. The Lebendun nappe, for example, consists mainly of tectonic slices of paragneisses and schists. The *Konglomeratgneis* of Heim (1922, p. 483) is in fact a partially granitized metaconglomerate. The augen gneiss is a feldspathized, thinly laminated, calcareous schist. The contact between the gneisses and the schists is commonly gradational. The Monte Leone "nappe" also does not have a basement core around which a well-defined sedimentary cover is wrapped. Southeast of the Antigorio Valley, the Monte Leone is separated from the Antigorio nappe by "an extremely heterogeneous zone containing streaks of cover-like marble and calcareous schist at various levels. The zone contains gneisses and schists of almost any composition and texture, as well as amphibolites and ultramafic pods and lenses" (Milnes 1974b, p. 336). Northwest of the Antigorio Valley, the same mélange, including a large exotic slab of the ultramafic body at Geisspfad Pass, forms part of this Monte Leone unit.

Pennine Paleogeography

The Aar massif is an autochthonous massif north of the Helvetic margin. The Gotthard massif is the basement under the Ultrahelvetic sediments; the Lower Jurassic dark gray shales on the south side of the Gotthard (Figure 7.5) are hemipelagic, typically Ultrahelvetic. The Penninic realms lay to the south of the Ultrahelvetic, and should thus have been located in the middle of the Tethys Ocean. Ocean floor is underlain by oceanic lithosphere, but the basement cores of the Penninic nappes consist mainly of granitic rocks derived from continent.

Figure 7.5. Structural cross section along the Lukmanier Pass across the Aar and Gotthard massifs (after Niggli and Nabholz 1967). The profiles show the valleys between the Aar massif (autochthon), the Tavetsch Zwischenmassif (basement of Helvetic nappes), and the Gotthard massif (basement of Ultrahelvetic nappes). The Penninic core nappes should have come from a region between the Ultrahelvetic and Austroalpine realms of sedimentation.

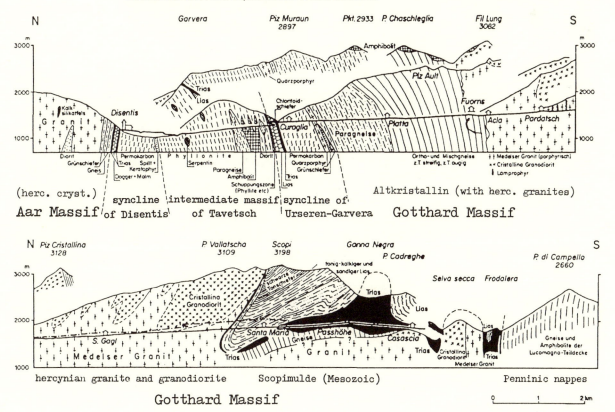

This paradox requires two explanations: (1) continental crust under island chains in the Tethys Ocean had been underthrust, then overthrust again to form the Penninic core nappes, and (2) ocean crust under the Tethys has been carried down the Benioff zone, except for a few slices preserved in ophiolite mélanges.

Where were those islands or island arcs in the Tethys Ocean and what has become of them?

Alpine geologists identified the Brianconnais Swell as the intra-oceanic island chain—a remnant arc—and the granitic basement of the swell as the core of the Bernard nappe. That the Bernard crystalline complex was the basement of an intra-oceanic swell has been a working hypothesis since the time of Argand (1911). The sedimentary cover of the basement is shallow marine and hemipelagic, such as that deposited on remnant arcs. Bernard was intra-oceanic, as indicated by the fact that the nappe is enveloped by *schistes lustrés*, metamorphosed hemipelagic sediments of the Tethys.

We have also discussed the fact that the stratigraphic sequence of the Median Prealps was the sedimentary cover of a remnant arc separating the Valais and Piedmont troughs, corresponding to that of the Brianconnais Swell in the French Alps, where the Brianconnais and Subbrianconnais units are sandwiched between the nappes of the Ultradauphinois (= Ultrahelvetic) and the Helminthoid Flysch (= Simmen nappe). The tectonic superposition of the Brianconnais was equivalent to that of the Bernard, and this fact is the basis of the postulate that the rocks of the nappes of the Median Prealps were the sedimentary cover above the Bernard basement. A comparison of the autochthonous Bernard cover and the sediments of the Brianconnais facies verified this interpretation.

The tectonic superposition on the left side of the Matter Valley, north of Zermatt, is as follows (Figure 7.6):

Dent Blanche nappe
Ophiolite Mélange (Saas Zermatt zone), with exotic radiolarite
Barrhorn Series
Schistes lustrés mélange, with Triassic exotics
Bernard nappe, with its sedimentary cover

The autochthonous sedimentary cover of the Bernard nappe was described by Iten (1948): In the Mettelhorn-Diablons region, the Carboniferous strata have been metamorphosed to form graphitic phyllite, graphitic quartzite, and chlorite/albite/epidote schist (Figure 7.7). The Permian consists of sericitic quartzite, metaconglomerate, and quartz porphyry, apparently a metamorphic equivalent of the Verrucano Formation, widespread in the Alpine region. The Lower Triassic includes quartzite, quartz-mica schist, and dolomitic phyllite, the Middle Triassic dolomite marble and Rauhwacke. Younger rocks at this locality have been sheared off. Elsewhere in the area of the Bernard nappe, the Permo-Carboniferous and Triassic strata are stratigraphically overlain by a predominantly carbonate, younger Mesozoic sequence (Laubscher and Bernoulli 1980; Sartori 1987; Escher 1988). The Barrhorn Series of the Mischabel region, for example, ranges from Permian to Eocene in age (Figure 7.8). The presence of Eocene Flysch, Cretaceous/Paleocene *couches rouges*, Upper Jurassic massive limestone (marble), and the insignificant development of Lower Cretaceous and Lower Jurassic leave little doubt that the Barrhorn sequence is correlative to the sedimentary cover of the Brianconnais Swell.

North of the Brianconnais Swell was the Valais Trough. The Antrona ophiolite is probably North Penninic, and the occurrence of Geisspfad ultramafic body in the Monte Leone mélange indicates that the Valais Trough was underlain in part by oceanic lithosphere. Farther north must have been the root zone for the Simplon nappes. The basement of the Antigorio nappe, the lowest, could be the "marginal basement high" or "Habkern Island" bounding the Ultrahelvetic on the south (Figure 6.10). The Lebendun and Monte Leone are not core nappes, but mélanges, and their exotic slabs

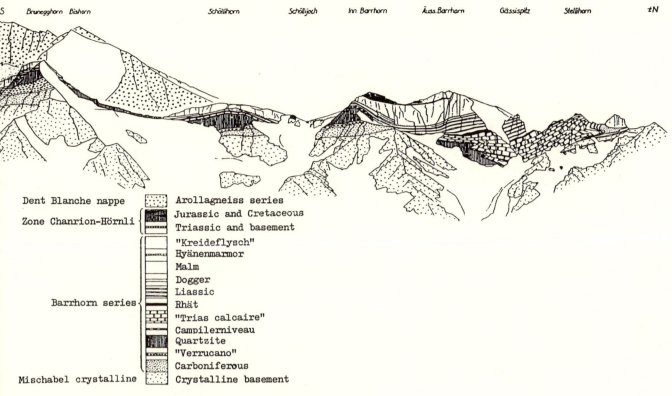

Dent Blanche nappe ▦ Arollagneiss series

Zone Chanrion-Hörnli { Jurassic and Cretaceous
                         Triassic and basement

                        "Kreideflysch"
                        Hyänenmarmor
                        Malm
                        Dogger
                        Liassic
Barrhorn series {   Rhät
                        "Trias calcaire"
                        Campilerniveau
                        Quartzite
                        "Verrucano"
                        Carboniferous

Mischabel crystalline ▨ Crystalline basement

Fig. 7.6. Barrhorn series between the Dent Blanche and Bernard nappes (after Staub 1958). Upper Paleozoic and Mesozoic sedimentary rocks between the Dent Blanche and Bernard nappes are the autochthonous and parautochthonous cover of the latter. These sediments have a facies development similar to that of the Brianconnais zone of the French Alps.

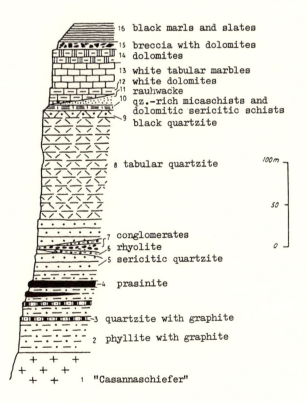

16 black marls and slates
15 breccia with dolomites
14 dolomites
13 white tabular marbles
12 white dolomites
11 rauhwacke
10 qz.-rich micaschists and dolomitic sericitic schists
9 black quartzite

8 tabular quartzite   100m

                                        50

7 conglomerates   0
6 rhyolite
5 sericitic quartzite

4 prasinite

3 quartzite with graphite
2 phyllite with graphite

1 "Casannaschiefer"

Fig. 7.7. Barrhorn series in the Mettelhorn-Diablons area (after Iten 1948). 1–4: Carboniferous; 5–7: Permian; 8–10: Lower Triassic; 11–14: Middle Triassic. The strata younger than Triassic evaporites have largely been sheared off to form the Median Prealpine nappe. Units 15 and 16 are allochthonous.

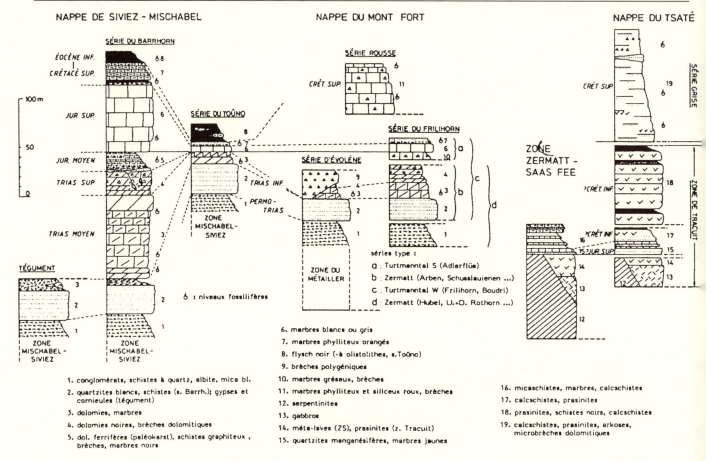

Fig. 7.8. The sedimentary sequences of the South Penninic realm (after Sartori 1987). Whereas normal sedimentary succession is still recognizable in the autochthonous cover (Barrhorn series) of the Bernard nappe, the succession of the Combin zone has been reconstructed on the basis of litho- and bio-stratigraphical correlations, because the present positions of the various slabs in this tectonic mélange do not necessarily correspond to their original stratigraphic superposition.

of sedimentary rocks, continental and/or oceanic basement, have all been derived from the North Penninic realm.

South of the Brianconnais Swell was a steep slope bounding an oceanic back-arc basin. The Jurassic breccias of the Breccia nappe were laid down at the foot of the escarpment, and the hemipelagic *schistes lustrés* were deposited farther out in the "Piedmont Ocean." The next basement unit above the Bernard is the Monte Rosa "nappe," sandwiched between mélanges of the Furgg and Saas/Zermatt zones. According to the simple rule of assigning a more internal paleogeographical position to a higher tectonic unit, the Monte Rosa gneiss should have been the basement of an island dividing the "Piedmont Ocean" into two troughs, corresponding to the Furgg and Saas/Zermatt ophiolites. There is, however, neither sedimentary nor stratigraphical evidence for the existence of such an island. Trümpy (1980, p. 63) stated that in the sedimentary "cover of the Monte Rosa nappe, relatively thin Triassic carbonates (dated by *Encrinus* sp.) are overlain by polygenic breccias which are reminiscent of the ultra-Brianconnais sediments" of the French Alps. Austroalpine margin is "ultra-Brianconnais," being situated in a position more internal than that of the Brianconnais. I have, therefore, postulated that the Monte Rosa unit is not a nappe, but an exotic slab detached from the Austroalpine margin and mixed in the South Penninic ophiolite mélanges (Hsü 1991a; see also Chapter 10).

The Combin zone contains the entire sequence of metamorphosed Mesozoic sediments and ophiolites, which are sandwiched between the Bernard/Monte Rosa and the Dent blanche Nappes (Argand 1911). Various tectonic interpretations have been proposed (e.g., Güller 1947; Iten 1948; Bearth 1976). It is generally recognized that these metamorphic rocks have not retained their stratigraphic coherence. Layers of marbles, quartzite, and other sedimentary rocks are exotic slabs in a schistose matrix of

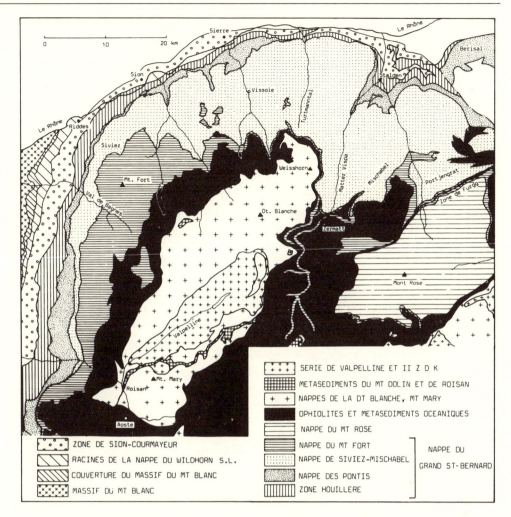

Fig. 7.9. Tectonic sketch map showing the distribution of the various Penninic elements in the area between Aosta and Berisal (after Escher 1988). Note that the Zermatt-Saas zone of Bearth and of Sartori is shown in this map as "*ophiolites et metasediments oceaniques*." Sartori's Mt. Fort "nappe" of the Combin zone is considered a part of the Bernard nappe, and his Barrhorn series a part of the *Nappe de Siviez-Mischabel*.

mélanges. Bearth recognized several thrust slices, such as the Upper Zermatt schuppen and Theodul-Rothorn zone. In a recent study of the Combin zone in the Zermatt region, Sartori (1987) proposed the following superposition of tectonic units (in descending order):

Tsaté "nappe"—calc-schists, metamorphosed Flysch, and ophiolites

Zermatt-Saas zone—metamorphosed ocean sediments (Jurassic and Cretaceous) and ophiolites

Mont Fort nappe—thrust slices, including

Rousse Series, Upper Cretaceous marble and microbreccia;

Frilihorn Series, quartzite, dolomite, and marble, ranging from Permian to Cretaceous in age;

Evolène Series, Mesozoic dolomite, marble, and breccia;

Metailler Series, Permo-Triassic metaconglomerate and metasandstone, Permo-Carboniferous gneiss, mica schist, glaucophane schist, basement gneiss

Barrhorn Series

Sartori's nappes are more or less equivalent to Bearth's schuppen zones; they are called nappes because Sartori was able to recognize southward vergent recumbent folds of sedimentary sequences (Mont Fort nappe) on the back of the Bernard nappe. The distribution of the various units in the area between Aosta and Berisal is shown in Figure 7.9.

The sedimentary rocks in the mélanges range from Permo-Carboniferous to Lower

Eocene. The Barrhorn Series is Brianconnais. The common occurrences in the Mont Fort nappe of breccias younger than Permo-Triassic suggest a correlation with the Breccia nappe. The presence of Flysch in the Tsaté "nappe" is indicative of deposition in a foredeep on the southern margin of the "Piedmont Ocean." The schists and ophiolites of the Zermatt-Saas mélange were obviously the sediments and seafloor of the ocean. The marbles and calc-schists of Tsaté and Mont Fort, representing metamorphosed shallow marine sediments, may have been derived from the Brianconnais or from an Austroalpine passive-margin sequence.

Origin of Core Nappes

Massive granite cannot be bent into folds by compression; it is deformed by faulting or ductile shear. The idea of forming core nappes was thus received with skepticism by students of rock deformation. Folds, and even recumbent folds, could develop, however, where massive rocks are bounded by a planar surface; the parautochthonous deformation of the contact between sedimentary cover and its underlying basement can thus serve as a model with which to reconstruct earlier stages of core-nappe formation (Figure 7.10).

The surfaces folded in the Pennine Alps are commonly not bedding surfaces, but shear planes or foliation. The earliest movement of the nappe complex seems to have been simple shear, or underthrusting. Continental and oceanic basements were sliced up, and discontinuous sedimentary layers (such as marble lenses) are separated from the enclosing schistose matrix by shear planes (Milnes 1974a). As the basement-nappe complex was underthrust to greater depth, the rocks became more ductile, resulting in the development of schistosity or foliation.

A northward vergence of the Penninic thrusting is generally assumed, and the north-dipping structures of the Bernard nappe have been explained by assuming of Tertiary back-thrusting. Yet Amstutz's unorthodox postulate of an earlier southward vergence has not been proven wrong, despite Bearth's argument concerning the correlation of the Antrona and Zermatt/Saas ophiolites. The presence of glaucophane schist of Cretaceous metamorphic age in the Bernard nappe led Hunziker *et al.* (1989) to conclude that the main subduction zone of Eo-Alpine deformation had to be north of the Brianconnais Swell. A north-dipping subduction zone could also account for the creation of the North Penninic Gurnigel Trench and the Ultrahelvetic Schlieren back-arc basin in the Ultrahelvetic (Hsü 1989). Baird and Dewey (1986) postulated Late Cretaceous (100–70 Ma) deformation of the Combin zone. The deformation of the Zermatt-Saas mélange could be related to Eo-Alpine underthrusting on an active margin south of the Piedmont Ocean. The presence of Upper Cretaceous *couches rouges* and Eocene Flysch in the Brianconnais realm argues against a Late Cretaceous deformation of the Bernard cover (Sartori 1987). I shall address the question of the kinematics of the Pennine deformation in Chapter 10.

The North Penninic Trough was eliminated when the Austroalpine nappes rode on top of the Helvetic margins after the early Tertiary Neo-Alpine collision. The Antigorio and Bernard nappes were overthrust northward as core nappes under the Austroalpine; the thrust slices and mélanges between them were folded into the recumbent antiforms of Lebendun and Monte Leone. The Austroalpine Dent Blanche nappe rode on a "cushion" of ophiolite (Zermatt-Saas zone) and eventually overtook all the Penninic nappes; it reached the Alpine front in the Miocene and then began to deliver detritus to the Molasse foreland basin.

Pennine Metamorphism

In the pioneer days of geology, most of the exposed metamorphic rocks of Europe and North America were considered Precambrian. Some British geologists even claimed that intense regional metamorphism had had no effect on Paleozoic or later rocks. Bailey, in his 1935 review of Alpine geology, thus expressed surprise that "Swiss

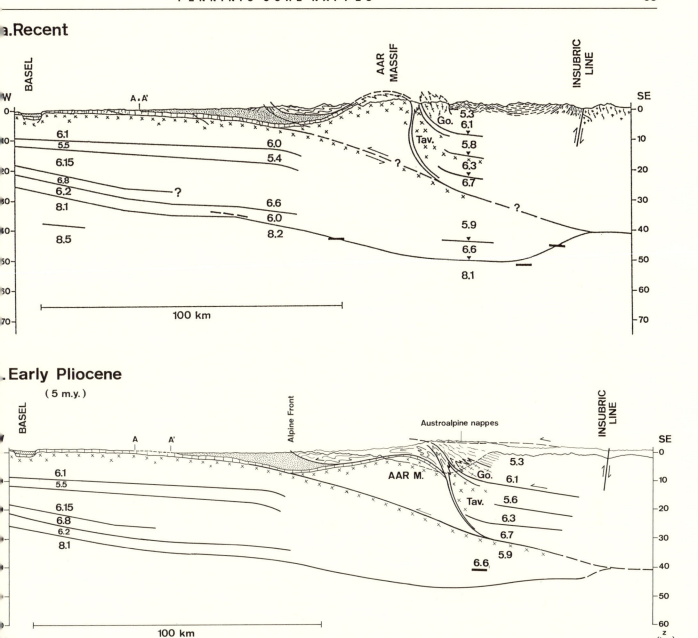

Fig. 7.10. Deformation of the Aar Massif as a model for the initiation of a core nappe. The two profiles are balanced crustal sections showing the Neo-Alpine deformation since the Early Pliocene (Hsü 1979). Crustal shorting has led to (1) the underthrusting of the Jura basement, (2) décollement folding of the Jura Mountains, and (3) compression of the Aar massif. The compression has led to the uplift of, and the folding of the thrust piles above, the Aar massif. Continued compression, in our opinion, could lead to the development of a recumbent core nappe like the Antigorio, while the thrust pile above the basement is folded into a recumbent antiform.

geologists ascribe much of their material to a Mesozoic date." Swiss pioneers were able to arrive at the correct conclusion because some of the Alpine metamorphic rocks are lithologically very similar to their sedimentary equivalents outside the Pennine Alps. The correlation was further verified by the few fossils that had been found in the metamorphic terrane (see Gerlach 1883).

The rocks of the Alps were subjected to various degrees of regional metamorphism during the Mesozoic and Tertiary times. The concept of episodic orogenesis led to the concept of deformational and metamorphic phases; the three phases of the Swiss Alps are Eo-Alpine (Cretaceous), Meso-Alpine (Paleogene), and Neo-Alpine (late Paleogene and Neogene). As we shall discuss later, there has been continuous deformation since the beginning of the Cretaceous. Hsü (1989) proposed, therefore, using the term Eo-Alpine for the deformation and metamorphism prior to the continental collision that eliminated the Piedmont Trough of the Tethys Ocean; the Eo-Alpine stage is mainly Cretaceous and early Paleogene. The term Neo-Alpine refers to the deformation and metamorphism (after the collision) when the Austroalpine nappes were overriding the Penninic and Helvetic nappes; the Neo-Alpine stage began during or after Late Eocene. The events that occurred between the Eo- and Neo-Alpine stages can be considered Meso-Alpine.

It is beyond the scope of the present elementary textbook to present the numerous recent studies that have dealt with the Alpine metamorphism. We shall, therefore, give a very brief summary based upon the short article by V. Trommsdorff (1980) in *Geology of Switzerland*. He recognized, on the basis of metamorphic grades, three domains of metamorphic rocks (p. 83):

1. An outer belt of zeolite and somewhat higher prehnite-pumpellyite grade metamorphism that has affected parts of the Helvetic and Upper Austroalpine nappes, plus the Aiguilles Rouges massif and parts of the Penninic units in the Prealps and in Graubünden.

2. A wide greenschist belt that covers the southern Helvetic, the Mont Blanc and Aar (and Gotthard) massifs, and also much of the Penninic and Lower Austroalpine nappes. This main (Neo-Alpine) greenschist facies overprints earlier (Meso- or Eo-Alpine) glaucophane schist and eclogite assemblages in the Penninic rocks of the Valais Alps as well as in middle and southern Graubünden.

3. The classic amphibolite facies belt that extends over the central Alps between Graubünden and the Valais. This main (Neo-Alpine) metamorphism locally overprints earlier eclogite and garnet-peridotite assemblages of earlier (Eo-Alpine?) age. Toward the east, the kyanite-sillimanite metamorphism of Barrowian type grades into an andalusite-sillimanite contact metamorphism around the Tertiary Val Bregaglia granitoid rocks.

The progressive increase in metamorphic grade southward in the Alps (Figure 7.11) gives a measure of the depth of underthrusting of the various segments of European continental crust below the overriding Austroalpine nappes.

Radiometric dates ranging from Early Cretaceous to Miocene have been reported for the metamorphic rocks of the Swiss Alps. Some are dates of emplacement of igneous rocks, others are dates of metamorphic ages, and still others are "cooling dates," signifying the time when the rock was cooled to such an extent that the escape of daughter product was effectively stopped. We shall refer to some of the key radiometric dates when we discuss the evolving history of the Alpine orogenesis in the last two chapters of this book.

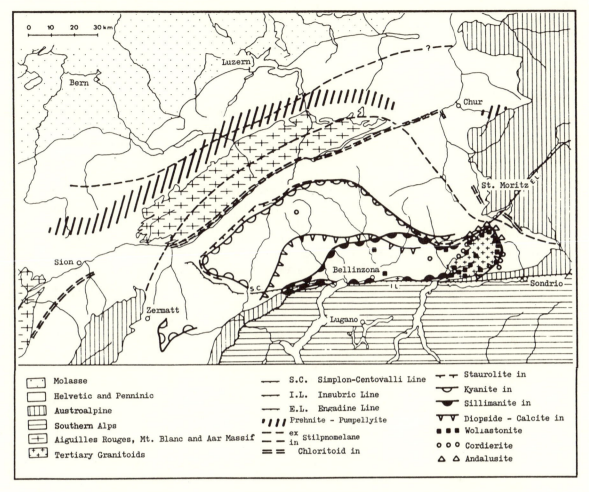

Fig. 7.11. Distribution of metamorphic grade in the Swiss Alps (after Frey and others, 1980).

8.

The Bündner-schiefer

Schists are found in many mountain belts, and they are seldom fossiliferous. The dichotomy between American and European geology can be traced back to the second half of the last century, when schists in mountains were considered basement in America, but ocean sediments on the other side of the Atlantic.

Portraits of Tethys

The Pennine Alps are underlain mainly by granitic rocks, as are the Helvetic and Austroalpine Alps. It was thought that the "Alpine Geosyncline" was ensialic, underlain by a continental crust. This impression seemed to be reinforced by the identification of Triassic quartzite and dolomite in the Alps. Not suspecting that an oceanic realm may once have existed, geologists, especially those in North America, tended to underestimate the amount of crustal shortening during the Alpine orogenesis. It was concluded, for example, that the deformation of the Swiss Alps was initiated by vertical movement uplifting the granite basement under the Pennine Alps (see Bucher 1933). The large horizontal displacement manifested by the Helvetic nappes was considered "secondary tectonics," induced by gravity sliding of the sedimentary cover above an uplifted basement (Haarmann 1930; van Bemmelen 1933).

Alpine geologists, however, seldom lost sight of the evidence for an Alpine Mesozoic ocean. Eduard Suess wrote as early as 1875 in his *Entstehung der Alpen* that some of the Alpine Triassic was deposited in a deep sea, even though much of the thick Triassic of the Austroalpine nappes is shallow marine or coastal sediments.

Suess's student Theo Fuchs published in 1877 a brilliant study on the origin of the Jurassic Aptychus Limestone. Aptychus is the calcitic part of an ammonite, probably a jaw bone. Aptychus Limestone has abundant Aptychus remains, but hardly any coiled ammonite shells. Fuchs knew that ammonite shells consist largely of aragonite and that aragonite is more soluble than calcite in sea water. He correctly concluded that the selective preservation of the calcitic part of an ammonite results from the preferential dissolution of aragonite. When the famous H.M.S. *Challenger* expedition found aragonite on ocean bottom as deep as 4000 m, Fuchs concluded that the Aptychus Limestone was deposited at great oceanic depth. At about the same time, W. Gümbel (1878) noted that the occurrence of nodules rich in iron and manganese in some Triassic red limestone of the Austrian Alps is comparable to the manganese nodules dredged up from ocean bottom floored by red clay. His observation supported Fuchs's postulate.

Another of Suess's students, Melchior Neumayr (1887), wrote on deep-sea sedimentation in the Alpine Tethys, after the results of the *Challenger* expedition were first published. He stressed the similarity of ammonitic limestones of the Alps to Recent oceanic deposits, and postulated that such red limestone with ferromanganese nodules was deposited at a depth between the accumulation levels of the globigerina ooze and the deep-sea red clay of the modern ocean.

British scientists then working on oceanic islands found radiolarian marls interbedded with globigerina chalk. This observation led H. A. Nicholson (1890) to conclude that radiolarian cherts, or radiolarites, and radiolarian marls in ancient rock sequences are in fact consolidated radiolarian oozes; this kind of pelagic sediment, consisting entirely of the siliceous skeletons of single-celled radiolarians, had been sampled on modern ocean floor (Murray and Renard 1891). Nicholson's opinion was enthusiastically endorsed by Gustav Steinmann (1905).

Steinmann noted the association of the Alpine oceanic sediments with ophiolites. These mafic and ultramafic igneous rocks were considered to be magmatic injections into deep-sea sediments. Suess (1909) agreed with Steinmann and considered the Alpine ophiolites part of raised ocean floor. The great resistance to acknowledging the oceanic nature of the Alpine ophiolite and radiolarite came from the English-speaking world. James Hall (1859) pointed out that the fossiliferous rocks of the Valley and Ridge province of the Appalachian mountains were the sediments of a shallow sea. The deep-water equivalent of those strata was largely metamorphic and thus considered the Precambrian basement. In Hall's "geosyncline," all sediments were deposited on shallow bottom.

The objection to the postulate of an oceanic realm in the Alpine Mesozoic was motivated by the dogma of the permanence of ocean and continent—the paradigm of geology during the first half of this century. Continent is underlain by an earth's crust with the composition of a granite, and ocean is underlain by a crust of basaltic composition. There is no way of changing continental crust into ocean crust or vice versa. The gneisses of the Pennine core nappes have a granitic composition, and the region was underlain by continental crust and could thus not have been very deep. All that talk by Austrian geologists about a deep Mesozoic ocean in the Alpine realm seemed to be idle speculation.

In the face of such prejudice, Murray and Renard, the authors of the *Challenger* report, rejected the idea of finding ocean sediments in areas that are now a part of continent. They wrote (1891, p. 189), "With some doubtful exceptions, it has been impossible to recognize in the rocks of the continents formations identical with these pelagic deposits [of modern oceans]."

Murray's influence was considerable. No less an authority than Johannes Walther (1897) expressed the view that pelagic sediments were not necessarily deep or oceanic; he stated flatly that the Alpine radiolarites were not ocean sediments. Another great sedimentologist during the first half of this century, A. W. Grabau, promulgated the same doctrine that pelagic sediments could be shallow marine (Grabau and O'Connell 1917).

A major hindrance to the acceptance of oceanic sedimentation in Mesozoic Alps was geologists' failure to recognize the true nature of the ophiolites and of their deformation. Coarse-grained igneous rocks, such as gabbro and peridotite (now serpentinized), were thought to be magmatic intrusions injected into the rocks around them. When the Alpine sedimentary sequence was considered to be deposits on a continental crust, ophiolites intrusive into the sediments must also have been "continental" in origin.

The hypothesis of continental drift revived the idea of a Mesozoic ocean: Tethys came into existence when Africa and Europe drifted apart, and the ocean was eliminated after the two continents came together and collided. Several Alpine geologists were attracted to Wegener's (1922) theory, because they found a ready explanation for the oceanic rocks they saw—they were the sediments and seafloor of the Piedmont and Valais troughs of the Tethys Ocean (e.g., Argand 1922; Staub 1928). At the same time, Wegener's idea met fierce opposition in the New World, because the North Americans had no place for ocean in their schemes of "geosynclinal" sedimentation.

After the triumph of the plate-tectonics theory in the early 1970s, the doctrine of immovable continents had to be discarded. We now believe that the intrusive rocks of the ophiolite suite, such as gabbros and peridotites, were injected into an ocean lithosphere. This suite, consisting of pillow basalts, pillow breccias, sheeted-dyke complex, mafic and ultramafic intrusions, was fragmented, and the broken slabs were mixed with pervasively sheared sediments, when the ocean plate was plunged into a Benioff zone. The Penninic mélanges were formed by such a subduction process during Eo-Alpine deformation. The Alpine gabbro and peridotite are not intrusive into sediments; they are embedded as exotic slabs in the schistose matrix of a mélange.

The theory of seafloor spreading explains the birth of a Penninic Ocean, and the theory of plate subduction accounts for its demise. In the ophiolites in Greece and on Cyprus, we have a model of ocean crust, and sampling of ocean sediments by the Deep Sea Drilling Project has yielded actualistic analogues of Penninic sediments of the

Alps. Investigations of the Franciscan Mélange of the California Coast Range have provided a basis for understanding Eo-Alpine deformation.

The knowledge that a Penninic Ocean once existed opens up new possibilities for interpreting the geological evolution of the Alps. Meanwhile, many new questions are raised. When did this ocean come into existence? Was the ocean floor underlain by a broad expanse of ocean crust, or was it a narrow trough? Was there one, or were there two or more deep-sea depressions in the Penninic domain? How did the ocean disappear? When did it begin to disappear? When did it disappear altogether? How was it deformed? Was the deformation always characterized by a northward vergence? If there was a swell continental crust separating the Penninic domain into two oceanic basins, what happened when the swell collided with the continents? Which collision came first, and which came later? What are the structures produced by those continental collisions? Where were the Penninic Flysch sediments laid down? How did they get deformed?

The Bündnerschiefer, the ophiolites, and the Flysch give a record of the geological evolution of the Pennine Alps. We shall attempt to find answers to these questions in this chapter and the next.

Bündnerschiefer and Schistes Lustrés

The rocks surrounding the Penninic core nappes are (1) *Bündnerschiefer* or *schistes lustrés*, (2) pelagic sediments such as radiolarite and *Calpionella* limestone, (3) ophiolites, and (4) Flysch.

The term *Bündnerschiefer* was used by Studer (1853) to designate the schistose formations of the Prättigau-Chur-Domleschg area, but was later extended to include all such rocks of northern and middle Graubünden. Studer (1872) gave a vivid description of its lithology:

> The *Bündnerschiefer* consist of gray and black shale, marl, and argillaceous limestone, which are hardened by silicification. Also intercalated are thick-bedded limestone and dark gray, well lithified sandstone and sandy shale layers. The clay minerals in the shale are locally recrystallized to sericite in phyllites and to muscovite in schists. [free translation by the author]

The rocks in western Switzerland correlative to the Bündnerschiefer are, on the whole, more highly metamorphosed. Lustrous mica schists in the French Alps were called *schistes lustrés*. Eventually the term was extended to include all the dominantly schistose rocks of Mesozoic age in the Penninic units of the Valais Alps. These rocks are a main constituent of the Combin zone of the Valais. Heim (1919–22) translated the term into German, and he considered *Glanzschiefer* a synonym of Bündnerschiefer. Aside from mica schists, Heim recognized the common presence of quartz schist, calcareous phyllite, and gray schist in the formation. The different types indicate the varying lithology of the original sediments and their different degrees of metamorphism.

The present usage of the term Bündnerschiefer, *sensu stricto*, is limited to the slightly metamorphosed, dominantly pelitic rocks in the Penninic units, which crop out, as a rule, in Canton Graubünden, whereas the term *schistes lustrés*, *sensu stricto*, is used for the mica schists of the Valais. Local formational names are given to different bodies (e.g., Ferret Schist, Prättigau Schist, etc.) to emphasize their various distinctive lithologies, paleogeographies, and degrees of metamorphism.

The Bündnerschiefer of Misox, Via Mala, and Domleschg

Returning from a vacation in Italy, one turns right at Bellinzona, near the "root zone" of the Penninic core nappes, and goes up the Misox Valley. One soon leaves the gneiss terrane behind. Exposed at roadcuts are mica schists and amphibolites, which are the metamorphosed sedimentary cover of a core nappe, the Adula. After crossing the San

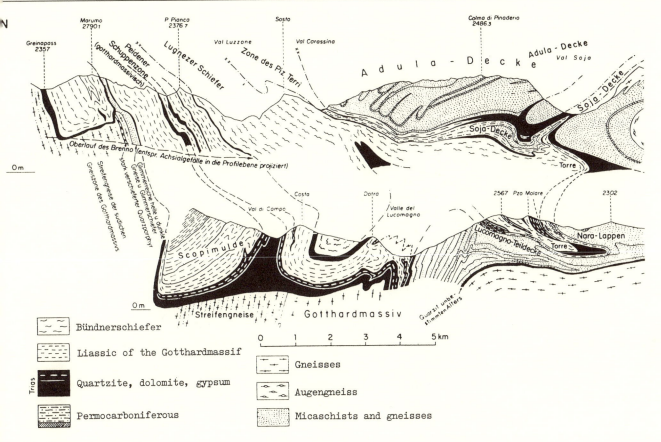

Fig. 8.1. Structural cross section of the east side of the Blenio Valley (above) and through the mountains between the Blenio and Leventina valleys (below) (after Niggli and Nabholz 1967).

Bernardino Pass and going down the country of the Schams Dilemma, one enters the gorge of Via Mala ("Bad Road"). The highway to Thusis has been cut into vertical cliffs of Bündnerschiefer (Figure 8.1). The road was "bad" because it was dangerous for travelers during the Middle Ages, and the road is "bad" because it requires constant repair. After Thusis, the valley opens up into the Domleschg, which is underlain by a "sea" of Bündnerschiefer. The highway turns to the right after the two arms of the Rhine meet; a freeway has been built in the broad Rhine Valley going north after Chur. The hilly country on the right bank of the Rhine is Prättigau, whence the Prättigau Schist derives its name (Figure 8.1). For a tourist, the two-hour drive traverses a breathtaking landscape of Switzerland. Geologists see a different landscape—they have descended the European continental margin at Misox, crossed the North Penninic Ocean through the Via Mala and Domleschg, gone up the Brianconnais Swell before arriving at Chur, back down to the deep sea of Prättigau again, before returning to the Helvetic (European) margin at Sargans. In the Bündnerschiefer of Misox, Via Mala, and Domleschg is a record of the history of the birth and demise of an ocean.

The Penninic core nappes of southeastern Switzerland and their correlatives in western Switzerland, according to Heim (1922), included the following (in descending order of tectonic superposition):

Margna nappe	Dent Blanche nappe
Tambo	Monte Rosa nappe
Suretta nappe	
Adula nappe	Bernard nappe
Tessin nappes	Lebendun nappe

The current consensus is that Tambo and Suretta are part of one nappe complex, the Tambo/Suretta, correlative to the Bernard/Monte Rosa of the west, and that the Adula

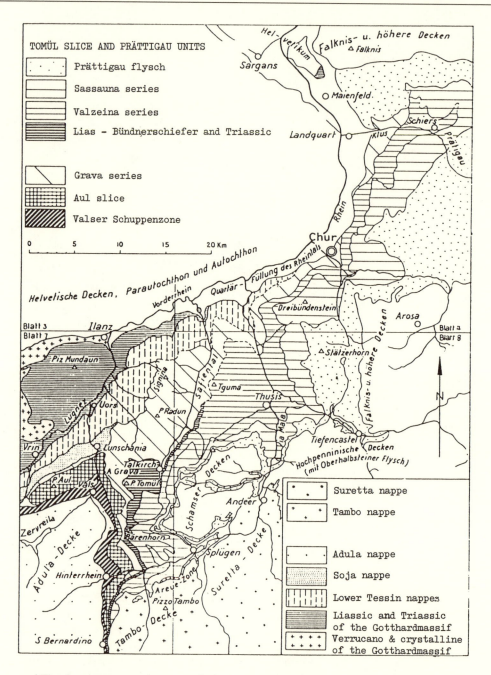

Fig. 8.2. Tectonic sketch map of the region between Prättigau and the Hinterrhein Valley (after Nabholz 1967). The tectonic units of the Misox zone, variously designated, are mélanges.

and Tessin nappes are thus correlative to the Simplon nappes. Hsü and Briegel (1991) proposed still another correlation:

Lower Austroalpine nappe	Dent Blanche nappe
Mélange ("Platta nappe")	Mélange (Tsaté/Zermatt-Saas)
Margna	Monte Rosa
Mélange (Avers)	Mélange (Furgg)
Tambo/Suretta/Schams	Bernard/Barrhorn Series
Ferret Schist/schistes lustrés	Mélange (Misox), Bündnerschiefer
Adula nappe	Simplon nappes

A profile across the Central Alps from Canton Tessin to Graubünden shows some similarity to the Simplon profile, although exact correlation of the core nappes and recumbent antiforms is neither possible nor sensible (Figure 8.2).

Fig. 8.3. Tectonic mixing of (1) calcareous schist, (2) ophiolite, and (3) granitic gneiss (after Gansser 1967). Note the scale of intimate mixing. × denotes shear surfaces.

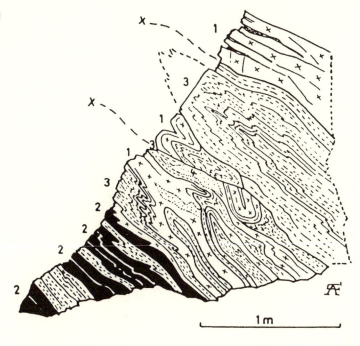

Fig. 8.4. Structural cross section of Giumella Grant (after Gansser 1967). The Ophiolite Mélange between the Adula and Tambo nappes consists of (1) Adula Gneiss, (2) gneiss slices, (3) dolomite and garnet schist, (4) calcareous schist, (5) pelitic schist, (6) marble, (7) ophiolite, (8) rauhwacke, and (9) Tambo Gneiss.

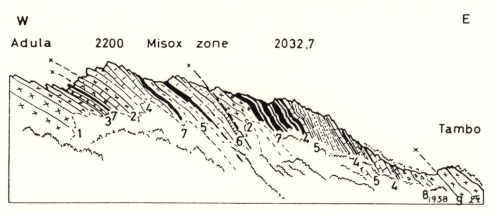

The metamorphic rocks of the Misox Valley belong to the Misox zone between the Adula and Tambo/Suretta nappes. They are a tectonic mixture of fragments and slabs of (a) sedimentary cover of the Adula basement, (b) ocean sediments, and (c) ocean floor. Nabholz (1945) recognized numerous tectonic slices, e.g., Tomül Sheet, Grava Sheet, Aul Sheet, and Upper and Lower Valserschuppen, in the Misox Valley (Figure 8.2).

Assuming normal stratigraphical superposition of the sedimentary strata in each sheet, and also that ophiolite rocks were intrusive into the sediments, Nabholz attempted stratigraphical reconstructions for each sheet. However, the interfingering of carbonate-orthoquartzite layers and Bündnerschiefer of hemipelagic origin does not make sense sedimentologically. We now realize that the Tomül, Grava, Aul, and Vals are neither schuppen nor thrust sheets of coherent packages of strata; they are mélange units, each characterized by a different assemblage of exotic slabs in a matrix of Bündnerschiefer. The intimate mixing of the ophiolites and sediments is illustrated on the scale of an outcrop (Figure 8.3) and that of a structural profile (Figure 8.4).

The Bündnerschiefer of Via Mala and Domleschg is a thick hemipelagic sequence of remarkably monotonous lithology. The dominant lithology was mud, but it has been metamorphosed into a shale or phyllite. The metamorphic grade of the rock is so low, however, that the only metamorphic mineral discernable in hand specimens is sericite.

m.y.	Period	Stage	Dinoflagellaten- & Palynomorphen-Zones	
92	CRETACEOUS	CENOMANIAN	Trithyrodinium suspectum	Complexiopollis [x]
100			Deflandrea echinoidea	Psilatricolporites
108		ALBIAN	Deflandrea vestita	Tricolpites minutus
115		APTIAN	Odontochitina [x] operculata	Clavatipollenites [x]
121		BARREMIAN		
126		HAUTERIVIAN	Druggidium rhabdoreticulatum	Ephedripites multicostatus
131		VALANGINIAN	Druggidium deflandrei / D. apicopaucicum	
141		BERRIASIAN	Biorbifera johnewingii	
	JURASSIC	PORTLANDIAN	Ctenidodinium panneum	
		KIMMERIDGIAN	Gonyaulacysta cladophora [x]	
161		OXFORDIAN	Gonyaulacysta jurassica	
		CALLOVIAN	Valensiella vermiculata	
170		BATHONIAN	Gonyaulacysta filapicata	
176		BAJOCIAN	Mancodinium semitabulatum	
180		AALENIAN / TOARCIAN	Nannoceratopsis gracilis [x]	
		PLIENSBACHIAN	Echinitosporites cf. iliacoides	
190		SINEMURIAN	Cycadopites subgranulosus	
195		HETTANGIAN		
	TR.	RHAETIAN	Corollina meyeriana	

(Retitricolpites geol. — marginal column between Albian/Cenomanian rows)

[x] also recognized in the Bündnerschiefer

PFÄFFIGRAT-Serie [*] TURONIAN
SASSAUNA-Serie MIDDLE CRETACEOUS (6) — VIA MALA [*] CENOMANIAN (8)
(5) VALZEINA-Serie LOWER CRETACEOUS (7) RHÄZÜNS [*] LOWER CRETACEOUS
KLUS-Serie (4) LOWER CRETACEOUS
BEDRETTO [*] (3) MIDDLE UPPER JURASSIC
SAN BERNARDINO II [*] (1) MIDDLE JURASSIC
SAN BERNARDINO I (2) LOWER MIDDLE JURASSIC
OPHIOLITHES
[*] NÄNNY, 1948
[#] PANTIĆ & GANSSER 1977

Fig. 8.5. Dating of Bündner-schiefer in the Hinterrhein Valley (after Pantic and Isler 1978). Note the facies change from carbonate to pelitic sediments at the end of the Triassic, signifying rapid subsidence during the rifting phase of the Alpine deformation.

The rock has about the same degree of lithification and metamorphism as a slate; it is not called slate, because the planar structure is not slaty cleavage, but bedding-plane shear, indicative of penetrative shearing. Intercalated in the shale are thin layers of siltstone, some of which show faint cross-lamination or graded bedding. The sedimentary association is typically that found on continental slope, continental rise, or distant abyssal plain in an open marine environment. The terrigenous component far outweighs the biogenic component of this sediment, indicating the existence of a nearby continent to supply fine detritus.

Very few megafossils have been found in the Bündnerschiefer, *sensu lato*. The sediments of the Misox zone had been considered Lower Jurassic, because Early Jurassic fossils were found in a schist. Staub (1937), Jäckli (1941), and Nabholz (1945) correlated the black pyritic shales (Nolla Tonschiefer) of the Tomül "sheet" to the black pyritic Aalenian (lower Middle Jurassic) shales of the Helvetic and Ultrahelvetic domains. Later Staub (1958) returned to the classical postulate of a comprehensive series ranging from Jurassic to Cretaceous in age. Trümpy (1960) favored the correlation of the Bündnerschiefer to the Ferret Schist; the two are lithologically similar, and they also have similar paleogeographical positions. Thanks to the palynological work by Pantic and his co-workers, the idea of a comprehensive series has prevailed (Pantic and Gansser 1977; Pantic and Isler 1978).

The Misox zone rocks in the Vals and Safien valleys (Figures 8.2 and 8.5) have yielded Jurassic pollen. Metamorphosed sediments of the carbonate-orthoquartzite association ("Gneisquarzit," "Stgir-Serie") are Lower Jurassic, but the schists of hemi

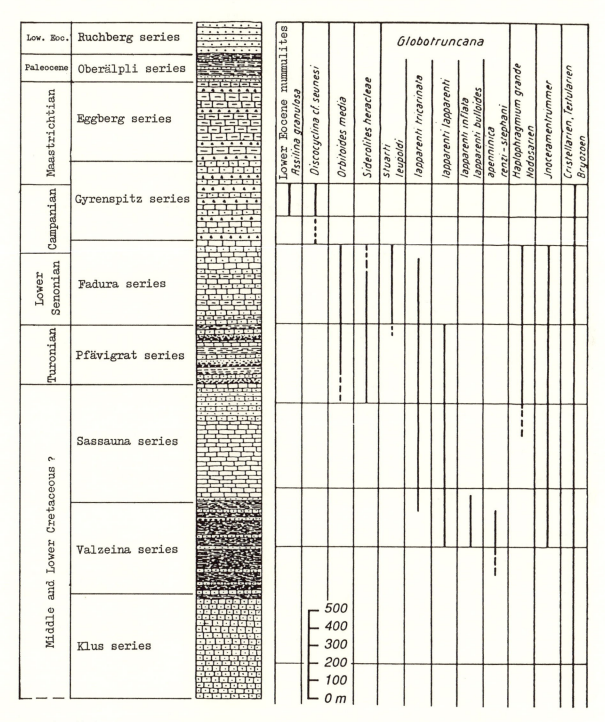

Fig. 8.6. Columnar stratigraphical
section of the Prättigau Schist
(after Nänny 1948).

pelagic origin are mainly Middle and Upper Jurassic (Pantic and Isler 1978). The Lias quartzite, like the ophiolite, occurs as exotic slabs in a Bündnerschiefer matrix.

South of San Bernardino, palynomorphs extracted from phyllite layers, which are intimately associated with ophiolite slabs in a mélange, are Middle to Late Jurassic in age. The Bündnerschiefer samples, obtained near Rhäzuns of Domleschg, are Lower Cretaceous. The Via Mala samples have yielded a Cenomanian palynoflora. Farther north, the Bündnerschiefer of the Prättigau Schist Group ranges from Early Cretaceous to early Upper Cretaceous (Turonian) in age. The stratigraphy of the group, as shown by the schematic columnar section (Figure 8.6), records a history of subsidence and continuous sedimentation during the Cretaceous.

The dogma that ophiolitic gabbro and peridotite are intrusive into the Bündner-schiefer persisted even to the time when Pantic and Isler concluded their significant contribution; these authors mistakenly suggested igneous (intrusive) emplacement of the ophiolitic rocks during the Middle Jurassic. We now recognize that the ophiolite are fragments of disrupted ocean floor; they must be older than the oldest Bündner-schiefer, but younger than the Liassic shallow marine deposits. The age of the ophio-lites is thus dated early Middle Jurassic.

Prättigau Flysch

The name *Prättigauschiefer*, or Prättigau Schist, refers to a thick sequence of Bündner-schiefer and Flysch of the Prättigau Valley in eastern Switzerland. These Mesozoic/Paleogene strata have been divided on the basis of sedimentological criteria into three groups (Trümpy 1960): the Bündnerschiefer, the pre-Flysch, and the Flysch.

The Bündnerschiefer, *sensu stricto*, is a monotonous alternation of shale, calcareous siltstone, and sandy limestone of Early and early Late Cretaceous (Cenomanian) age. Microbreccia beds are present, but uncommon. The pre-Flysch, ranging from Late Cretaceous (Turonian) to Paleocene in age, consists of sandy, silty, and argillaceous limestones, calcareous sandstone, and marly shale, all slightly metamorphosed. Breccia and conglomerate interbeds are not uncommon. The lithology is very similar to that of the Niesen Flysch. Rudolf Trümpy (1960) introduced the term pre-Flysch to describe its transitional lithology between that of the Bündnerschiefer, *sensu stricto*, and that of the Schlieren-Gurnigel type of sandstone flysch. The pre-Flysch sediments were laid down on continental rise or in a deep-sea basin by contour currents, or by dilute turbidity currents. The typical flysch of Prättigau, or the Ruchberg Series, is Early Eocene. The interbedding of thick arkosic sandstone or breccia beds with shale is a Schlieren-Gurnigel lithology. Carbonate strata are rare or absent.

The Prättigau Schist is a North Penninic unit, commonly correlated to the Ferret Schist/Niesen Flysch in the western Alps. The sedimentary sequence shows a progres-sive increase in topographic relief in the Prättigau domain during the Cretaceous, cul-minating in paleogeographic conditions favorable for the deposition of coarse flysch turbidites.

North Penninic Bündnerschiefer

The sedimentary cover on the south side of the Gotthard massif ranges from Permo-Carboniferous to Jurassic in age (Figure 8.7). The schistes lustrés or Bündnerschiefer, *sensu lato*, form several tectonic units. They have been considered schuppen zone, but may in fact be mélange units. Assuming that all the rocks within each unit constitute

Fig. 8.7. Stratigraphy of the Mesozoic sediments in northern Penninic zones (after Bolli *et al*. 1980). The Bündnerschiefer south of the Gotthard massif consists mainly of hemipelagic sediments.

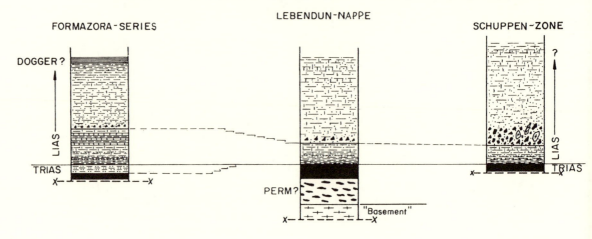

one sedimentary sequence, Bolli *et al.* (1980) reconstructed the stratigraphy of various units (Figure 8.7). The superposition of Lower Jurassic hemipelagic sediments above Triassic quartzite and dolomite records a subsidence event that led to the birth of the Tethys Ocean. The originally hemipelagic shales have been metamorphosed and became garnet-mica schists, yet megafossils have been found in those high-grade rocks, indicating their Early Jurassic age. Palynological studies have revealed the presence of Upper Jurassic hemipelagic deposits in the Formazora Series near the San Giacomo Pass (Bolli *et al.* 1980). Cretaceous strata have not been identified in the Gotthard region; they have apparently been sheared off to form the décollement nappes of the Ultrahelvetic nappes.

Ferret Schist

The schistose rocks between the Helvetic and the Bernard nappes south of the Rhône River crop out in a belt that extends from Sion in Switzerland to Courmayeur in Italy. They form the nappes of the Sion-Courmayeur zone between the Middle Penninic Subbrianconnais/Brianconnais and the Ultrahelvetic nappes (Figure 8.8). They are named Ferret Schist, with the type locality in the Ferret Valley of western Switzerland.

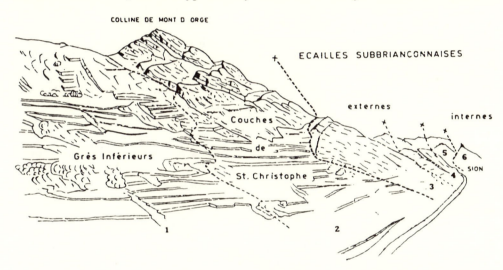

Fig. 8.8. The Ferret Schist on Mont d'Orge, west of Sion (after Burri 1967). The Ferret Schist (units 1 and 2) is thrust above the Ultrahelvetic. The thrust slices of Subbrianconnais rocks (units 3 to 6) are thrust above the Ferret Schist.

The Ferret Schist is formed by low-grade metamorphism of fine-grained detrital sediments. Quartz, sericite, chlorite, albite, and calcite are the most common constituents. The pelitic rocks of this monotonous sequence were once considered schistes lustrés, even though they resemble far more the Bündnerschiefer, *sensu stricto*, in view of their low grade of metamorphism. Lithostratigraphical subdivisions have been attempted (Trümpy 1954), but these "formations" are rarely recognizable far beyond their type locality. The schist overlies or is overthrust above a Triassic formation, which is apparently the detachment layer of strip sheets. Above a conglomerate and schist at the base, the lower division of the Ferret Schist consists mainly of slightly metamorphosed sandstone, calcareous siltstone, and calcareous shale. The middle division is mainly a radiolarian shale, with intercalations of calcareous siltstone and sandstone. The unit is locally 400 m thick, but is thin or missing near Sion north of the Rhône (Burri 1958). The upper division, the St. Christophe Beds of Trümpy, consists of interbedded micaceous and calcareous quartzite and silty shale. Fossils are very rare in the Ferret Schist. The age of the few microfossils found in the schist is Cretaceous (Trümpy 1954; Burri 1958). It is probable, however, that the Ferret Schist is a comprehensive series of Jurassic and Cretaceous hemipelagic sediments of the North Penninic domain, like the Bündnerschiefer of Graubünden.

The tectonic superposition indicates that the Ferret Schist was deposited in a basin north of the Brianconnais Swell. In contrast with the rapid facies changes on the northern slope of the swell, as manifested by the sediments of the Subbrianconnais facies, the monotonous lithology of the Ferret Schists is typical of hemipelagic sedimentation in a distant offshore basin. Turbidite and micro-breccia beds have been described, but are not common (Fricker 1960).

The North Penninic position of the Ferret Schist suggests a close relation to the Niesen Flysch. The latter is mainly Upper Cretaceous and is thus younger. The Niesen is also distinguished by the interbedded coarse turbidite beds, indicative of more rugged topographic relief at the time of their deposition. The site of Early Cretaceous hemipelagic sedimentation (Ferret) apparently evolved into a Late Cretaceous Flysch basin (Niesen).

Evolution of the Penninic Domain

The Penninic Domain is defined as the region between the Helvetic/Ultrahelvetic margin on the north and the Austroalpine margin on the South. The region was underlain by continental crust and was a domain of shallow marine and/or subaerial sedimentation until the Triassic. Subsidence began during the latest Triassic (Rhät) or Early Jurassic (Lias). The Penninic Ocean was separated into two troughs, the Piedmont and the Valais, by the Brianconnais Swell, which was underlain by continental lithosphere. The North Penninic Bündnerschiefer and the Ferret Schist gave evidence that deep-sea sedimentation began in early Jurassic time. Hemipelagic or "pre-flysch" sedimentation continued then throughout the Jurassic and Cretaceous periods. The mélanges of the Misox zone and the occurrence of North Penninic ophiolites in western Switzerland indicate that the deep-sea floor north of the Brianconnais was floored in part by ocean crust. The Piedmont Ocean south of the Brianconnais is an oceanic depression formed by Jurassic extension. Jurassic and Early Cretaceous pelagic sediments are present in regions distant from detrital sources, while breccias and hemipelagic shales were deposited closer to shore.

The Penninic region became deeply submerged at the beginning of the Late Cretaceous, when *couches rouges* were the hemipelagic deposit on the Brianconnais Swell. Hemipelagic shales and Flysch turbidites were laid down in South and North Penninic trenches bordering active margins. We shall return to the theme of the changing Pennine paleogeography when we discuss the history of the Penninic deformation (see Chapter 10).

9.

The Ophiolite Mélanges

Various Austroalpine nappes are underlain by ophiolite mélanges. They have been known by local names, such as the Zermatt-Saas zone, Arosa schuppen zone, and "Platta nappe" (Figures 7.9 and 9.1). Invoking the geosyncline theory to interpret Alpine geology, the sedimentary and volcanic rocks of the mélange were considered to constitute a continuously deposited sedimentary sequence, the metamorphic and granitic rocks its basement, and the plutonic ophiolites intrusives; there seemed to have been no oceanic realm between the Mesozoic Europe and Africa (e.g., Trümpy 1960). It is therefore of paramount importance to make a brief resume of the factual information leading to the recognition of the mélange nature of the sedimentary, igneous, and metamorphic rocks under the Austroalpine nappes, because the new interpretation is the critical argument for the plate-tectonic interpretation of Alpine geology.

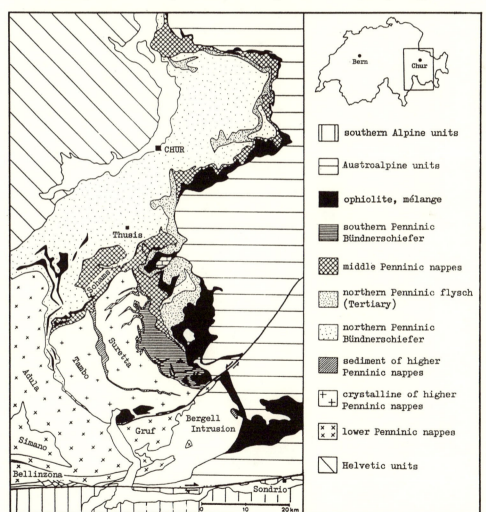

Fig. 9.1. Geological sketch map of the Penninic units in southeastern Switzerland (modification of tectonic map of Switzerland by P. Rück [1990]). The "Platta nappe," "Arosa schuppen zone" and other ophiolitic mélanges are all grouped under the heading "ophiolite-bearing elements of the Penninic Alps." This unit corresponds more or less to what was called the Rhätic nappe before Staub (1917) began to differentiate the various tectonic units between the Middle Penninic (Suretta, Schams, Falknis/Sulzfluh) and the Austroalpine nappes.

Legend:
- southern Alpine units
- Austroalpine units
- ophiolite, mélange
- southern Penninic Bündnerschiefer
- middle Penninic nappes
- northern Penninic flysch (Tertiary)
- northern Penninic Bündnerschiefer
- sediment of higher Penninic nappes
- crystalline of higher Penninic nappes
- lower Penninic nappes
- Helvetic units

"Platta Nappe"

Assuming that all the rocks between the Margna nappe and the lowermost Austroalpine nappe originally constituted the coherent sequence of one nappe, Cornelius (1935) reconstructed the sedimentary succession of the "Platta nappe" as follows:

Upper Jurassic: radiolarite
Middle Jurassic: Aptychus Limestone
Lower Jurassic: Liasschiefer, Kieselkalk, Spatkalk
Noric: Hauptdolomit
Carnic: dolomite, dolomite breccia, schist
Carboniferous: Flix Beds
Basement: Phyllite, Augengneiss

Coarse-grained ophiolitic rocks in the area were considered igneous intrusives and thus not a part of the stratified sequence; pillow basalts were supposedly emplaced by extrusion onto a Jurassic or Cretaceous ocean floor. Cornelius's conclusion reflects the consensus of his time and a paradigm of classic Alpine geology. As it was not recognized that the Triassic and upper Paleozoic rocks are exotic blocks from Austroalpine nappes, the reconstructed stratigraphy for the "Platta nappe" implies a sedimentary history not much different from that for the Austroalpine domain. The "Platta sequence" was supposedly laid down on a continental crust, because the "Platta basement" seemed to consist of gneiss and phyllite.

After the Helvetic, Austroalpine and Penninic core nappes were proven to be derived from continental crust, the interpretation of a continental basement for Platta left no place for a Mesozoic/Tertiary ocean between Europe and Italy/Africa: the Alpine Tethys had to be underlain everywhere by continental crust. This establishment tradition may have retarded for a decade or two the acceptance of the theories of seafloor spreading and of plate tectonics by Alpine geologists.

That the various formations of the "Platta nappe" cannot be traced from one outcrop to another is obvious (Figure 9.2). Unfortunately, the concept of olistostrome was innovated to interpret tectonic mélanges as sedimentary deposit (Flores 1955). After the consensus was reached that the Platta ophiolites were not intrusive, the advocates for a sedimentary origin for mélanges could still insist that they were exotic blocks in a sedimentary formation, having been mixed with the matrix mud during downslope mass transport. Mélanges of tectonic origin were called "olistostromes," and were considered sedimentary deposit in a normal stratigraphic sequence. Denying the obvious intensive shearing parallel or subparallel to bedding planes, a pile of mélanges was interpreted as imbricated thrusts of sedimentary cover peeled off from granitic basement. The invention of "olistostromes" rescued the idea of a "Platta nappe" and rendered the theory of plate tectonics unnecessary.

The notion that all the rocks in the "Platta nappe" constituted one sedimentary sequence was a last argument for the permanence of oceans and continents. The "fixists" could accept the idea of a deep-sea origin for the Jurassic radiolarites, but these were considered sediments of a sunken continent; the pelagic sediments were laid down above shallow marine Triassic carbonates and the granite basement of a continental crust. The postulate of a "Platta nappe" would thus make the Alpine orogenesis an intracontinental process, rather than the result of continental collision.

Ophiolite Mélanges

Slabs and matrix in a mélange are not normally superposed (Hsü 1968). One cannot even assume that they have all been detached from one and the same rock sequence. The "Platta nappe" is a mélange, like the rocks of the Misox zone. Slabs of ocean crust (ophiolite) and the matrix, consisting mainly of ocean sediments (Bündnerschiefer), were derived from the Piedmont Ocean; they are Penninic. Slices of continental crust (gneiss) and its sedimentary cover (quartzite, dolomite) were derived from the southern continent; they are Austroalpine. These rocks of distinctly different origins are mixed in a mélange now found in the suture zone between the Brianconnais and the Austroalpine nappes. But the mélange is neither a Penninic nor an Austroalpine nappe; a mélange is a mélange is a mélange, to paraphrase Gertrude Stein. Suture mélanges

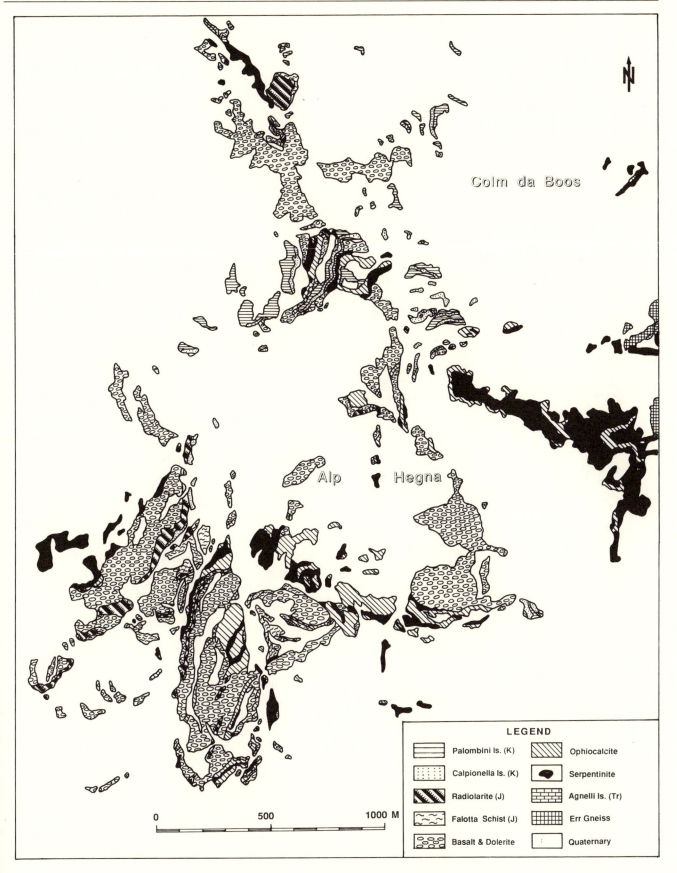

Fig. 9.2. Geological map of Alp Flix (after Leemann 1988). The map shows the lack of lateral continuity of the various stratified rock formations in a tectonic mélange.

correlative to the "Platta nappe" are the "Arosa schuppen zone" in eastern Switzerland and the Zermatt-Saas zone in the Valais.

"Arosa Schuppen Zone"

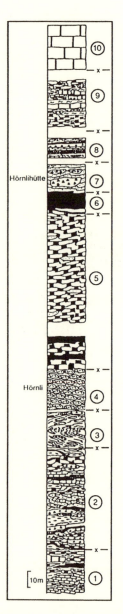

Fig. 9.3. Hörnligrat profile (after Lüdin 1987). This columnar section shows the lack of normal succession in a non-Smithian stratigraphic sequence. See text for explanation.

An ocean once existed, but was consumed; this fact is recorded by the ophiolite mélange that has received the misnomer nappe. That the "Platta nappe" is a mélange was obvious to me during a field course for students at Alp Flix in 1969. I tried in vain to call the attention of my colleagues to the fact that "blocks of granite and Triassic shelf carbonates . . . found in the ophiolite zone of . . . Switzerland, are exotic slabs in a tectonic mélange" (Hsü 1973, p. 76). Nevertheless, the Platta was still considered a nappe, or a tectonic unit with a coherent sedimentary sequence, and it took a daring young undergraduate to throw the first stone at the establishment.

Weissert (1974) did field mapping for a master's thesis in the Davos area of eastern Switzerland. The "strange bedfellows" in the "Arosa schuppen zone" near Davos were well known, rocks derived from the Austroalpine having been found in close proximity to rocks from the Upper Penninic (Figure 9.3). The term schuppen zone was used in place of nappe, because no recumbent folds were recognizable (Streckeisen 1948). Cadisch and Streckeisen (1967) described a structural profile near Davos as follows (the description applies to Figure 9.4 as well):

> A footpath leads from the aerial cable car station Weissfluh at 2685.5 m elevation northward to a water divide at 2625 m elevation. The path crosses a shear zone, which includes numerous slabs of Triassic dolomite, Totalp Serpentine, and other rocks. From south to north, the outcrops are (1) Triassic dolomite at 2622 m elevation, (2) brecciated Rauhwacke and ophiolite (serpentinite, etc.), (3) Triassic dolomite, (4) a shear zone of serpentinite (similar to 2), (5) limestone, Rhätian, (6) calcareous schist, Liassic, (7) green and black schist and arkose (Verrucano formation?), (8) limestone, Rhätian, (9) schist, Liassic (similar to 5 and 6, respectively), (10) red and green schists and radiolarite, (11) Calpionella limestone, thick-bedded, (12) breccia (similar to 2), (13) Triassic dolomite at the base of the Weissfluh. [free translation by the author]

With the spread of the plate-tectonic theory in the early 1970s, the time was finally ripe for the mélange concept to be imported to Switzerland. Weissert wrote:

> The thrust slices in the eastern part of the "Gotschnagrat Slice" consist of radiolarite, Aptychus Limestone, and Calpionella Limestone and a subordinate amount of ophicalcite. Yet the serpentinite and ophicalcite are the dominant rock types in the western part of the zone; they form the matrix of the Parsenn Mélange. Since the Parsenn Mélange cannot be separated tectonically from other rocks of the unit, the question can be asked whether we should designate the whole "Gotschnagrat Slice" as one single mélange. In other words, we have to apply the concept of mélange, not only to the Parsenn Mélange, but also to designate the less pervasively sheared "Gotschnagrat Slice." At any rate, the traditional interpretation of a normal stratigraphic sequence of an "olistostrome" overlain by a radiolarite is not admissible, because the Parsenn Mélange contains radiolarite slabs; the radiolarite can thus not be younger than the mélange (see Hsü 1968). [free translation by the author]

It did not take much imagination, but it did require courage to state that the sediments such as Triassic quartzites and dolomites are Austroalpine rocks derived from higher tectonic units (Gruenhorn Schuppe, Casanna Schuppe), and the whole pile is overriden by the Austroalpine Silvretta nappe. The separation of the Penninic ophiolites from the Austroalpine granite in this "Platta mixture" is the "egg of Columbus" that opened the way for a plate-tectonic synthesis of Alpine geology.

The geology of the "Platta nappe" and of the "Arosa schuppen zone" has been the theme of numerous theses in Swiss universities during the last two decades. Contacts between rocks of different lithologies are commonly identified as shear planes. An

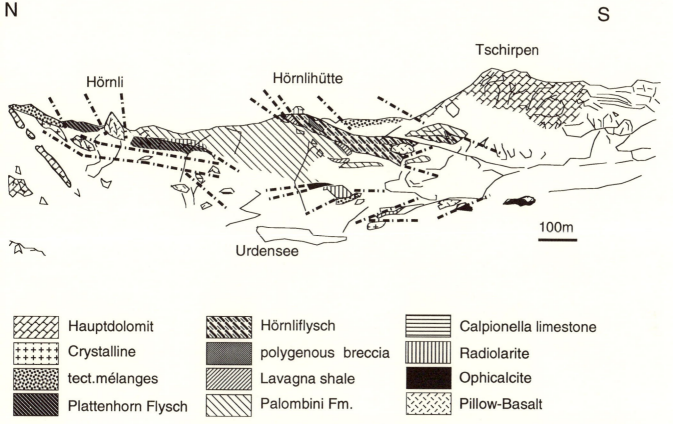

Fig. 9.4. Panoramic view of the Platten-Hörnli area (after Lüdin 1987). This view shows the chaotic distribution of various exotic slabs of the "Arosa schuppen zone." The profile extends from a point north of the Hörnli to the peak south of the Hörnlihütte (the Tschirpen). The gneiss outcrop north of the Hörnli is an exotic slab in a mélange (Lüdin's unit 2). The pillow lava under the Hörnli is part of an ophiolite slab (unit 4). Exotic blocks of ophicalcite are found south of the Hörnlihütte (unit 9). Other slabs have been detached from Austroalpine or from Penninic units.

example of the intimate mixing of rocks of different origins is shown by the Hörnligrat profile near Plattenhorn (Figures 9.4 and 9.5). Unit 1 at the base of the "column" is an intercalation of slices of Cretaceous Calpionella Limestone, Palombini Limestone, and Lavagna Black Shale. Unit 2 is a tectonic mélange, consisting of slabs, blocks, and fragments of granitic gneisses, quartzite, Hauptdolomit, serpentinite, radiolarite, Calpionella Limestone, Palombini Limestone, and sandstones. Unit 3 is a slab of isoclinally folded phyllites and sandstones; they contain abundant detritus derived from serpentinites and other ophiolites. Unit 4 is an ophiolite slab, including diabases, pillow basalts, pillow breccias, and basaltic tuffs. Above that is unit 5, a slab of folded Palombini Limestone sequence, about 80 m thick; a few sandstone lenses are interbedded in the limestone. Unit 6 is the Lavagna Shale slab, and unit 7 is Hörnli Flysch, with tectonically mixed blocks of ophiolites and breccias. Unit 8 is a mélange of rocks of various lithologies, including ophicalcite, radiolarite, Calpionella Limestone, pillow basalt, phyllite, sandstone, etc. Unit 9 is another tectonic mélange of mica schist, dolomite, calcarenite, black shale, pillow basalt, diabase, serpentinite, and Palombini Limestone.

This shear zone, or tectonic mixture, of deep-sea sediments and ocean crust from one terrane and of shallow marine sediments and continental crust from another, is more than 200 m thick (Lüdin 1987). The Lower Jurassic, Triassic, upper Paleozoic, and gneissic basement are derived from an Austroalpine margin; only the oceanic sediments and ophiolites are originally South Penninic (Figure 9.6).

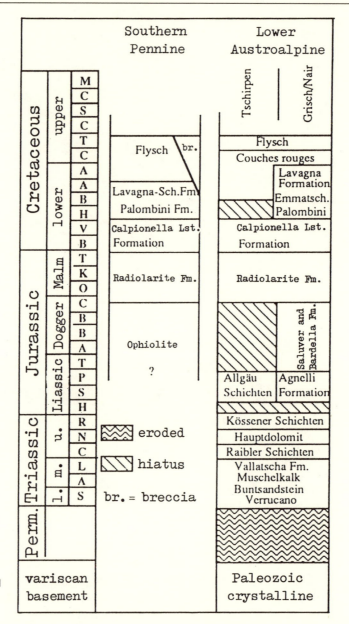

Fig. 9.5. Stratigraphic reconstruction of the Penninic and Austroalpine sequences (after Lüdin 1987).

Fig. 9.6. "Arosa schuppen zone" near Davos (after Cadisch and Streckeisen 1967). G Sch: Schiahorn; Sl: Schafläger; W: Weissfluhjoch; (1) total serpentinite; (2) Dorfberg nappes of Davos; 2a: Mittelgrat Series; 2b: gabbro; 2c: gneiss; (3) Schafläger sediments; 3a: Permo-Werfenian; 3b: Rhät limestone; 3c: Lias calcareous schist; (4) Silvretta basement rocks; 4a: mixed gneisses; 4b: paragneiss; 4c: orthogneiss; 4d: amphibolite; (5) dolomite.

"Margna Nappe"

The term Margna nappe, in a broad sense, was used by Staub (1917) to designate the complex shear zone between the Lower Austroalpine Err/Bernina nappes and the Penninic Tambo/Suretta nappes. This shear zone includes not only what was once considered a core nappe ("Margna nappe," *sensu stricto*), but also the "Platta nappe" (ophiolite mélange) and the Bündnerschiefer of Avers. The "Margna nappe," *sensu stricto*, was considered a core nappe enveloped by its deformed sedimentary cover; the tectonic superposition in the Maloja area of eastern Switzerland is thus

Lower Austroalpine nappes (Err/Bernina)
"Platta nappe" (ophiolite mélange)
"Margna nappe," *sensu stricto*
Melange of the Bündnerschiefer of Avers/Malenco Gabbro
Tambo/Suretta nappes

The basement rocks of the "Margna nappe" are mainly augengneiss (Trommsdorff 1980). The sedimentary cover of the "nappe" is, in descending order (Cornelius 1935),

Jurassic: radiolarite, Aptychus Limestone, calc-schist, breccia
Triassic: dolomite (with crinoids), Rauhwacke, marble, quartzite
Permian: conglomerate, porphyry

The facies development of the Margna sediments is thus typically Lower Austroalpine. Yet the nappe was considered Upper Penninic by Staub (1937; 1942; 1958) and by Cadisch (1953), because the Margna rocks underlie the "Platta nappe." Furthermore, slabs of serpentinite are present within the "Margna nappe." Having identified three stratigraphical sequences in the area, Staub (1958, p. 142) postulated three deep-sea troughs south of the Brianconnais Swell, i.e., Avers/Malenco, Margna, and Platta. The former existence of a South Penninic Piedmont Trough (Platta) is now generally accepted, and the possible presence of an Avers Trough has not been ruled out (see the next chapter). There is, however, no evidence for a third "Margna Trough." The Margna sediments are Lower Austroalpine, and the augengneiss was the continental basement of the southern Tethyan margin. The ophiolites in the "Margna nappe" were not the seafloor of a "Margna Trough," but exotic slabs derived from an ocean lithosphere elsewhere.

The classic rule of using nappe superposition to reconstruct paleogeography assumed (1) unidirectional vergence, and (2) a coherent sedimentary sequence in each nappe. The vergence of the various Austroalpine and Upper Penninic units is unidirectional; it is directed from south to north or east to west. On the other hand, the contact between the Austroalpine and the Upper Penninic is not a simple, clear-cut surface of dislocation between two coherent packages of rocks. I have pointed out the presence of numerous Austroalpine elements in the ophiolite mélange that has been called the "Platta nappe," and I have cited field evidence for the tectonic mixing of the Austroalpine and Upper Penninic components in the "Arosa schuppen zone." Both of these are mélanges, and the internal structure of the "Margna nappe, *sensu stricto*" is not much different from the two mélange units.

A reading of a geological map of the Crap da Chüern area (Figure 9.7) near Sils in the Upper Engadine Valley shows that the Margna unit consists of an interlayering of exotic slabs (Heierli 1967). When one makes a traverse from Splüga toward the northeast, one encounters (1) graphitic phyllite (Margna basement?), (2) Triassic dolomite, (3) Liassic limestone, (4) Triassic dolomite, (5) graphitic phyllite (like 1), (6) Liassic breccia, (7) Liassic limestone (like 3), (8) gneiss and mylonite (Margna basement), (9) Liassic limestone (like 3), (10) Jurassic radiolarite, (11) Triassic dolomite, (12) gneiss and mylonite (like 8), (13) Liassic limestone (like 3), (14) serpentinite (Penninic ophiolite), (15) Liassic limestone (like 3), (16) Jurassic radiolarite (like 10), (17) Liassic limestone (like 3), and (18) serpentinite (like 14).

The rocks at Crap da Chüern thus constitute a mélange like those in the "Arosa

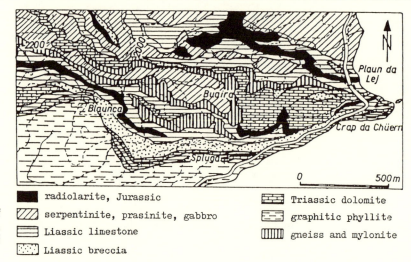

Fig. 9.7. Geological sketch map of the Crap da Chüern area, near Sils, Upper Engadine Valley (after Heierli 1967).

Legend:

■ radiolarite, Jurassic
▨ serpentinite, prasinite, gabbro
▤ Liassic limestone
⋯ Liassic breccia

▥ Triassic dolomite
⊟ graphitic phyllite
▥ gneiss and mylonite

schuppen zone" near Davos, although more slabs of Liassic limestones and of the Margna basement are present in one area and more slabs of Triassic dolomites and Jurassic serpentinite in another. That one should be considered an Upper Penninic "Margna nappe" while the other is called a Lower Austroalpine "Platta nappe" (or "Arosa schuppen zone") is a purely arbitrary interpretation, resulting from an erroneous understanding of Alpine tectonics.

The geology of the transition between the Austroalpine rigid-basement nappes and the Penninic core nappes is typified by shear zones and mélanges. In a tectonic mélange, the original paleogeographical positions of the various slabs are not necessarily those suggested by their tectonic superposition; a lower unit was not necessarily located north of the overlying unit. Invoking the plate-tectonic theory, I suggest that the Austroalpine basement and its passive-margin cover had been cut into numerous tectonic slices, and that they formed an accretionary wedge under the Austroalpine margin when the subduction of the Piedmont Ocean started. They were tectonically mixed with exotic blocks of sedimentary and igneous rocks derived from an ocean lithosphere. Where smaller slices of Austroalpine rocks are present as exotics in a predominantly ophiolitic mélange, the shear zone is called the "Platta nappe" or the "Arosa schuppen zone." Where a large pile of tectonic slices of Austroalpine origin is mixed with Penninic serpentinite and Bündnerschiefer, the unit is called the "Margna nappe." We thus have the paradoxical puzzle that the predominantly ophiolitic "Platta nappe" lies on top of the predominantly Austroalpine "Margna nappe."

"Monte Rosa Nappe"

We discussed in Chapter 7 the different opinions on the paleogeographical position of the Monte Rosa. The Monte Rosa, underlain by continental crust, was considered Middle Penninic by Argand and Staub, but Austroalpine by Amstutz and Milnes. The possibility that the Monte Rosa nappe is Austroalpine has also been proposed by Laubscher and Bernoulli (1982).

Accepting the conclusion that the ophiolites in the Furgg zone separating the Monte Rosa and Bernard nappes are the remnant of the Piedmont Ocean, we can proceed to the postulate that the Monte Rosa and its tectonic correlative of Gran Paradiso/Dora Maira basement were originally situated on the southern Tethyan margin (Hsü 1990). Like the "Margna nappe," the continental basement is now enveloped in ophiolite mélanges. Evidence of high-pressure metamorphism indicates that this slice of continental crust was underthrust to a depth of 100 km during the Eo-Alpine deformation, but was again overthrust above the Bernard nappe after the collision of the southern Tethyan margin with the Brianconnais Swell (see the next chapter). The Monte Rosa/

Gran Paradiso/Dora Maira rocks were tectonically mixed with other rocks of the Benioff zone during its upward journey. This interpretation has placed the "Monte Rosa nappe" into a paleogeographical and tectonic position similar to that postulated for the "Margna nappe."

Paleogeography of the South Penninic Domain

The Penninic Ocean, as we indicated in the last chapter, was separated into two troughs, the Piedmont and the Valais, by the Brianconnais Swell. A thin sequence of radiolarite and pelagic limestones was laid down on the seafloor of the Jurassic Piedmont Ocean. The deposition of exclusively biogenic sediments signifies that terrigenous detritus was largely trapped in the Ultrahelvetic and North Penninic basins north of the Brianconnais Swell, while the southern continent was largely submerged under the sea during the Middle and Late Jurassic. Pelagic sedimentation continued until earliest Cretaceous; the Calpionella limestone is as young as Berriasian. Rare pollen and microfossils indicate that the South Penninic Bündnerschiefer is largely Cretaceous. The influx of terrigenous materials into the Piedmont Ocean signifies a major change in the paleogeography at the beginning of the Cretaceous. The Brianconnais Swell subsided and became a site of hemipelagic sedimentation in the Late Cretaceous, when flysch sediments were deposited in South and North Penninic troughs.

10.

A History of the Pennine Deformation

Elie de Beaumont (1831) postulated periodic catastrophes to explain the transgressions and regressions of seas. A corollary was the scheme of dividing geological formations by mountain-building "events." The concept was simple and beguiling. With the introduction of the geosynclinal theory by James Hall, the periodicity idea became the paradigm. Geological history, it was thought, was comparable to human history: there were long ages of tranquility, alternating with comparatively brief epochs of revolutions (Dana 1873). This concept continued to dominate the thinking of geologists during the first half of this century: Bucher (1933) and Stille (1924) postulated globally synchronous orogenic episodes, or "phases," punctuating long periods of inactivity.

James Gilluly (1949), in his presidential address at the Geological Society of America entitled *The distribution of mountain-building in geologic time*, brought uniformitarianism back to the question of the timing of mountain-building. He envisioned a steady-state process of orogeny, and pointed out that the postulate of orogenic "phases" had been based upon meager evidence and poor logic. The controversy between Stille and Bucher on one side and Gilluly on the other is a well-known episode in the history of geology. The innovation of the plate-tectonic theory has settled the dispute: orogenic movements are a consequence of the displacement of lithospheric plates, which move at linear rates for long intervals of time. There can be no global orogenic revolutions as Elie de Beaumont or Stille once envisioned.

Time and Place of the Alpine Orogenesis

While we all agree that the edifice of the Swiss Alps was not formed in one catastrophic upheaval, many believed, however, that compressional deformations took place during distinct, relatively short periods; uplift and erosion followed only after a significant delay (see Trümpy 1973). It was thought that the main deformation in the Alps occurred at the end of the Eocene and/or the beginning of the Oligocene. Indications for earlier and later orogenic activities were considered evidence for the Eo-Alpine or the Neo-Alpine "event." "The present morphology, especially of the western Alps, is the product of Pleistocene uplift and erosion" (Trümpy 1960, p. 898).

The traditional thinking on "orogenic events" or "tectonic phases" is contradictory to the theory of plate tectonics. There were no orogenic episodes in the evolution of the Alps, but rather 130 million years of compressional deformation since the beginning of the Cretaceous (Hsü 1989). Nor were there epeirogenic episodes; uplift is the vertical component of the forward movement of a wedge-shaped overriding plate (Hsü 1991a).

The Swiss Alps have been formed by continental collision, and the moment of collision when one continent began to touch another was an event. But the Alps were not formed by this event; orogenic deformation took place continuously long before and long after the event. Using the time of collision as a reference datum we can recognize at least two stages of compressional deformation. Prior to the collision, the deformation was continent-ocean interaction and should have involved the subduction of oceanic lithosphere; this can be considered the stage of Eo-Alpine deformation. The orogenesis after the collision was a continent-continent interaction, when one continental block was thrust under another. Deeply buried continental basement and part of its sedimentary cover wass stripped off and pushed onto the foreland as décollement nappes. The deformations after the collision can thus be considered Neo-Alpine (Hsü and Briegel 1991).

A twofold subdivision of orogenic chronology assumes one event marker for a single continental collision. In fact, the Mid-Penninic Brianconnais Swell was underlain by continental crust and was, in our opinion, separated from both the Helvetic and Austroalpine margins by oceanic basins, narrow as they may have been. There should have been three or more continental collisions during the Alpine orogenesis: one event was marked by the collision of the Brianconnais Swell and the European margin, another by the collision of the Brianconnais and the Austroalpine margin, and a third between the Austroalpine and European margins, after the Brianconnais was accreted to one or another of the plates. In the scheme of multiple continental collisions, one can use the term Meso-Alpine to designate deformations by continent-continent interactions after an initial continental collision (involving the Brianconnais Swell) and before the final suturing of the Austroalpine and European margins. Used in this sense, the terms Eo-, Meso-, and Neo-Alpine stages refer to the nature of the deformation, rather than definitive time intervals. Subduction may have been going on at one plate margin while collision was happening at another. A Late Cretaceous Eo-Alpine deformation on the South Penninic margin may thus have been taking place at the same time as Europe and the Brianconnais Swell were being sutured by a Meso-Alpine collision.

Eo-Alpine South Penninic Deformation

The Eo-Alpine deformation was an ocean-continent interaction, when the ocean lithosphere under the Penninic domain was subducted under the Austroalpine margin. The margin had been a passive margin during the Jurassic seafloor-spreading stage, when the Pennine Ocean was widened. When compression prevailed, the boundary between the Pennine and Austroalpine became an active plate-margin, and the deformation in the subduction zone at the active margin should be typified by penetrative shearing, forming accretionary prisms and ophiolite mélanges. The downward plunge of a cold slab to great depth produces, furthermore, an immediate increase in pressure, while there is a time lag in thermal equilibrium. Consequently, the metamorphic reactions resulting from subduction processes are typically high-pressure, low-temperature metamorphism, forming rocks of the glaucophane schist facies. Reestablishment of thermal steady state at great depth leads to high-pressure, high-temperature eclogite-facies metamorphism. The evidence for Eo-Alpine deformation is thus to be found in the tectonic mélanges and in high-pressure metamorphisms. Later, when tectonic mélanges on active plate-margin were deformed in a suture zone after continental collision, high-pressure metamorphic rocks could have been recrystallized at medium/low P/T range, erasing in part the evidence of the Eo-Alpine deformation.

Stratigraphical evidence, such as the occurrence of coarse detritus in some Alpine Cretaceous formations, suggested Eo-Alpine deformation (see Trümpy 1960). The evidence became irrefutable after Dal Piaz (1972) and Hunziker (1974) found a Rb-Sr whole rock age of 130 Ma for granitoids in eclogite facies from the Sesia-Lanzo zone; there was high-pressure metamorphism in the Cretaceous prior to the Neo-Alpine regional metamorphism in the Tertiary. Additional Cretaceous dates have since been reported from the Alps, and the pattern that has emerged is one of continuous subduction and high-pressure metamorphism during the Eo-Alpine deformation: Deutsch (1973) found 110–70 Ma and Phillip (1982) found 90–70 Ma metamorphic ages for alkali amphiboles in the mélange of the "Platta nappe"; Bearth (1973) cited 90–60 Ma and Bocquet et al. (1974) obtained 100–80 Ma ages for the glaucophane schists and eclogites in ophiolitic mélanges of the Western Alps.

The high-pressure, low-temperature metamorphic rocks were originally ocean crust and ocean sediments: the glaucophane schists of Zermatt were pillow basalt and pillow breccia, whereas the radiolarite with a Sr-Rb isochrone age of 85 Ma was a Jurassic sediment (Oxfordian, 165 Ma) in the Piedmont Ocean. The eclogite facies granitoid of the Sesia zone was, on the other hand, a part of the continental basement under an Austroalpine sequence, but this basement must have been underthrust to great depth during the Eo-Alpine deformation. Elsewhere, the Austroalpine basement has been locally subjected to greenschist/amphibolite metamorphism of Cretaceous age. Satir

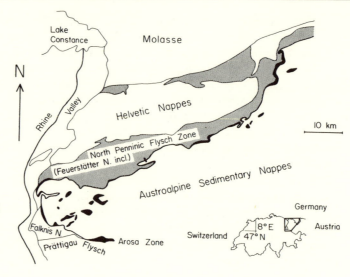

Fig. 10.1. Tectonic framework and section locality of the Walsertal zone (after Winkler and Bernoulli 1986). This zone is considered the eastern tectonic equivalent of the Arosa zone in Switzerland.

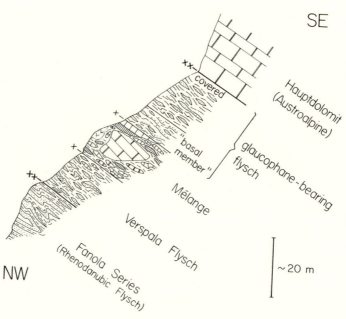

Fig. 10.2. Section showing the thrusting of the Walsertal zone under the Austroalpine Hauptdolomit (after Winkler and Bernoulli 1986).

(1975) suggested an onset of metamorphism during the Early Cretaceous (124 Ma), whereas Thöni (1983) obtained Late Cretaceous (100–85 Ma) dates for rocks of Upper Austroalpine nappes.

The radiometric ages of metamorphism suggest that the margin between the Penninic and Austroalpine domains was changed from a passive or transform margin to an active margin at or near the beginning of the Cretaceous Period. Continental crust at the Austroalpine margin went down first, along a south-dipping Benioff zone under the southern continent, to a depth of 50–100 km, where eclogite was formed. Subduction of the ocean lithosphere ensued. This deformation of the southern margin progressed as a steady-state phenomenon from the beginning to the end of the Eo-Alpine stage. Kinematic considerations permit an estimate of a subduction rate of about 1 cm/yr, and some 700 km of the Penninic Ocean should thus have been consumed (Hsü 1989).

The rocks metamorphosed under high pressure were soon uplifted to deliver detritus to Flysch basins on the active margin (Figures 10.1 and 10.2). This is indicated by the finding of ophiolite detritus in Penninic Flysch of Aptian age in Austria and of glaucophane and lawsonite detritus in the South Penninic Flysch of Turonian/Coniacian age in eastern Switzerland; these sediments of 110–85 Ma age were deposited not very long after the start of the Eo-Alpine metamorphism (Flügel *et al.* 1987; Winkler and Bernoulli 1986). In western Switzerland, the Simmen nappe consists of slices of ophiolites, pelagic sediments, and Simmen Flysch. The Flysch conglomerates include clasts from ophiolitic mélange, as well as detritus from an Austroalpine provenance. The Flysch sandstones, ranging from late Cenomanian to Maastrichtian (95–70 Ma) in age, are distinguished by the detritus derived from ophiolites, such as chrome spinel (Wildi 1985). The South Penninic Flysch of eastern Switzerland and the Simmen Flysch of the Prealps had equivalent paleogeographic positions: these turbidites were laid down in foredeep or on abyssal plain north of an Eo-Alpine subduction zone.

The Eo-Alpine ocean-continent interaction is best exemplified by the Circum-Pacific orogenesis, and an actualistic analogue is the deformation of the Pacific Northwest of North America: the Eocene ophiolite complex that once lay under the Eocene ocean sediments has been uplifted and is exposed on Vancouver Island (Massey 1986). The ocean sediments form a thick pile of accretional wedges, and the continuous underthrusting of younger accretional wedges has jacked up ophiolites, which occur as exotic slabs in older mélanges. It took less than 30–40 Ma for a subducted ocean floor to be uplifted and exposed in a coast range. It is thus not surprising that the mélanges formed during Early Cretaceous subduction were exposed to furnish detritus for Late Cretaceous South Penninic Flysch. An older and even better documented case history is the Franciscan deformation of the California Coast Ranges (Hsü 1971a), where the Farallon and Pacific plates were thrust under the North American plate, from the Jurassic to the early Tertiary, at rates of centimeters per year.

The Schams Dilemma

The basement under the Brianconnais Swell was deformed prior to its burial under the advancing Austroalpine nappes; evidence for such deformation is the Cretaceous metamorphic ages in the Middle or Upper Penninic core nappes (Hunziker 1974). A postulate of a Meso-Alpine collision of the Brianconnais and the European margins is a working hypothesis to explain the complex history of the tectonic evolution of the Schams nappes (Hsü and Briegel 1991).

The Schams nappes are several strip sheets of detached sedimentary cover enveloping the Tambo-Suretta nappe of eastern Switzerland (Figure 10.3). The Schams nappes are rootless, and the Schams sediments are allochthonous. The three major Schams thrust slices are sandwiched between a mélange zone (Areua-Bruschghorn) and the Tambo-Suretta nappes (Streiff 1962). The tectonic superposition above the core nappe, in descending order, is (Figure 10.4)

> Flysch
> Areua-Bruschghorn zone (ophiolite mélange)
> Gelbhorn nappe (basement/Triassic/Jurassic/Cretaceous)
> Tschera nappe (U. Triassic to Cretaceous or Paleogene)
> Kalkberg nappe (Middle Triassic)
> Suretta nappe (Rofna Gneiss/Verrucano/Triassic)

The Tambo-Suretta core nappe, enveloped by the Schams nappes and the overlying mélange, was overthrust as a recumbent antiform northward above the Adula nappe during the Neo-Alpine orogenesis, but the Schams nappes are distinguished by structural features indicative of a southerly vergence. This apparent contradiction has been called the "Schams Dilemma" by Trümpy (1980).

The facies similarity between the Schams nappes and the Falknis and Sulzfluh nappes east of the Rhine Valley is well known. The thick marble beds of the Tschera nappe (Marmor zone) have been compared with the Upper Jurassic Sulzfluh Limestone. The Triassic to Cretaceous sequence of the Gelbhorn nappe (with its Vizan Breccia) is correlative to the Falknis sequence (with its Falknis Breccia). To the west,

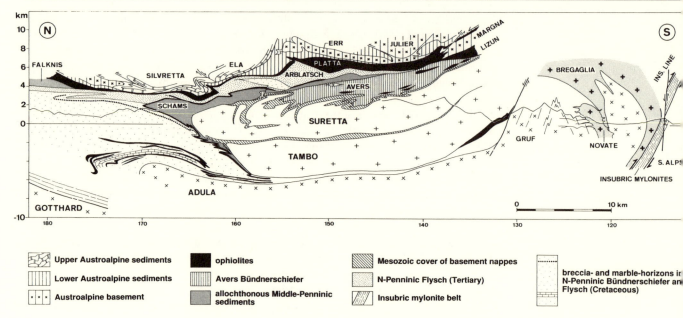

	Upper Austroalpine sediments		ophiolites		Mesozoic cover of basement nappes		breccia- and marble-horizons in N-Penninic Bündnerschiefer and Flysch (Cretaceous)
	Lower Austroalpine sediments		Avers Bündnerschiefer		N-Penninic Flysch (Tertiary)		
	Austroalpine basement		allochthonous Middle-Penninic sediments		Insubric mylonite belt		

Fig. 10.3. A north-south profile across eastern Switzerland (from Schmid *et al*. 1990).

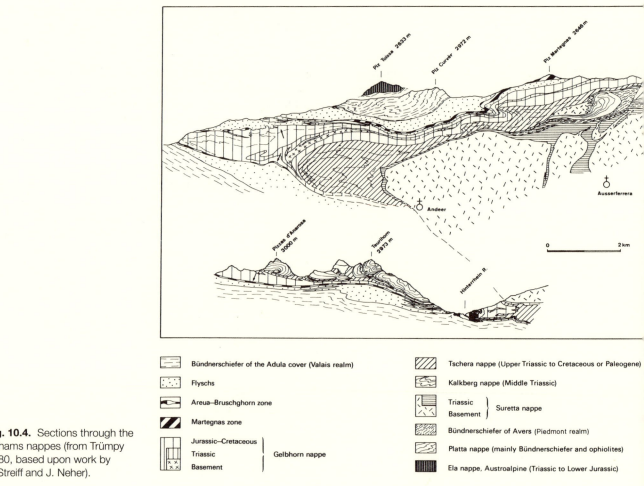

	Bündnerschiefer of the Adula cover (Valais realm)		Tschera nappe (Upper Triassic to Cretaceous or Paleogene)
	Flyschs		Kalkberg nappe (Middle Triassic)
	Areua–Bruschghorn zone		Triassic / Basement } Suretta nappe
	Martegnas zone		Bündnerschiefer of Avers (Piedmont realm)
	Jurassic–Cretaceous / Triassic / Basement } Gelbhorn nappe		Platta nappe (mainly Bündnerschiefer and ophiolites)
			Ela nappe, Austroalpine (Triassic to Lower Jurassic)

Fig. 10.4. Sections through the Schams nappes (from Trümpy 1980, based upon work by V. Streiff and J. Neher).

those sequences found their equivalent in the Rigid and Plastic Median Prealps and in the Brianconnais and Subbrianconnais.

The core nappe Tambo-Suretta has been considered an equivalent of the Bernard/ Monte Rosa or, in my opinion, the Bernard nappe (Hsü and Briegel 1991). The sediments of the Schams nappes cannot, however, be the cover of the Suretta basement, because it underlies its own sedimentary cover of Verrucano and Triassic (Streiff 1962). The Schams sediments now superposed on the back of the Suretta nappe have been thrust above the Suretta cover.

The idea that the Schams strata were the sedimentary cover of the Tambo basement was first proposed by Emil Haug in 1925. Several generations of Alpine geologists followed his lead and concluded that the Schams nappes were stripped off from a swell north of the Suretta and were thrust southward on top of the Suretta basement. This swell was the eastern extension of the Brianconnais Swell, and the basement under the swell should be the core of the Tambo nappe.

That the cover should have been moved backward while the core marched forward was a postulated mechanism difficult for many geologists to accept. An alternative hypothesis was to deny the southerly vergence of the Schams deformation and to find a home for the Schams sediments in the Austroalpine domain (Staub 1958). In this scheme, an Austroalpine sequence was pushed northward to a position above the Suretta, before the Suretta basement and its cover of strip sheets advanced farther northward as a recumbent fold (Staub 1958; Milnes and Schmutz 1978).

The idea of a unidirectional vergence in the Alps has taken root in the geological thinking in Alpine geology since Bertrand proposed the northward thrusting of the Glarus Overthrust. Under such an assumption, the tectonic displacement of a strip sheet northward is an overthrust, but a southward movement can take place only in the form of underthrusting. Therefore, the two alternative hypotheses on the tectonic evolution of the Schams nappe have also been called "*solution supra*" and "*solution infra*." The "supra" solution, which suggests northward movement of thrust slices, is more elegant from a tectonic viewpoint, because a south-to-north displacement of cover thrusts is the norm in the Alps. The "infra" solution would require that the Schams sediments had been squeezed out from under the Tambo-Suretta nappe and moved southward on top of the core nappe while the latter was overthrust northward. The stratigraphical evidence is unequivocal; however, the facies correlation of the Schams sediments with the Brianconnais/Subbrianconnais is so well established that an Austroalpine home for them is all but impossible (Trümpy 1980).

It should be recalled that the south-to-north vergence is not a physical law, but an empirical observation. In fact, this vergence is an exception rather than a rule in the Tethyan system of mountains. Suess (1909) puzzled long over the fact that the vergence of the Himalayas is southerly, whereas that of the Alps is northerly. Hsü and Schlanger (1971), in their interpretation of the paleogeography of the Ultrahelvetic/ North Penninic Flysch, had to postulate a Late Cretaceous/Paleocene north-dipping Benioff zone under the European continental margin; deformations related to such a zone should have a southerly vergence of overthrusting. North-dippling reflectors, implying southerly vergence, were found by seismic sounding during a geotraverse of eastern Switzerland (Frei *et al.* 1989). The evidence clearly shows a southerly vergent deformation prior to the northerly vergent Neo-Alpine orogenesis (Figure 10.5).

The ocean floor of the North Penninic Valais Trough was bounded on the north by the granitic basement of the European margin and on the south by the Brianconnais Swell; the former now forms the core of the Adula nappe and the latter the core of Tambo-Suretta. When this North Penninic Ocean was consumed, the margin above the Benioff zone was the active plate-margin, while the other could be a passive margin. The Brianconnais/Subbrianconnais facies are typically passive-margin or transform-margin sediments, whereas the occurrence of North Penninic Flysch suggests an active margin to the north. This paleogeographical reconstruction is consistent with the postulate by Hsü and Schlanger (1971) that the Late Cretaceous subduction of the ophio-

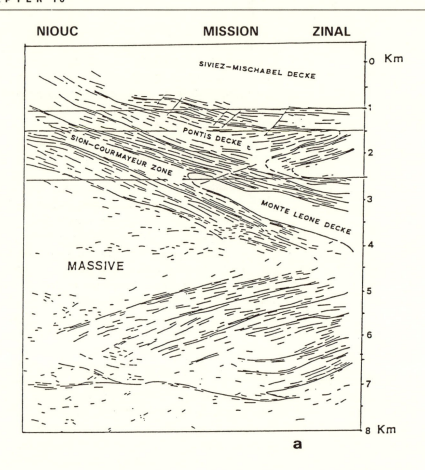

NIOUC   **MISSION**   **ZINAL**

a

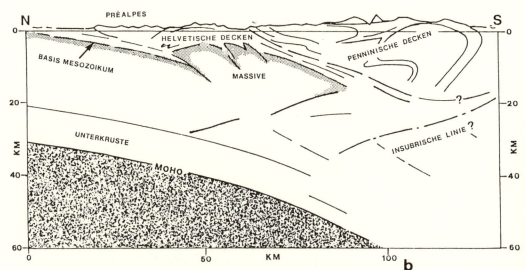

b

Fig. 10.5. Crustal structure under the Alps (from Frei *et al.*, 1989). (a) Sketch showing the results of deep seismic profiling. Note the presence of north-dipping reflectors. (b) Sketch showing the interpretation of deep seismic profiling. The north-dipping reflectors are truncated by northerly vergent structures of the Neo-Alpine deformation. They may well have been formed during the Meso-Alpine collision of Europe and Brianconnais. The décollement deformation of the sedimentary cover of the Brianconnais Swell should thus have a southerly vergence, as is shown by the vergence of the Schams nappes.

lite crust under the Valais Ocean took place along a north-dipping Benioff zone. The Adula basement and the Brianconnais Swell would have to collide when this North Penninic Ocean was completely consumed. The overthrusting of cover nappes that resulted from such a collision should have had a southerly vergence.

Thrust slices such as the Schams nappes are characteristic tectonic elements of a collision-type orogen (Hsü 1989). Recognizing that the Schams nappes are cover thrusts, I came to the conclusion that the Schams strata were stripped off from that part of the Tambo basement that now assumes a "sub-Tambo" position (Hsü and Briegel

1991). The southerly vergent overthrusting did not take place during the Neo-Alpine, but during a Meso-Alpine collision of the European margin and the Brianconnais Swell. This hypothesis of two stages of deformation (with opposite vergences) made unnecessary the awkward postulate by Schmid *et al.* (1990) that the Schams strata were squeezed out from a position below the Tambo nappe by backthrusting during Neo-Alpine orogenesis.

The suture zone of the Meso-Alpine collision is marked by the Areua-Bruschghorn zone of ophiolitic mélange, which has been superposed tectonically above the décollement nappes derived from the Brianconnais/Subbrianconnais passive margin. The North Penninic Bündnerschiefer and the mélange were "obducted" and the Schams cover nappes overthrust onto the Tambo-Suretta Swell under the "plunger action" of the Adula basement. In other words, the Adula nappe was overthrust southward, during the Meso-Alpine orogenesis; this *traineau écraseur* moved toward the Brianconnais Swell, pushing the Bündnerschiefer, mélange, and Schams nappes onto the Tambo-Suretta during the Meso-Alpine orogenesis. Later, after the collision of the southern Tethyan margin and the mid-oceanic Brianconnais Swell, the Austroalpine nappes moved northward as the Neo-Alpine *traineau écraseur*. The Brianconnais (then autochthonous) and its tectonically superposed cover were folded into a giant recumbent fold called Tambo-Suretta; the basement core was enveloped by the cover thrusts and mélanges in the mode of *Deckeneinwicklung* of the Helvetic by the Ultrahelvetic described in Chapter 5.

This hypothesis of two stages of Schams deformation is based upon the recognition of the Schams structures as the Meso-Alpine cover thrusts. It is a "supra" solution from the viewpoint of tectonics, and this tectonic interpretation has been well documented by detailed investigations on the basis of structural geology by Milnes and Schmutz (1978). Their conclusion has not been well received, because they have adopted the conventional approach of assuming a southerly origin for the Schams sediments. Now nobody debates the fact that the Schams sediments belong to the Brianconnais/Subbrianconnais facies. They are not Austroalpine, and they could not have come from the south. Now that we assume a northerly source for the Schams nappes, and a Cretaceous "supra" position for the Schams nappes prior to the Tertiary recumbent folding of the Tambo-Suretta, the negative stratigraphical arguments are not applicable and the positive tectonic arguments are reassuring.

The postulate of Meso-Alpine collision of the Brianconnais and Adula implies Eo-Alpine subduction of the Valais Trough. Thanks to the excellent dissertations on structural geology and metamorphism by several students, the evidence for the Cretaceous deformation of the Adula nappe is now well documented.

Eo-Alpine and Meso-Alpine North Penninic Deformations

The Adula nappe is a Lower Penninic core nappe of eastern Switzerland. The sedimentary cover of the Adula nappe consists of Triassic shallow water strata and Jurassic/Cretaceous Bündnerschiefer. The continental basement of the Adula was separated from the Tambo-Suretta Swell by an ocean crust, the remnants of which are represented by the ophiolite slabs in the Misox zone and by the ophiolites of Chiavenna.

Analyzing the structural geology data of the Adula basement core, Löw (1987) identified four stages of metamorphism (Figure 10.6):

1. Thrusting faulting and shearing of the Adula basement and of its Mesozoic cover (Sorreda stage);
2. Intensive shearing and isoclinal folding of the earlier planar structures, namely the shearing surfaces between the basement and its cover (Zapport stage);
3. Recumbent folding when the Adula basement formed the core of a nappe enveloped by its sheared sedimentary cover (Leis stage);
4. Bending of the Neo-Alpine Adula nappe (Lepontine metamorphism).

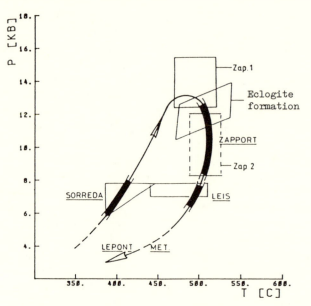

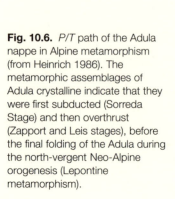

Fig. 10.6. *P/T* path of the Adula nappe in Alpine metamorphism (from Heinrich 1986). The metamorphic assemblages of Adula crystalline indicate that they were first subducted (Sorreda Stage) and then overthrust (Zapport and Leis stages), before the final folding of the Adula during the north-vergent Neo-Alpine orogenesis (Lepontine metamorphism).

The metamorphism accompanying the first two stages of deformation was characterized by high-pressure reactions: kyanite, zoisite, and chloritoid were formed during the first stage, and minerals of the eclogite facies during the second (Löw 1987). Löw further emphasized the continuing nature of the deformation and metamorphism, especially during the first three stages. Reconstruction of the *P/T* path during the progression of metamorphism indicates the rapid displacement of Adula rocks, forming high-pressure, low-temperature mineral assemblages at first (Sorreda stage), and leading ultimately to the crystallization of eclogite assemblages (Zapport stage) at a depth with a pressure of 12–15 kb (Löw 1987; Heinrich 1986).

The continuous downward plunge of a cold lithospheric plate is a manifestation of subduction. Such a descent of the Adula basement and its sedimentary cover down to 40–50 km could take place only when this segment of continental crust was taken down the Benioff zone during the subduction of an ocean lithosphere, or when the North Penninic Ocean between the European margin and the Brianconnais Swell was being consumed. The subsequent stages of Adula deformation and metamorphism, as shown by the *P/T* path diagram, are decompressional, or overthrusting. This indicates that the previously subducted Adula Slab rose gradually up a Benioff zone when Meso-Alpine subduction took place more oceanward. A similar history of initial descent and subsequent rise of a subducted continental margin is recorded by the deformation of the Monte Rosa and Margna nappes (see Chapter 9).

Deformations by thrusting and by penetrative shearing are typical of subduction tectonics. The studies by Löw and Heinrich indicate that the Adula basement, its sedimentary cover, and slabs of ocean lithosphere were first deformed by simple shear, apparently in a Benioff zone during the Eo-Alpine subduction. The shear surfaces were then isoclinally folded, probably during the Meso-Alpine collision. The Adula rocks were overthrust upward prior to and after the collision. Such a postulate agrees with the record of the *P/T* path indicated by metamorphic assemblages. Löw (1987) found that the hinge of the isoclinal fold faces south, verifying the postulate of a southerly tectonic transport during the Meso-Alpine collision between the Adula and Tambo-Suretta continental basements.

The last stage (Lepontine) of Adula metamorphism is Neo-Alpine: the Adula basement, with its deformed and metamorphosed tectonic cover (Misox zone), was overthrust northward (and/or northeastward) to form a core nappe (under the Tambo-Suretta).

Paleogeography of the North Penninic Domain

The South Penninic Piedmont and the North Penninic Valais troughs were separated by the Brianconnais Swell. The Valais oceanic crust was bounded on the north by the Adula continental crust. The presence of breccia beds in the North Penninic Bündner-schiefer south of the Gotthard massif suggests that the Adula was a Jurassic submarine swell situated south of the Ultrahelvetic realm (Bolli *et al.* 1980). The Flysch paleo-geography suggests the presence of a Late Cretaceous island arc—the Habkern Island of Hsü and Schlanger (1971) or the Ultrahelvetic Marginal High of Homewood (1974). This island chain, located between the Ultrahelvetic and the North Penninic realms, furnished detritus to both the Ultrahelvetic and North Penninic Flysch. Whereas the Ultrahelvetic Schlieren Flysch was laid down in a back-arc basin, the trench (Gurnigel and Wägital) south of the island arc should have been the surface expression of a north-dipping subduction zone. It seems logical to postulate that this Late Cretaceous island arc has evolved from the Jurassic Adula Swell. The subduction that caused the Eo-Alpine deformation and metamorphism of the Adula rocks (described by Löw and Heinrich) found its surface expression in the trenches of Gurnigel/Wägital Flysch sedimentation.

The evidence for a southerly vergence during the Meso-Alpine collision between the European margin and the Brianconnais Swell opens up the question of the significance of northerly dipping shear surfaces between the Monte Rosa and Bernard nappes in the Zermatt region. Is it indeed Neo-Alpine back-thrusting, or could it be a Meso-Alpine structure of southerly vergence?

The rocks of the Combin zone have been folded by ductile deformation with a southerly vergence (Sartori 1987, p. 801). Baird and Dewey (1986) suggested a Late Cretaceous age for their deformation, but their conclusion was refuted by Sartori on the evidence that the youngest sediments of the zone are Eocene Flysch. On the other hand, the possibility cannot be excluded that the Flysch was deposited unconformably upon a deformed Brianconnais sequence before its final Neo-Alpine deformation.

We cannot resolve these questions with our present knowledge. At the moment, the geology of the Barrhorn Series/Mont Fort nappe in the west is not comparable to the Schams geology in the east, because there is no evidence for a Meso-Alpine collision of Europe and Brianconnais in western Switzerland. On the other hand, we should not let our admiration for Haug, Argand, and other masters of Alpine geology blind our vision. The possibility of a southerly vergent Meso-Alpine deformation of the western Swiss Alps should be explored.

Deformation of the Southern Margin of the Brianconnais Swell

In western Switzerland, the sediments deposited on the southern margin of the Brian-connais are distinguished by the Jurassic breccia facies. The only Jurassic breccia in eastern Switzerland is the Vizän Breccia of the Gelbhorn unit of the Schams nappes, which was the sedimentary cover above the Tambo basement; the Suretta cover contains no breccia deposits correlative to those of the Breccia nappe of the Prealps. Superposed tectonically above the Suretta Triassic are the Avers Bündnerschiefer. Like the Misox Bündnerschiefer, the Avers is also a tectonic mélange, including exotic blocks of ophiolites in a schistose matrix.

Tectonic investigations of the Suretta by Milnes and Schmutz (1978) led to the recognition of four stages of tectonic deformation:

1. Thrusting and penetrative shearing of the Avers (A stage);
2. Early isoclinal folding with southerly vergence (F stage);
3. Tectonic emplacement of the Schams strip sheets (S stage);
4. Late isoclinal folding (N stage).

The crystalline basement of the Suretta, Rofna gneiss, was deformed and metamorphosed during the second stage, which has been dated as Cretaceous, 110–80 Ma

(Hanson *et al.* 1969). This date implies that the earlier penetrative shearing of the Avers mélange must have taken place during the Early Cretaceous. The Cretaceous Meso-Alpine deformation has, according to Milnes and Schmutz (1978), a southerly vergence. The tectonic superposition of the Mid-Penninic nappes would thus place the Avers Bündnerschiefer in a paleogeographic position between the Tambo and Suretta nappes. One thus has to assume the existence of such a narrow trough and that its consumption was during Early Cretaceous deformation (first stage of Milnes and Schmutz). The Bündnerschiefer was squeezed out of, whereas the Triassic rocks were squeezed in between, the Tambo and Suretta basement cores during a southerly vergent Meso-Alpine collision (second stage of Milnes and Schmutz). The Schams cover thrusts were emplaced on top of the sutured Tambo-Suretta in the Late Cretaceous (third stage of Milnes and Schmutz) after the Adula/Tambo collision. The last stage of deformation was the Neo-Alpine recumbent folding, with a northerly vergence, of the Tambo-Suretta nappe.

Meso-Alpine and Neo-Alpine Penninic Deformations

Passive-margin sequence was stripped off, after a continental collision, to form décollement nappes. The age of the youngest sediments in cover nappes thus places a constraint on the age of collision.

The youngest formation in the Schams nappes is the *oberer Hyänenmarmor*—a metamorphosed pelagic limestone correlative to the *couches rouges* of the Falknis nappe (Streiff 1962). These hemipelagic sediments range in age from Late Cretaceous to Paleocene. The collision between Adula and Tambo-Suretta probably took place during the Late Cretaceous, but postcollisional southerly vergent deformation may have continued until the early Tertiary.

Flysch is present above the Schams nappes in tectonic superposition, and the oldest Flysch of the Oberhalbstein nappes is Maastrichtian (W. H. Ziegler 1956). The age relations suggest, therefore, that a Penninic foredeep existed during late Late Cretaceous time. On the north side was the Adula-Tambo-Suretta arc, now sutured together by an arc-arc collision. On the south was the Austroalpine margin. The Oberhalbstein Flysch, the Prättigau Flysch, and other Upper Cretaceous North Penninic Flysch were deposited in this foredeep, in front of the advancing Austroalpine nappes.

The collision between the Brianconnais and the Austroalpine took place after the beginning of the Tertiary, after the deposition of the *couches rouges* formation, which is as young as Paleocene. The Eocene Flysch formations in these nappes may have been deposited in a deep-sea trough, on this active margin of suturing, shortly after the collision.

The collision between the Austroalpine and the European margins took place during the Late Eocene; the Gurnigel/Schlieren Flysch, which should have been deposited before this collision, includes sediments as young as Middle Eocene. The Austroalpine front advanced far to the north toward the end of the Eocene, when the first flysch deposits were laid down in a foredeep close to the autochthonous realm.

The age of continental collision can also be constrained by the metamorphic dates of Penninic core nappes: the southward underthrusting of the Penninic basement could have taken place only after the Austroalpine-Brianconnais collision. Hunziker (1974) suggested the Late Eocene (38 Ma) for the beginning of the Neo-Alpine regional metamorphism. The mobilized basement nappes were first "tucked" under the Austroalpine, before they moved up again as crustal wedges during the Neo-Alpine orogenesis (see Figure 7.10).

11.

The Austroalpine Nappes

The basement and sedimentary cover of the Austroalpine nappes were those of a continent which was originally situated south of the Tethys Ocean. The nappes were thrust onto the European margin during the Cenozoic, after the Tethys was consumed by the subduction of the ocean lithosphere, and the Austroalpine nappes are separated from the Helvetic by the Penninic nappes and the ophiolite mélanges.

Engadine Window

Driving with a Chinese geologist friend from the Lower Engadine Valley up to the Swiss National Park, we made our first stop at the roadside outside of Zernez and saw basement rocks of an Austroalpine nappe. The road then winds its way up toward the Ofenpass, and we entered the Triassic terrain of limestones and dolomites. Our final stop was at the summit of the pass, where we observed the sedimentary structures exhibited by the Hauptdolomit of Late Triassic (Norian) age. The Triassic strata are deformed, but there are no spectacular recumbent folds like those we had seen in the High Limestone Alps. My friend was somewhat disappointed. "Where are the nappes?" he asked; it seemed that we had seen only the Triassic sedimentary cover of an autochthonous massif. To convince him that the nappes do exist, I drove him down to the valley again and to a serpentinite outcrop near the village of Schuls. The nappes exist because we have found a window of Penninic ophiolite and Bündnerschiefer—the Engadine Window—beneath the Austroalpine rocks.

The story of the discovery dates back to the years shortly after the beginning of the century, and the hero was Pierre Termier. Maurice Lugeon (1902) had just published his grand synthesis of the nappe tectonics of the Western Alps, and the leaders of geology were gathered at Vienna for the 1903 International Geological Congress. On that occasion, Termier joined his Austrian colleagues for an excursion to the Tauern region in western Austria.

The rocks of the Hohe Tauern seemed to be old basement rocks. A gneiss core is wrapped around by a *Schieferhülle*, or schist wrapper, and between the two are discontinuous layers of quartzites and marbles. Overlying the *Schieferhülle* are the thick Mesozoic strata of the Eastern Alps, and fossiliferous Paleozoic rocks are present locally. Austrian geologists considered the schists the Paleozoic basement of the sedimentary strata. Coming from the west, Termier immediately recognized the similarity between the *Schieferhülle* and the *schistes lustrés* of the Western Alps. He was told that Charles Lory, who first described the *schistes lustrés*, had expressed the same opinion some 30 years earlier. The correlation was not possible, the Austrians believed, because Devonian fossils had been found in some schist of the region. Yet the similarity to the Western Alps was too impressive to be ignored. The Devonian fossils could have come from an older formation; the bulk of the *Schieferhülle* has to be a correlative of the Mesozoic *schistes lustrés*. Carrying the hypothesis a step further, Termier (1903) tried to persuade his host that the Eastern Alps, like the Western Alps, are underlain by a pile of nappes.

The western edge of the Eastern Alps is exposed on the right side of the Rhine Valley (Figure 11.1). The tectonic superposition of the basement rocks of the Upper Austroalpine and North Limestone Alps on the Prättigau Schist is unmistakable (Figure 9.2). Rothpletz (1900) considered it evidence for westward thrusting of the Eastern Alps over the Western Alps. J. Blaas (1902), in a geological guide of eastern Switzerland and western Austria, had in fact found the Engadine Window before Termier discovered the Tauern Window. The *schistes lustrés*, which extend to eastern Switzer-

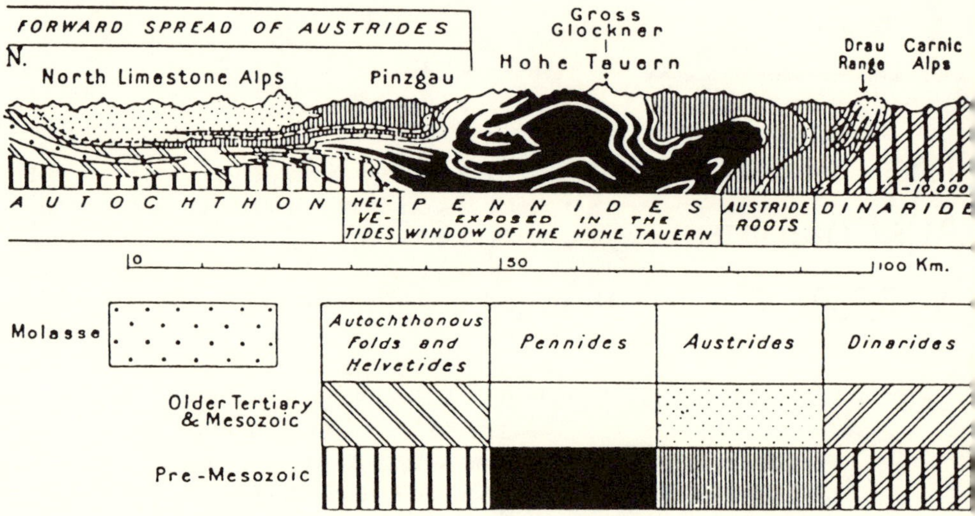

Fig. 11.1. The Tauern Window (after Bailey 1935). Termier correlated the schists (Pennides, in black) of the Hohe Tauern in Austria to the Mesozoic *schistes lustrés* of the Swiss Alps; the formations yielding Paleozoic and Mesozoic fossils above the schists are thus recognized to constitute a thrust complex, now called the Austroalpine nappes (Austrides in profile).

land under the name of Bündnerschiefer, plunge in Prättigau under the nappes of the Rhäticon (Falknis and Sulzfluh), only to reappear again in the Lower Engadine Valley.

The geological significance of the two immense windows (the Lower Engadine Window and the Tauern Window) was, however, not appreciated until Termier's (1903) elaborate synthesis. Termier's insight was so brilliant and his arguments so persuasive that Eduard Suess (1904), the great master of Alpine geology, had to admit it a year later: the Lower Engadine is a structural window!

The Inn River has cut deep enough through the tectonic edifice of the Austroalpine nappes that the Bündnerschiefer is exposed on both sides of the Lower Engadine (Lower Inn) Valley of southeastern Switzerland (Figure 11.2). Cadisch *et al.* (1963), invoking the nappe concept, suggested the following tectonic superposition (in descending order):

Silvretta and Oetztal, rigid-basement nappes
Hauptdolomit schuppen zone
Tasna nappe
Champatsch zone
Ophiolite
Bündnerschiefer

The Silvretta and Oetztal are Upper Austroalpine nappes, and the Hauptdolomit schuppen zone consists of strip sheets of the Upper Austroalpine sedimentary cover. The Tasna sequence is, in descending order,

Flysch
Upper Cretaceous, *couches rouges* pelagic limestone
Lower and Middle Cretaceous, limestone and glauconite sandstone
Upper Jurassic, thick-bedded limestone
Lower Jurassic, limestone
Triassic, dolomite, gypsum and anhydrite

The Champatsch zone has been considered Upper Penninic; Cadisch *et al.* (1963, p. 5) wrote:

Cropping out on the southwestern border of the Engadine Window are blocks or slabs of diabase, Triassic dolomite, Liassic limestone, and Jurassic radiolarite, embedded in a matrix of South Penninic schist. This group of rocks and the ophiolites on Piz Nair nearby constitute the Champatsch zone. [free translation by the author]

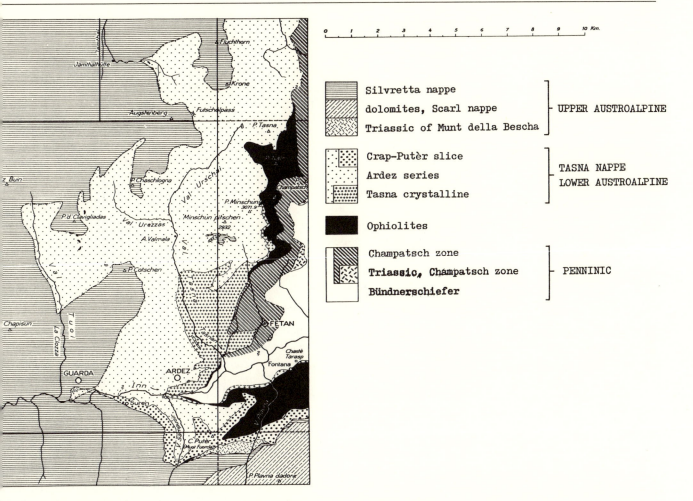

Fig. 11.2. Tectonic sketch map of the southwest end of the Lower Engadine Window (after Cadisch *et al.* 1941). Between the Upper Austroalpine nappes and the basal Bündnerschiefer are the tectonic mélange (Lower Engadine Mélange) containing rocks of both Austroalpine and Penninic origin. Note that ophiolites (in black) and basement rocks form slabs of very limited lateral extent.

The ophiolite unit described by Cadisch *et al.* consists mainly of large slabs of serpentinite, ophicalcite, diabase, spilite, and green schists, like the ophiolites of the "Platta nappe." We agree, therefore, with the assignment by Cadisch *et al.* of an Upper Penninic position to the Champatsch rocks, but the "Champatsch zone," like the "Arosa schuppen zone" or the "Platta nappe," is a tectonic mélange. The exotic components in Champatsch are derived from two different terranes—Austroalpine (dolomite, limestone) and South Penninic (ophiolite, radiolarite, Bündnerschiefer). Subjacent to the Champatsch Mélange are the Bündnerschiefer of the Engadine Window; they are metamorphosed calcareous and sandy shales and siltstone similar to those of North Penninic origin cropping out at the Via Mala Gorge.

Applying the classic concept that a higher nappe had a more southerly paleogeographic position, Staub (1958) concluded that the Tasna was Lower Austroalpine, because the "nappe" lies above the Upper Penninic Champatsch zone. To extend the correlation a step further, the nappes of Falknis/Sulzfluh, of Schams, and of the Median Prealps were all considered Lower Austroalpine. Now that Falknis/Sulzfluh and Schams, as well as the Klippen nappe, are all considered Mid-Penninic, their correlative sequence of Brianconnais/Subbrianconnais facies in the Engadine Window can hardly be considered Lower Austroalpine; the Tasna sequence was the sedimentary cover of the Brianconnais Swell. The apparently reversed tectonic superposition of the Upper and Middle Penninic units in the window can be understood if one accepts the interpretation that the various Austroalpine, Upper and Middle Penninic elements in the window constitute a tectonic mélange—the Lower Engadine Mélange. The tectonic superposition is, in descending order, as follows:

Upper Austroalpine nappes
    Basement and its parautochthonous cover
    Hauptdolomit schuppen zone
Lower Engadine Mélange, including the "Tasna nappe," "Champatsch zone," ophiolite,
    and Bündnerschiefer (Upper and Middle Penninic)
Bündnerschiefer (Lower Penninic)

In this scheme, the presence of Brianconnais rocks (Tasna) above those of Upper Penninic (Champatsch) in the window can be explained by assuming a differential displacement of the various components in a mélange; the tectonic superposition may have been in part reversed during the penetrative shearing of Eo-Alpine subduction and Neo-Alpine overthrusting.

That the Lower Engadine Mélange crops out in a structural window under the Austroalpine nappes is now a foregone conclusion. These rocks have been broken into thrust sheets, called nappes in Alpine literature. The nature of the Austroalpine deformation is, however, significantly different from that of the Helvetic and Penninic: the original nappes, such as the Morcles (Helvetic) or the Antigorio (Penninic), are large recumbent folds with an overturned limb, but such an inversed limb is hardly ever found in Austroalpine nappes. Cornelius (1940) wrote:

> It is a source of much misunderstanding that some Alpine geologists emphasized the recumbent folding of nappes. Opponents to the nappe theory correctly argued that overturned limbs are rarely found, but they concluded wrongly that there are no "nappes." They overlooked the work by Albert Heim, who had pointed out long ago that the lower limb of a recumbent anticline is reduced by overthrusting to become a thin shear zone such as the *Lochseitenkalk*; one should not expect at all to find a complete recumbent fold in a nappe. [free translation by the author]

The Austroalpine nappes are mostly rigid-basement thrusts (Trümpy 1960); the sedimentary cover is stripped off in places, but in other instances the cover is still found on top of its basement as autochthonous or parautochthonous cover. Strip sheets, such as the Hauptdolomit schuppen zone around the Engadine Window, are thrust slices or cover thrusts, but they rarely form recumbent folds such as those in the Helvetic domain.

The Alpine deformation of the rigid-basement thrusts is restricted to an intensively sheared zone at the base of the thrusts. The deformation at the base of the Silvretta indicates a long history of Eo-Alpine deformation and metamorphism before its overthrust onto the Penninic units of the Engadine Window.

The physical problem of moving large overthrusts over long distances was first pointed out by Smoluchowski (1909): the friction at the base of overthrusts is, under normal circumstances, so large that a large slab should break before the friction can be overcome. Hubbert and Rubey (1959) maintained that bodily displacement of large overthrust sheets is made mechanically feasible because of a reduction of sliding friction when the pore pressure at the base of such thrusts is abnormally high. This hypothesis can be applied to explain the mechanics of overthrusting by frictional sliding along a brittle fracture, but the postulate is not applicable if the rocks in the shear zone have a finite cohesive strength (Hsü 1969a). In considering the mechanics of the Glarus Overthrust, I pointed out that the overthrusting took place along shear zones of flowage. The maximum shear stress, allowing for the stability of the Glarus thrust-slab, is less than 1 kb, and the deformation took place at a strain rate of about 10^{-10} to 10^{-13} per second and a temperature of 300–400 °C. Later Schmid (1975) obtained a maximum temperature of 390 °C for the thrusting. Using experimental data, the strain rate at this temperature should be 10^{-10} to 10^{-11} per second for the Glarus overthrust (Hsü 1969b). Studies of the limestone mylonite at its base indicate overthrust deformation by superplastic flow, with a stress as low as a hundred bars (Briegel and Goetze 1978).

The absence of a crystallographic preferred orientation indicates that grain-boundary sliding was a dominant strain-producing mechanism during the mylonitization (Briegel, personal communication).

The mechanics of rigid-basement thrusting has not yet been systematically investigated. The size of such slabs is on the order of dozens of kilometers long and 10–20 km thick, and their stability suggests deformation under low shearing stress. Field observations revealed that the Upper Austroalpine Silvretta and Oetztal nappes are marked by a thin shear zone at their base. The rocks in the shear zone were subjected to temperatures of up to 500 °C in the Early Cretaceous (130–100 Ma), and up to about 300 °C in the Late Cretaceous, before they were cooled down to less than 200 °C during the Tertiary (Flisch 1986). The rate of the vertically upward component was estimated on the basis of those dates and of fission-track cooling ages to be about 0.15 mm per year (Flisch 1986). The displacement rate for the relative motion between central Europe and Italy should have been 0.7 mm/yr during the Cretaceous and early Tertiary (Hsü 1989). The 0.15 mm/yr "uplift" rate corresponds to a 0.3 mm/yr convergence rate, if the Upper Austroalpine nappes were thrust up a ramp with a 30-degree inclination. It seems that about half of the displacement between the two plates was taken up by the shear zone between the Lower and Upper Austroalpine nappes.

Observations of the shear zone at the base of the Silvretta indicate that garnet-mica schist and quartz-feldspar gneiss were formed during an early stage of deformation (probably Early Cretaceous) when amphibolite facies metamorphism prevailed at a depth of 15–20 km; the thrusting movement took place along an intracrustal shear-zone between the Lower and Upper Austroalpine. As the Upper Austroalpine basement moved higher up the ramp and was thrust northward, the temperature of deformation decreased. Consequently, the strain rate should have slowed down, assuming that the stress remained constant. Under such conditions, probably during the Late Cretaceous, mylonitic rocks, including ultramylonite, were formed at the base of the Silvretta nappe. Further movement up the ramp after the Neo-Alpine collision caused the Tertiary overthrusting, at relatively low temperature and along a brittle fracture surface; the Silvretta basement was finally emplaced tectonically above the ophiolite mélange and Bündnerschiefer of the Engadine Window.

The last deformation of the Engadine Window region was Eocene or younger, because Maastrichtian microfossils are found in the schists of the Champatsch zone (Cadisch *et al.* 1963), and Paleocene/Eocene microfossils in the Lower Engadine Flysch (Oberhauser 1983). The Penninic Engadine was still a site of deep-water sedimentation in the Early Eocene before the arrival of the Lower and Upper Austroalpine nappes.

The Lower Engadine Mélange, involving the fragmentation of the Tasna, the ophiolites, and the Champatsch rocks and their mixing in a Bündnerschiefer matrix, owed its genesis to the Eo-Alpine subduction, but the mélange became the suture between the Austroalpine and the Helvetic elements after the Neo-Alpine collision.

Tectonic Superposition of Austroalpine Nappes

The Eastern Alps of Switzerland, Liechtenstein, and western Austria are a complex edifice of nappes of great coherent masses. Trümpy (1980) proposed a threefold division of the complex in the region:

Central Austroalpine complex
   Silvretta-Oetztal nappes, etc.
   Campo, Languard, Ela/Ortler nappes, etc.
Northern Calcareous Alps
Lower Austroalpine nappes, Err-Bernina nappes, etc.

The Lower Austroalpine nappes consist largely of the Err-Bernina basement; the Lower Austroalpine sedimentary cover has been sliced into numerous thrust sheets, which are sandwiched between the basement. The Languard and Campo are rigid-basement nappes; a small remnant of sedimentary cover has been retained on top of the Languard, whereas the Campo basement is overlain by the sedimentary cover (Permian to Paleocene) of the Ortler zone. The Silvretta-Oetztal nappes also consist largely of Paleozoic basement. Along its western margin the Silvretta basement overlies lenses of Triassic dolomite and the Ela nappe and comes into contact with the nappes of the Northern Calcareous Alps. The relative position between the two has been a subject of debate between Swiss and Austrian geologists: Tollmann (1977) considered the Northern Calcareous Alps an Upper Austroalpine element and the Silvretta-Oetztal a Middle Austroalpine element. The mainly Mesozoic sedimentary cover of the former was considered to have been emplaced from a home south or east of the Silvretta-Oetztal basement. Trümpy (1980) suggested, however, that sediments of the Northern Calcareous Alps once formed the cover of the Languard basement; he wrote (p. 80):

> The Northern Calcareous Alps of the Rhätikon mountains (between Graubünden, Liechtenstein, and Vorarlberg) are constituted by a series of imbricate slices, with apparent movement to the northwest. The underlying incompetent rocks of the Arosa zone are squeezed into thrust zones separating these slices . . . In the middle Montafon Valley (north of Sulzfluh) the Permian and Triassic rocks of part of the Northern Calcareous Alps seem to be in near-stratigraphic contact with the "Phyllitgneis zone," a narrow band of partly phyllonitic basement rocks along the northern margin of the Silvretta block. At their northwestern edge, the Silvretta override the phyllite gneisses and also inverted slabs of Triassic rocks pertaining to the Northern Calcareous Alps.

Whereas all seemed to agree that the "Phyllitgneis zone" is a slice of the basement under the Mesozoic of the Northern Calcareous Alps, this basement was considered by Tollman (1977) ultra-Oetztal, and by Trümpy (1980) sub-Silvretta, i.e., at sites south and north of the Oetztal-Silvretta basement, respectively. Stratigraphy cannot yield a unique answer; palinspastic reconstructions can produce various scenarios of Triassic submarine topography, all equally reasonable. Theoretical considerations of mountain-building mechanism give, however, strong support to the Trümpy alternative. He suggested that the sedimentary sequence of the Northern Calcareous Alps has been stripped off from the uppermost Lower Austroalpine Languard nappe. They are cover thrusts and pushed northward by the plunger action of the Silvretta nappe. We should recall that a folded tablecloth (cover thrusts) needs to be pushed by a hand (plunger) from behind. Silvretta could be such a "hand"; the Languard cover under the Silvretta could easily have been stripped off and moved onto foreland under the "plunger action" of this great Upper Austroalpine rigid-basement nappe. If, however, those sediments should be ultra-Oetztal in origin, it would be difficult to move them across the Central Austroalpine basement nappes; there seems to be no appropriate "hand" to push the "tablecloth" from one end of the table to the other.

Sedimentation on Austroalpine Margin

The Austroalpine Permian and Triassic sediments are parts of the shallow-water sedimentary cover of the supercontinent Pangea. The Austroalpine, the Apulia, and other microcontinents may have had the Andean type of margin, where the Paleotethys Ocean was subducted under Europe along a northwestern-dipping Benioff zone (see Chapters 12 and 16). The extensional rifting leading to the genesis of the Tethys between the European (Helvetic) and Austroalpine margins had its beginning during the Late Triassic. The Norian Hauptdolomit, widespread in the Eastern Alps, was laid down in lagoons or on tidal flats. The Rhätian formation consists mainly of shallow marine limestone and calcareous shale, deposited on a subsiding margin. The sediments formed during the Jurassic rifting stage of the Tethys Ocean are mainly re-sedimented deposits, laid down when the Austroalpine margin was actively subsiding, and when the ocean-continent transition was characterized by very rugged relief. The first ocean crust of the Tethys was emplaced at the beginning of the Middle Jurassic.

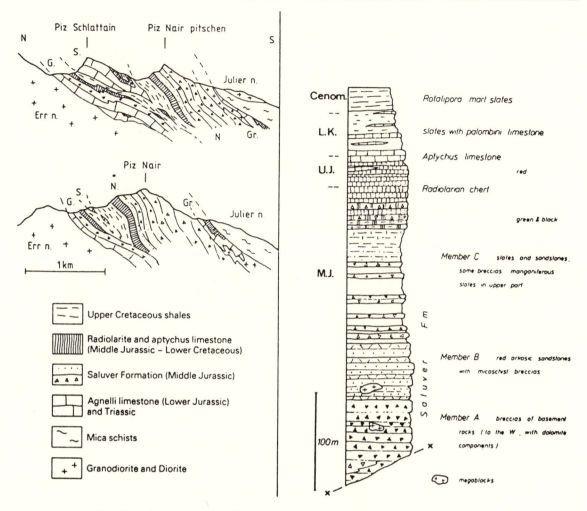

Fig. 11.3. Structural and stratigraphic sections of the Piz Nair area (after Finger 1978).

The subsidence history in the Lower Austroalpine domain is recorded by the imbricated sedimentary cover on top of the Err nappe in the mountains north of St. Moritz in the Upper Engadine Valley (Figure 11.3). The Lower Jurassic is mainly a limestone, with little evidence of resedimentation. The initiation of the rifting is manifested by a thick Middle Jurassic breccia deposit (Saluver Formation). These submarine-slide deposits have clasts of basement rocks, although dolomite fragments are also common. The breccias are overlain by resedimented arkosic sandstone and microbreccia beds, and by manganiferous pelitic sediments. There was less tectonic activity and the submarine relief reduced toward the end of the Middle Jurassic, when the first radiolarian chert was deposited. The increased influx of detritus to the Austroalpine domain during the Early Cretaceous changed the sedimentation pattern; the pelagic deposits of radiolarite and Aptychus Limestone grade upward into hemipelagic sediments, which have been lithified into slates or phyllites. Occasional pelagic deposits occurring as intercalations have been known as limestone of the Palombini type. The youngest hemipelagic deposits in the area are Cenomanian (Finger 1978).

Rifting was not confined to the Lower Austroalpine margin. The presence of megabreccias and other resedimented deposits indicates extensional tectonics toward the continental interior in the depositional realms of the Central Austroalpine complex. The Allgäu Formation represents sedimentary prisms, which accumulated in Upper Austroalpine basins formed by Jurassic rifting. The formation consists of marls and limestones interbedded with different types of redeposited carbonates. Eberli (1988) recognized the following associations of resedimented deposits (Figure 11.4):

(1) Talus Association. Thick chaotic breccia beds, up to 100 m thick, are the deposits at the base of fault scarp. These are blocks that have fallen down from submarine cliff and/or were laid down by catastrophic debris-flow. The largest blocks are several meters across. The breccias are clast supported, and the matrix consists mainly of broken debris of dolomite.

(2) Basal Breccia Association. This is an association of megabreccias, thin-bedded turbidites, marls, and limestones. Sediments of this association are commonly the basal sediments in Jurassic basins of Upper Austroalpine domains.

(3) Thick Turbidite Association. This is an association of thick-bedded turbidite sandstones, conglomerates, and interbedded marls, deposited not far from the base of the slope.

(4) Thin Turbidite Association. This is an association of thin-bedded turbidite beds and marls, deposited, in most instances, far away from the base of the slope.

(5) Hemipelagic Association. This association is dominantly hemipelagic marl and argillaceous limestone. Turbidite layers are not uncommon in basin-plain deposits, but are rare in sediments deposited on the opposite slope.

(6) Area of nondeposition.

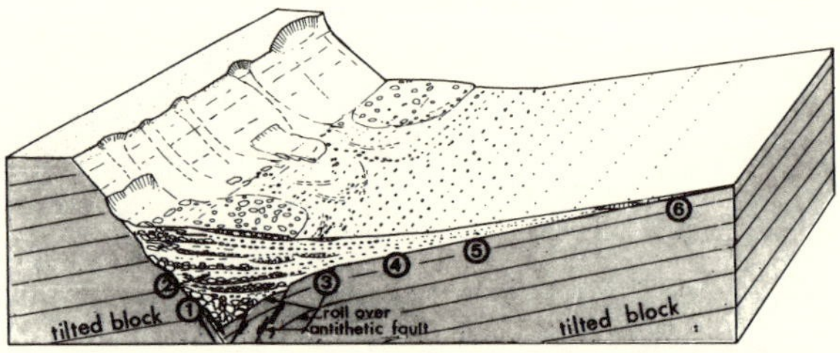

Fig. 11.4. Model of sedimentary facies of the Allgäu Formation (after Eberli 1987). (See text for explanation of sediment types 1–6.)

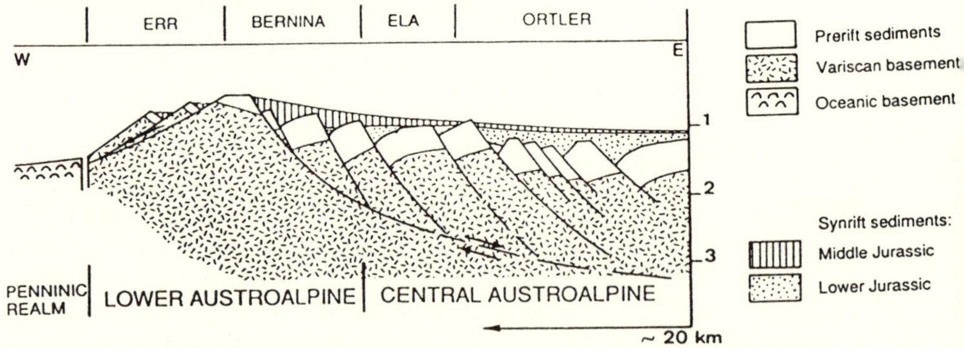

Fig. 11.5. Middle Jurassic Austroalpine passive margin (after Eberli 1987). The Tethys Ocean came into existence owing to seafloor spreading west of the Err terrane. The various types of sediments of the Allgau Formation were deposited in basins east of Err.

The distribution of the various associations within a basin is shown diagrammatically in Figure 11.4. The Middle Jurassic paleogeography reconstructed by Eberli (Figure 11.5) shows that a series of N-S-trending half-graben basins were formed by extensional rifting. The continental margin was fragmented progressively northward or westward. The Upper Austroalpine basins had their origin in the Early Jurassic whereas the Lower Austroalpine (Bernina, Err) basins were first rifted during the Middle Jurassic. There is also a preferential dip of the faults toward the interior of the continent. The rifting is not symmetrical, and cannot be explained by the simple extension model of McKenzie (1978). A lack of symmetry has been predicted by Wernicke (1985); his model postulated crustal thinning as a consequence of extension along a major detachment fault (Figure 11.6). The axis of maximum crustal thinning migrated northward, and Africa was finally detached from Europe in the Middle Jurassic along an axis north or west of the Lower Austroalpine domain (Figure 11.7)

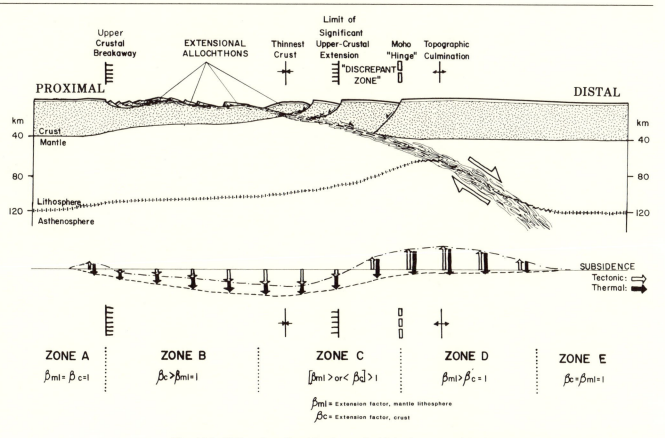

Fig. 11.6. Rifting of continent leading to the birth of an ocean (after Wernicke 1985). The rifting of the southern Tethyan margin is best explained by Wernicke's model of crustal extension, except that his model shows half-grabens with the downthrown side oceanward, whereas Eberli postulates rift grabens with the downthrown side mostly toward the continental interior.

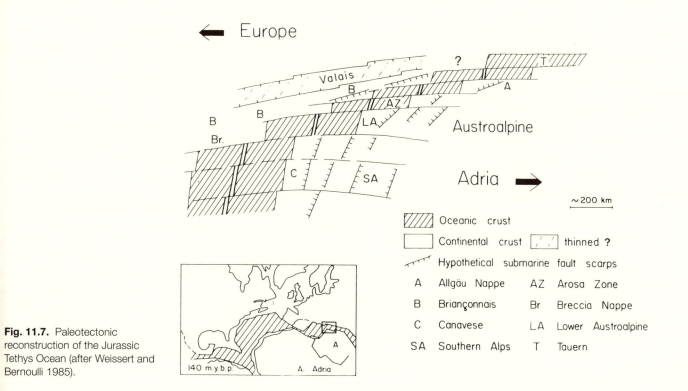

Fig. 11.7. Paleotectonic reconstruction of the Jurassic Tethys Ocean (after Weissert and Bernoulli 1985).

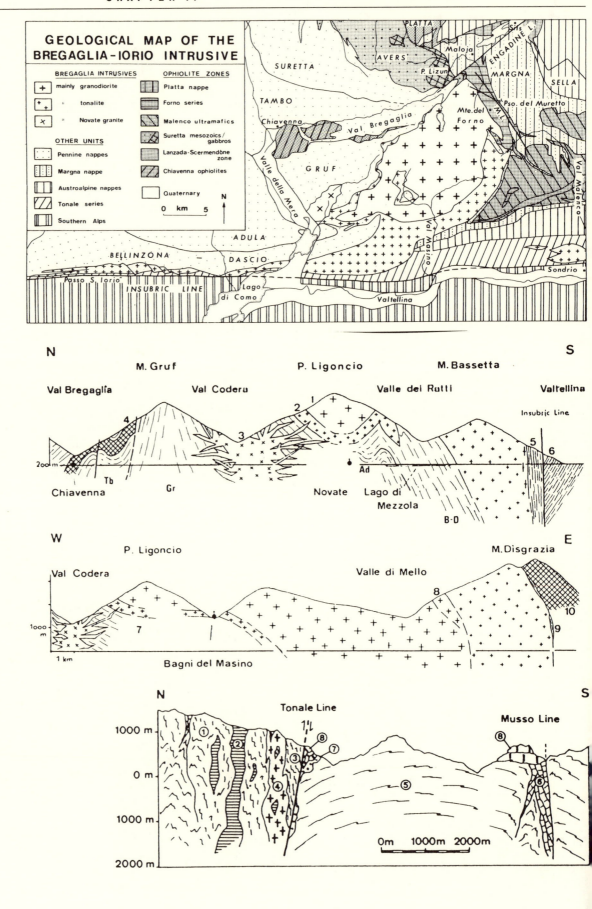

GEOLOGICAL MAP OF THE
BREGAGLIA-IORIO INTRUSIVE

BREGAGLIA INTRUSIVES
- mainly granodiorite
- " tonalite
- " Novate granite

OTHER UNITS
- Pennine nappes
- Margna nappe
- Austroalpine nappes
- Tonale series
- Southern Alps

OPHIOLITE ZONES
- Platta nappe
- Forno series
- Malenco ultramafics
- Suretta mesozoics / gabbros
- Lanzada-Scermendòne zone
- Chiavenna ophiolites
- Quaternary

0 km 5

Bergell Granites

In contrast to other mountain belts, the Alps are distinguished by a paucity of granitic intrusions. The Bergell granites are the sole plutonic bodies of the Swiss Alps. They are intruded into the region between Maloja and Mera Valley and are bounded by three late Tertiary faults, namely the faults along the Insubric, Engadine, and Muretto lines (Figure 11.8). The superposition of the tectonic units in the surrounding terrane, is in descending order,

"Platta nappe"
"Margna nappe"
Forno Series/Malenco Gabbro
Suretta nappe
Tambo nappe
Chiavenna Ophiolites
Gruf unit/Adula nappe

Three types of calc-alkaline rocks constitute three distinctive intrusive bodies: they are the tonalite, the granodiorite, and the leucogranite. The tonalite, or quartz diorite, is medium-grained, biotite-hornblende quartz diorite. This rock type is present mainly on the southern border of the intrusive complex (Figure 11.9). Toward the west the tonalite becomes a strongly foliated and steeply dipping sheet (Figure 11.10), trending parallel to the Tonale Line. Ductile deformation of the tonalite at more than 500 °C is manifested by the deformation of aplite dykes. The southern border of the tonalite is a young augengneiss. Vogler and Voll (1976) postulated that the tonalite underwent ductile deformation during the Neo-Alpine orogenesis, when the San Jorio zone was flattened to one-twentieth of its original width; it is now only 5 km wide. The tonalite is also present on the northern border of the Bergell complex (Figures 11.8 and 11.9). Alignment of hornblende crystals produces a lineation parallel to the contact with the surrounding Gruf unit, and no intrusive relation can be observed at this contact (Wenk 1973). In the Bagni del Masino Window area (Figure 11.9), the tonalite appears under schistose granodiorite, and the Bergell-Jorio complex seems to have been emplaced as a thrust plate above a Penninic core nappe.

The tonalite is absent in the east, where a biotite granodiorite is intrusive into the country rock. Cross-cutting relationships indicate that massive granodiorite was intruded into the Monte del Forno complex after those rocks were pervasively sheared and metamorphosed (Figure 11.8). A zone of contact metamorphism along this eastern border was noted by Cornelius as early as 1913, and the contact aureole is about 2 km wide.

Andesite-basalt dykes have been found near the eastern border of the Bergell intrusive (Nievergelt and Dietrich 1977). The dykes have been subjected to contact metamorphism caused by the intrusion of the Bergell granodiorite. Those found at more distant sites are neither deformed nor metamorphosed, and they cross-cut the Malenco serpentinite and the metamorphic rocks of the "Margna nappe." Staub (1920) suggested that volcanoes in the Bergell area were the source of the volcanic detritus of the Late Eocene Taveyannaz Sandstone in the Helvetic domain. Preliminary data on the age, chemical composition, and location of the dykes seem to verify this postulate (Trommsdorff and Nievergelt 1983).

The third type of Bergell intrusive, a leucogranite, is younger, mostly foliated, medium-grained, inhomogeneous, locally garnet-bearing, two-mica aplite-granite. An intrusive body (Novate Leucogranite) cuts cross all structures and older Bergell intrusions (Figure 11.9); this discordant intrusive may not be genetically related to the tonalite or to the granodiorite (Gulson and Krogh 1973).

The tectonic significance of the Bergell granites has been a matter of controversy. Citing the evidence of an intrusive relationship on the eastern border, Staub (1924) considered the Bergell a textbook example of a post-tectonic intrusion. Emphasizing the concordant structural relationship on the northwestern and southern borders, Wenk

ON OPPOSITE PAGE:

Fig. 11.8. Tectonic map of southeastern Switzerland and the geology of the Bergell intrusive (after Trommsdorff and Nievergelt 1983). Note the separation of the Swiss Alps from the Southern Alps by the Insubric Line.

Fig. 11.9. Schematic profiles across the Bergell intrusive (after Trommsdorff and Nievergelt 1983). Tb: Tambo nappe; Gr: Gruf unit; Ad: Adula nappe; B-D: Bellinzona-Dascio zone. (1) Granodiorite; (2) tonalite; (3) Novate Leucogranite; (4) Chiavenna Ophiolite; (5) Tonal Series; (6) Southern Alps; (7) Penninic basement, *sensu lato*; (8) Cameraccio Granodiorite; (9) Suretta nappe; (10) Malenco ultramafics. Note the window at Bagni del Masino, where the Bergell granodiorite is thrust on top of Penninic basement, *sensu lato*.

Fig. 11.10. Section through the boundary between the Swiss Alps and the Southern Alps, north of Lake Como (after Fumasoli 1974). (1) Gneisses of the root zone; (2) amphibolites; (3) gneiss, mica schist, and marble of the Tonale Series; (4) Tertiary tonalite; (5) basement of the Southern Alps; (6) marble; (7) Permian sandstone and conglomerate; (8) Triassic dolomite.

(1973) proposed that the Bergell igneous rocks were emplaced as a nappe concordantly in the stack of higher Pennine nappes. Trommsdorff and Nievergelt, in their 1983 review, stated that Wenk's interpretation had been proven wrong, because the Bergell granodiorite clearly cuts across nappe boundaries. Gautschi and Montrasio (1978, p. 329) also considered the Bergell post-tectonic, because "the contact metamorphism of the Bergell intrusives postdates the alpine regional metamorphism in the eastern part of the complex." Bucher (1977) pointed out, however, that the Bergell granodiorite and tonalite may have been emplaced between the two stages of deformation and metamorphism.

The tectonic units east of the Bergell pluton are (1) the Suretta Nappe, (2) the Monte del Forno complex, (3) Malenco Serpentinite, and (4) the "Margna nappe." Units 2, 3, and 4 constitute one big mélange unit, and the various components in the mélange include ocean sediments and ocean floor of the South Penninic domain, as well as Austroalpine basement and cover. The Suretta rocks consist of gneiss basement, quartzites, marbles, and Bündnerschiefer. They had been thrust under the mélange, and all these rocks were penetratively deformed and regionally metamorphosed before the intrusion of the Bergell granodiorite. Citing an isotope date of 70 Ma for Margna metamorphism (Jäger 1973), Bücher recognized the Eo-Alpine age of the mélange deformation. The intrusion of the Bergell must then have taken place after this deformation. Correlating the andesite-basalt dykes near the eastern border of the Bergell pluton with the Taveyannaz volcanism, the date of intrusion would be subsequent to the Late Eocene collision of the Austroalpine and the Helvetic margins. In other words, the Bergell is a Neo-Alpine granite. The age of the intrusion has also been estimated on the basis of isotope dating and of stratigraphy. U/Pb analysis of zircons from Bergell granites indicates an age of crystallization of 30 Ma (Grünenfelder and Stern 1960; Gulsen and Krogh 1973). Such an age is consistent with the observation that pebbles derived from the Bergell pluton are found in the Late Oligocene/Early Miocene Molasse deposit of the Como district.

In conclusion, the evidence shows that the Austroalpine and Penninic nappes in the east had already been superimposed and deformed when the Bergell intrusion was emplaced. In the western Bergell, however, high-grade regional metamorphism and ductile deformation persisted even beyond the time of tonalite and granodiorite intrusions (Trommsdorff and Nievergelt 1983). This could explain the concordant margin and the schistose appearance of the granodiorite in the west. The thrust movement at the base of the Bergell took place while partial melting at the top produced the discordant intrusives of tonalite and granodiorite. Dating of micas has yielded isotope ages ranging from 22 to 25 Ma for the granodiorite and tonalite, and even younger ages for the Novate granite. The pluton was then uplifted, and ductile deformation was replaced by brittle faulting, which was to produce the Engadine, Insubric, and Muretto lines.

12.

Geological Evolution of Switzerland

Paleogeographical reconstructions have shown that Europe and Africa were both parts of Pangea during the Permian Period. Bullard *et al.* (1965) reconstructed the continents around the Atlantic and Indian oceans without having to disturb the rigidity of the continent. An implication of this is the presence of a wedge-shaped gap in Pangea that opens eastward (Figure 12.1). This giant ocean separating Gondwanaland from Laurasia was the Paleotethys, and remnants of it are now found in the Cimmeride orogenic belt, which extends from Dobrugea, to Crimea and Turkey, to the Caucasus, to Iran and Afghanistan, and from there through Tibet, Yünnan, and Indo-China to the Southwest Pacific (Sengör 1984).

Pangea began to be rifted apart during the Triassic. Eurasia separated from Gondwanaland in the early Jurassic, and the Neo-Tethys was born. The Mediterranean regions of Italy and the Balkans constituted a northern promontory of Africa, but they became a separate Mediterranean plate during the early Cretaceous (Hsü 1989). The northward march of the Mediterranean plate caused the Eo-Alpine consumption of the Neo-Tethys, before the final Neo-Alpine continental collision (Figure 12.2). Switzerland sits on a suture zone, and the geological history of Switzerland is a record of the separation and suturing of the two plates.

Fig. 12.1. Paleogeographic reconstruction of the Jurassic Tethys (after Dewey 1988). The Paleotethys (T_1) was consumed after the Cimmerian continent (Qiangtang) was split off from Gondwanaland and collided with Laurasia in the early Jurassic. The Neo-Tethys (T_2,T_3) was consumed during the Alpine-Himalayan orogenesis, causing the suturing of the Lhasa block and India to Laurasia. (1), (2), (3), (4) denote the sequential order of the splits.

Hercynian Europe

The geology of Europe gives evidence of Paleozoic mountain-building processes, leading to continental collision. The Hercynides of central Europe embrace all areas where compressional deformations came to an end during the late Paleozoic. Hercynian is the term used in Romance languages; its German equivalent is *variszisch*, or Variscan. The tectonic divisions of the central European Hercynides were established by the classic paper of Kossmat (1927); they are, from north to south, the Rhenohercynian, Saxothuringian, and Moldanubian zones, equivalent more or less to the Mo-

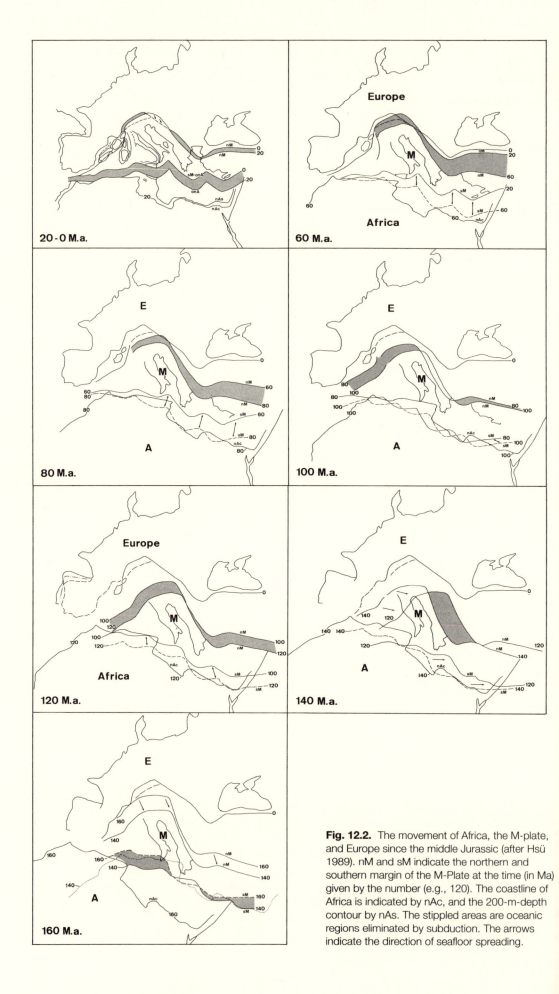

Fig. 12.2. The movement of Africa, the M-plate, and Europe since the middle Jurassic (after Hsü 1989). nM and sM indicate the northern and southern margin of the M-Plate at the time (in Ma) given by the number (e.g., 120). The coastline of Africa is indicated by nAc, and the 200-m-depth contour by nAs. The stippled areas are oceanic regions eliminated by subduction. The arrows indicate the direction of seafloor spreading.

NW

SUBVARISCAN
FOREDEEP

Lahn - Dill
Syncline

SE

NORTHERN MID-GERMAN
PHYLLITE CRYSTALLINE
ZONE RISE

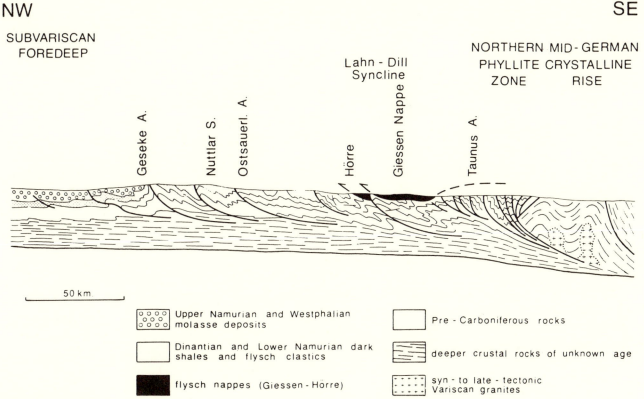

50 km

○○○○ Upper Namurian and Westphalian
○○○○ molasse deposits

Dinantian and Lower Namurian dark
shales and flysch clastics

flysch nappes (Giessen-Hörre)

Pre - Carboniferous rocks

deeper crustal rocks of unknown age

syn- to late - tectonic
Variscan granites

Fig. 12.3. Diagrammatic cross section through the Rhenohercynian zone (after Franke 1989).

lasse/Helvetic, Penninic, and Austroalpine units of the Alps. South of the Belgian coalfields (foreland-basin molasse) is the Rhenohercynian foreland deformed belt, which extends from the Ardennes to and beyond eastern Germany; the Devonian and Carboniferous sediments of the area constituted a passive-margin sequence. The predominantly carbonate sequence grades southward to the slates, phyllites, and turbidites that have been thrust northward as the flysch nappes of the Hercynides (Figure 12.3). Farther south was the "Saxothuringian Ocean," now manifested by the rocks of the Saxothuringian zone (Franke 1989). These metamorphic and igneous rocks, deformed by mid-Paleozoic subduction and late Paleozoic suturing, are found in the core nappes and mélanges of the central European Hercynides. The Moldanubian zone of Precambrian basement and late Paleozoic granite is comparable to the Austroalpine rigid-basement nappes and the Bergell intrusives in the Alps (Finger and Steyrer 1990); the advance of this overriding wedge ultimately caused the foreland deformation of the sediments in the Rhenohercynian zone.

Classical ideas on the orogenic chronology of the Hercynides are expressions of Stille's dogma of episodic orogeny. Stille postulated, on the basis of local unconformable relations, the following "orogenic episodes," namely

Pfälzic (post-Permian, pre-Triassic)
Saalic (Early Permian)
Asturic (late Late Carboniferous)
Erzgebirgic (Late Carboniferous)
Sudetic (post–Early Carboniferous, pre–Late Carboniferous)
Bretonic (post–Devonian, pre-Carboniferous)

The Sudetic was supposed to be the main orogenic phase, because Upper Carboniferous deposits are commonly found in rifted basins in regions underlain by Hercynian igneous and metamorphic rocks. This hypothesis suggested that a phase of deformation, metamorphism, and igneous intrusion took place during the mid-Carboniferous,

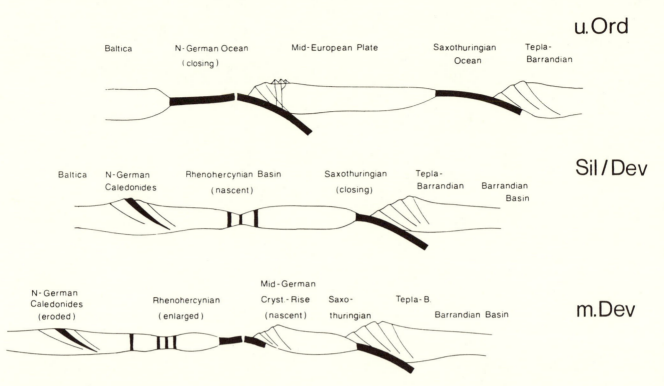

u.Ord

Sil/Dev

m.Dev

Fig. 12.4. Plate-tectonic model for the northern flank of the Variscan belt (after Franke 1989).

and that the metamorphic and igneous rocks were then quickly unroofed to furnish detritus for sediments deposited in Late Carboniferous rift valleys. The late Carboniferous/Permian episodes of deformation in France, Belgium, and the Ruhr region of Germany were considered subsidiary events after the climax, like aftershocks after a main event. Now that the igneous and metamorphic rocks in the Saxothuringian zone have been found by radiometric dating to be as old as 480 Ma, the evidence indicates continuous deformation since the Late Ordovician. The Hercynides have a record of Eo-Hercynian subduction prior to Neo-Hercynian collision (Franke 1989), and the mountain belt had an orogenic history similar to that of the Eo- and Neo-Alpine deformations of the Alps. Adopting the model of Alpine orogenesis, I have to conclude that Stille's postulate of Hercynian deformation during several late Paleozoic episodes is incorrect. The compressive deformation, involving the subduction of continental crust and ocean lithosphere on an active plate-margin, started in the Late Ordovician or Silurian ("Caledonian"). The plate movement resulted in the consumption of the Saxothuringian Ocean in the Devonian and ended in the foreland deformation of Rhenohercynian in the Permo-Carboniferous ("Hercynian"). This history of orogenic deformation by plate interaction was shown by Franke (1989) and is reproduced in Figure 12.4.

The theory of plate tectonics provides a rational explanation for the observation that different styles of deformation take place in different parts of an orogenic belt during successive stages. The Alps serve as a model: the Penninic rocks of Liguria were deformed, uplifted, and eroded prior to the Oligocene Molasse deposition, whereas the Helvetic strip sheets of Switzerland were not folded until the Miocene (Trümpy 1973). Using this Alpine analogue, I envision that the metamorphic rocks of the Hercynides were deformed by subduction and collision long before the "Sudetic phase." These "Pennides" were unroofed and became the basement upon which Upper Carboniferous sediments were laid down unconformably. The unconformity thus signifies only the termination of a long history of "Pennine" deformation; it is not evidence for an epi-sode of "paroxysmal deformation" (Sudetic) during the middle Carboniferous. Nor are the subsequent deformations "episodes of aftershocks"; the unconformities are frag-mentary evidence of continuing deformation in the Late Carboniferous and Early Per-mian. Stille's six late Paleozoic "orogenic phases" are, in my opinion, artifacts, be-cause he invoked the geosynclinal theory of orogeny to interpret the record of a long and continuous deformation.

The Intra-Alpine Hercynides

The basement rocks of the Austroalpine Alps consist mainly of (1) an early Paleozoic volcanogenic sequence, (2) a Paleozoic marine sequence, and (3) Carboniferous and Permian granitic intrusions. In contrast to the northerly or northwesterly vergence of the European Hercynides, the vergence of late Paleozoic deformation in the region south of the Moldanubian zone, *sensu stricto*, and the Intra-Alpine zone is southerly or southeasterly. The Intra-Alpine Hercynian granites have been called I-type, typically formed by ocean-continent interactions on an active margin of the circum-Pacific re-gion. Finger and Steyrer (1990) concluded, therefore, that the late Paleozoic magma-tism manifested by the Austroalpine granites took place along the Eurasian margin of the Paleotethys. The genesis of this magmatic arc was related to the subduction of the Paleotethyan ocean floor along a northwest-dipping Benioff zone. According to this scenario, the late Paleozoic deformation of the region between the Moldanubian and the Intra-Alpine was caused by a continent-arc collision subsequent to the collapsing of a back-arc basin between the two.

Carboniferous and Permian

The Upper Carboniferous and Permian sedimentary formations of Switzerland are commonly considered "post-orogenic," because these formations lie unconformably on the pre–Upper Carboniferous basement complex. The basement rocks include both metamorphic and granitoid rocks, and the Hercynides of the Alpine autochthonous massifs and Austroalpine nappes have remained more or less intact during the Alpine deformation. Amphibolite-facies metamorphism during Hercynian deformation is most common, although rocks belonging to the greenschist or the granulite facies are also present. The granitoid rocks are both Carboniferous and Permian, ranging in age from 360 to 260 Ma (Trümpy 1980; Finger and Steyrer 1990).

The geosynclinal theory of orogeny divided the magmatism into sharply defined orogenic episodes. The Carboniferous granitoids were called "Sudetic"; it was thought that the emplacement of those plutons took place during the so-called Sudetic Phase of orogenic deformation. The Permian granitoids were called "Saalic," and were suppos-edly emplaced during the middle Permian episode. The theory of plate tectonics postu-lates continuous magmatic activity at active plate-margins, producing I-type granites where subduction of ocean lithosphere is the dominant process, and S-type where col-lision has taken place (Pitcher 1982). Finger and Steyrer (1990) found the former in the Alps and the latter in the Moldanubian zone; the I-type granites were considered the product of magmatism along a Cordilleran-type plate margin where the Paleotethyan ocean lithosphere was thrust under, along a NW-dipping subduction zone, the Hercyn-ian European margin.

The oldest continental deposits, namely sediments of the molasse facies, of the Her-cynides were laid down long before the plutonic activities at depth ceased. The Alpine Molasse is not post-orogenic; the Oligocene Molasse was deposited in foreland basins long before the Miocene deformation of the Helvetic strip sheets. Similarly, the Upper Carboniferous molasse sediments of Switzerland are not post-orogenic; they were laid down in narrow faulted troughs, while orogenic and deformational activities were going on elsewhere.

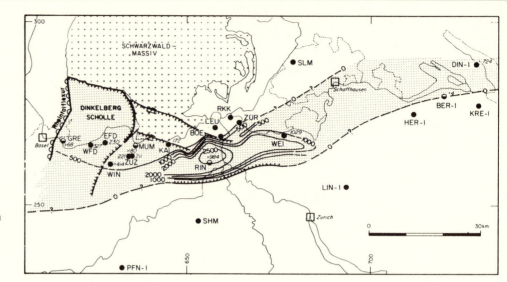

Fig. 12.5. The Permo-Carboniferous Trough of northern Switzerland (after Matter 1987). Numbers indicate thickness in meters, and dots show the locations of boreholes.

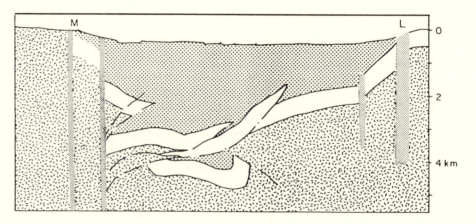

Fig. 12.6. Geological interpretation of the Permo-Carboniferous Trough on the basis of seismic information (after Laubscher 1987). Mesozoic sediments are shown in white. The basin is bounded by vertically dipping strike-slip faults.

The geology of the Permian and Carboniferous strata is obscured by Alpine deformation, except for the relatively undeformed sequence found by drilling under the décollement folds of the Jura Mountains (Figure 12.5). This Permo-Carboniferous Konstanz-Frick Trough is at least 80 km long, some 10 km wide, and more than 4 km deep (Figure 12.6). Such long, straight, and deep faulted basins typically owe their genesis to strike-slip faulting, and the Konstanz-Frick Trough has thus been described as a transpressive (Laubscher 1987). The Upper Carboniferous and Permian sediments of the Konstanz-Frick Trough, more than 1 km thick, have been penetrated by drilling (Diebold and Müller 1985) (Figure 12.7). The Carboniferous consists mainly of sandstone, siltstone, and shale; several coal seams are present. The Carboniferous cross-bedded sandstone, with grain size fining upward, is an alluvial channel deposit, whereas siltstone and mudstone are river-bank and swamp sediments. Coal beds, up to 6 m thick, have been penetrated, and the coal may have been deposited in broad, flat river valleys similar to those in New Zealand where peat is forming; the peat accumulates where the supply of terrigenous detritus is negligible, because the surrounding hills are overgrown with dense vegetation, and the meandering streams have a very gentle gradient.

The Permian consists of bituminous shale, sandstone, siltstone, and conglomerate. The sequence records of a history of alluvial, lacustrine, and desert sedimentation (Matter 1987). Volcanism was active, as evidenced by the intercalated ash beds in the sequence. A large lake came into existence in the area during the Early Permian. The lacustrine deposits include bituminous mudstone and turbidite siltstone, as well as

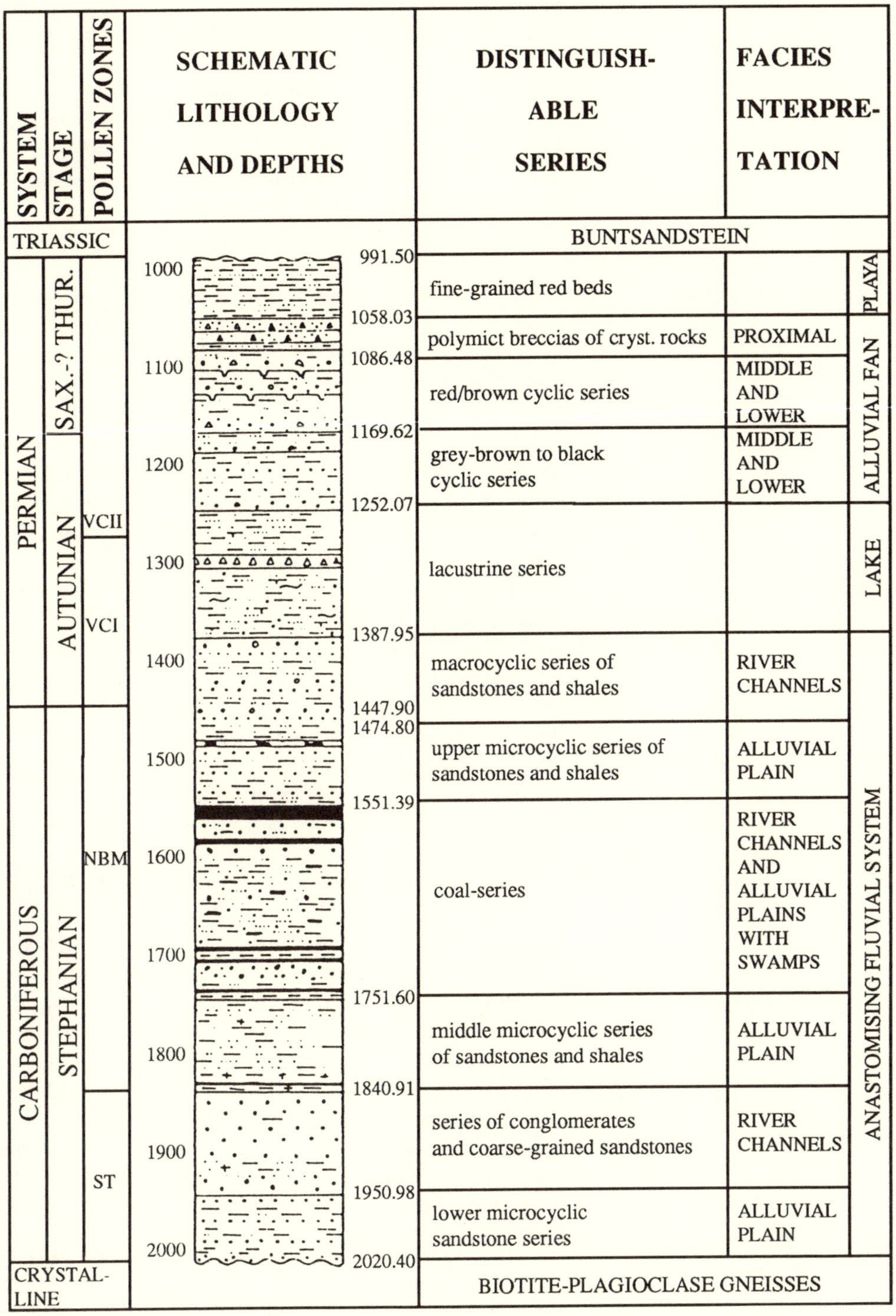

SYSTEM	STAGE	POLLEN ZONES	SCHEMATIC LITHOLOGY AND DEPTHS	DISTINGUISHABLE SERIES	FACIES INTERPRETATION
TRIASSIC			991.50	BUNTSANDSTEIN	
PERMIAN	SAX.-? THUR.		1000 — 1058.03	fine-grained red beds	PLAYA
			1058.03 — 1086.48	polymict breccias of cryst. rocks	PROXIMAL / ALLUVIAL FAN
			1100 — 1169.62	red/brown cyclic series	MIDDLE AND LOWER
	AUTUNIAN	VCII	1200 — 1252.07	grey-brown to black cyclic series	MIDDLE AND LOWER
			1300	lacustrine series	LAKE
		VCI	1400 — 1387.95		
			1387.95 — 1447.90	macrocyclic series of sandstones and shales	RIVER CHANNELS
CARBONIFEROUS	STEPHANIAN	NBM	1474.80 — 1500	upper microcyclic series of sandstones and shales	ALLUVIAL PLAIN
			1551.39 — 1600 — 1700	coal-series	RIVER CHANNELS AND ALLUVIAL PLAINS WITH SWAMPS
			1751.60 — 1800	middle microcyclic series of sandstones and shales	ALLUVIAL PLAIN
		ST	1840.91 — 1900	series of conglomerates and coarse-grained sandstones	RIVER CHANNELS
			1950.98 — 2000 — 2020.40	lower microcyclic sandstone series	ALLUVIAL PLAIN
CRYSTALLINE				BIOTITE-PLAGIOCLASE GNEISSES	

(Right-hand column "ANASTOMISING FLUVIAL SYSTEM" spans the lower portion from macrocyclic series downwards.)

Fig. 12.7. Columnar stratigraphical section of the Weiach borehole (after Matter 1987).

sandstone and conglomerate deposited on lacustrine deltas. The climate became more arid during the Late Permian. The Upper Rotliegende consists of playa mudstone and alluvial-fan conglomerate.

The Upper Carboniferous and Permian formations elsewhere in Switzerland are also intramontane-basin sediments. The Upper Carboniferous of the autochthonous massifs consists of arkosic sandstone, silty shale, and conglomerate, with anthracite coal seams; these alluvial and lacustrine deposits were laid down when the climate was humid. The Verrucano of Glarus, mainly Permian, is a group of conglomerate, arkosic sandstone, variegated shales, freshwater limestone, red claystone, etc.; these were sediments of more arid regions. Volcanic rocks are common, including rhyolite, rhyodacite, andesite, and albitized basalt ("spilite").

Anthracite-bearing slate and meta-sandstone are present in the *Zone Houillère* on the northern margin of the Bernard nappe. They and the "Upper Casanna Schist" of the Bernard nappe have also been subjected to Alpine metamorphism, and this volcano-genic-sedimentary formation is chronostratigraphically equivalent to the Verrucano Formation of Canton Glarus.

Germanic Triassic

The tectonic regime changed from compressional to extensional toward the end of the Paleozoic. Trümpy (1980) suggested that the mid-Permian was the turning point, marking the end of the Hercynian and the beginning of the Alpine cycle. Sedimentological evidence points to a reduction of relief on the continent and an increasing aridity of climate.

The Triassic is named after its threefold division in Germany, Buntsandstein, Muschelkalk, and Keuper (see Chapter 1). The Triassic formations of the Jura and the Helvetic are typically Germanic Triassic. The Buntsandstein is encountered by drilling in the Jura Mountains. The red arkosic sandstone formation is less than 100 m thick; it thins in a southeasterly direction, and is probably absent under the Molasse Basin. The Triassic either overlies the Permian red beds of the Konstanz-Frick Trough, or lies directly on basement rocks (Ramseyer 1987). The lowermost Triassic formation in the Aar massif is a quartz sandstone. The Mels Sandstone of the Helvetic Alps is more feldspathic. These basal transgressive deposits overlie either the Permian Verrucano or granite basement; they are coeval to the Buntsandstein or to the basal Muschelkalk of the German Triassic (Trümpy 1959). The red beds owe their color to hematite formed by ground-water diagenesis in an arid environment. Other diagenetic minerals filling the pore space of the sandstones include nodular calcite (caliche), illitic clay derived from the decomposition of detrital potash feldspar, and ferruginous carbonates (Blüm 1987).

The Muschelkalk records a major marine transgression of Europe. The base of the Muschelkalk in eastern France and southern Germany is a sandstone—*Grès coquillier*. The Middle Muschelkalk is the main evaporite; the detachment surface of the Jura décollement folds lies within this "Anhydrite Group." Sulfates—anhydrite and gypsum—are the dominant evaporite minerals, and they occur in nodular or selenitic forms, as detrital deposits (arenite and rudite), or as vein minerals (Dronkert 1987). Halite is present locally. Associated with the sulfates is stromatolitic dolomite, formed in a tidal-flat environment. Black clays and chert layers are intercalated.

The Muschelkalk in the Aar massif and the Helvetic regions is the Röti Dolomite; it consists of rauhwacke and dolomite in the west, and of dolomite only in the east. Trümpy (1959) discussed the relation of the Triassic facies of northern Switzerland to the tectonics of the Jura Mountains; he noted (p. 430):

> The boundary limiting the distribution of the anhydrite-bearing Muschelkalk in Switzerland crops out on the northern margin of the Aar massif near Innertkirchen (Canton Bern). This facies change may have a fundamental significance in tectonic development, because the detachment surface of Jura folding lies within the "Anhydrite Group." The pinching out of anhydrite can explain why the chain of Jura Mountains ends near Dielsdorf in Canton Aargau. The contrasting style of the Molasse folding in western and central Switzerland and the more brittle deformation of the Molasse in eastern Switzerland and Bavaria can also be explained by this facies change in the Middle Triassic. One can, therefore, postulate that the evaporite of the Middle Muschelkalk lies to the west of the line Innertkirchen-Winterthur. [free translation by the author]

The Germanic Keuper consists of variegated shales, sandstone, dolomite, and evaporite. The so-called *Gipskeuper* of the Jura Mountains is more than 100 m thick, and the occurrence of red shales indicates an increased influx of detrital sediments during the Late Triassic.

The Keuper of the autochthonous massifs and of the Helvetic Alps has been called Quarten Shale, named after its type locality on the south side of the Lake of Walenstadt (Cadisch 1953). This dominantly continental deposit consists of variegated shales, dolomite, quartz sandstone, cherty shale, and breccias. Evaporites are absent.

Halite and sulfate deposits are present in the Ultrahelvetic formations at Bex in western Switzerland and in the Gotthard region (Cadisch 1953); they were apparently deposited in a local basin. The Triassic of the North Penninic realm has also been divided into three units; they are, in descending order,

Quarten Shale (Keuper)
Rauhwacke and dolomite (Muschelkalk)
Quartzite (Buntsandstein)

The Penninic Triassic is thus similar to the Germanic Triassic of the Helvetic region. The Triassic quartzite is, however, commonly absent in Tessin nappes (Cadisch 1953).

Brianconnais/ Subbrianconnais Triassic

The Triassic sequence of the nappe of the Rigid Median Prealps is reduced in thickness and consists of dolomite, rauhwacke, gypsum, and a basal quartzite unit, whereas the Triassic of the Plastic Median Prealps is more complete (Cadisch 1953). There is not much facies variation of the Triassic in the Brianconnais/Subbrianconnais realm. Above the Lower Triassic quartzite, the Middle Triassic (Anisian and Ladinian) can be subdivided into eight units, and the Upper Triassic (Carnian, Norian, and Rhätian) can be subdivided into four units (Baud 1972). These units can be traced from one tectonic slice to another, from one nappe to another. These Triassic formations, however, underwent erosion before the Middle Jurassic transgression. The succession in the Subbrianconnais realm is relatively complete, and the affinity to the Austroalpine Triassic is obvious. The thickness of the Brianconnais Triassic is much reduced because of erosion; all of the Upper Triassic and much of the Middle Triassic were removed prior to the deposition of the Mytilus beds (Figure 6.3). The similarity of the Brianconnais Triassic to the Germanic facies, an observation cited by Cadisch (1953), is thus more apparent than real. The Upper Triassic dolomite of the Subbrianconnais is not very thick, and the sequence of interbedded carbonates, variegated shales, and sandstone belongs to a facies similar to the Keuper of Germany, yet sufficiently different to be referred to as Carpathian Keuper.

Alpine Triassic

The Triassic of the Austroalpine realm and the Southern Alps is characterized by a thick carbonate sequence. Detrital sedimentation continued from Permian to Early Triassic time, and the first carbonate sediments were laid down in the Austroalpine realm toward the end of the Early Triassic.

The Anisian and lower Ladinian formations, both Middle Triassic, are mainly limestone, with silicious shale intercalations in the upper part. The upper Ladinian and lower Carnian carbonates are dolomite, and the common occurrence of dolomite distinguishes the Central Austroalpine Triassic. Farther south, in the region of the Southern Alps, depressions on marine bottom were surrounded on all sides by carbonate banks, and the bottom water was stagnant. Middle Triassic thin-bedded carbonate and bituminous shale, deposited in one such euxenic basin, have yielded a rich marine reptilian fauna.

The Carnian is distinguished by dolomite beds, weathered yellow, with intercalations of gypsum, rauhwacke, shale, and sandstone. The gypsiferous Carnian is one of the major detachment horizons of the Austroalpine realm in Switzerland. The Norian is commonly represented by a dolomite formation, up to more than 1000 m thick. Its sedimentary structures, such as stromatolite, oncolite, and edgewise conglomerate, are typical of sedimentation on tidal flats. The sediments were lime muds, and they were dolomitized in a supratidal environment shortly after their deposition. This Norian Hauptdolomit is a distinguishing feature of the Alpine Triassic.

The uppermost Triassic, the Rhätian, and the overlying Lower Jurassic are mainly limestone formations, deposited during a subsidence stage, when Africa and Europe were being separated by extensional rifting.

Triassic Rifting and Jurassic Seafloor Spreading

Several models have been proposed for the origin of passive margins by lithospheric stretching. The simplest model of crustal thinning by stretching was first suggested by Walter Bucher (1933), using the ductile "necking" of metals before failure as an analogue. A modern "pure-shear" model to evaluate the subsidence history of passive margins has been presented by D. P. McKenzie (1978).

The consequence of crustal thinning by lithospheric stretching is twofold. The replacement of light crust by denser asthenosphere materials causes isostatic subsidence, which has also been referred to as "subsidence immediately after stretching" (Le Pichon and Sibuet 1981). The second consequence results from the exponential decay of a thermal anomaly introduced by the emplacement of hot asthenosphere material (Oxburgh 1982). This thermal subsidence under a stretched continental crust is similar to that experienced by an oceanic lithosphere as it moves away from the axis of seafloor spreading (Sclater *et al.* 1971).

The pure-shear model implies a symmetry of the two rifted margins. Also, the site of initial subsidence, the site of maximum initial subsidence, and the site of final separation of the extended continental margins should coincide; the locus of the initial flaw determines the locus of the final failure. The rifting should propagate in the direction away from the site of eventual failure. The steep side of rifted half-grabens should face toward the site of initial rifting.

The history of the rifting leading to the birth of the Tethys provides a test case for the applicability of the "necking" model. We should recall that the Norian dolomite beds were all deposited near sea level, both the oldest and the youngest. The thickness of the dolomite thus gives a measure of the total subsidence during the Norian time, with greater dolomite thickness indicating greater subsidence. The Hauptdolomit of the Upper Austroalpine nappes is more than 1000 m thick (Dössegger and Müller 1976). Its thickness in the Lower Austroalpine realm is only a few hundred meters, and is less than 100 m in the Subbrianconnais realm, near the site of the final separation of Europe and Africa, and even thinner in the Helvetic region. Thus the rifting was not symmetrical. Also, the axis of maximum subsidence during the Late Triassic lay within the Austroalpine realm, a site considerably more southerly than the site of eventual crustal separation in the Penninic.

A more sophisticated model of lithospheric stretching has been called the simple-shear model (Wernicke 1985). The theory postulates that the crust under extension yields by simple shear along a low-angle crustal fault (Figure 12.8). The high-angle normal faults bounding the rifted basins near the surface are secondary structures induced by the master shear-fracture. Under these circumstances, the rifting is not symmetrical, and the site of maximum earlier subsidence does not mark the site of eventual crustal separation.

The complicated rifting structures on the Austroalpine margins have been reconstructed on the basis of sedimentological evidence (Eberli 1988). The actual rifting history may be even more complicated than that deduced on the basis of the simple-shear, one master-fault model. Frotzheim and Eberli (1990) postulated a model modified after that of Wernicke, assuming simple shear along two low-angle normal faults, one under the Lower Austroalpine and the other under the Upper Austroalpine realm. They have observed near-surface rifting structures related to such low-angle normal shear zones.

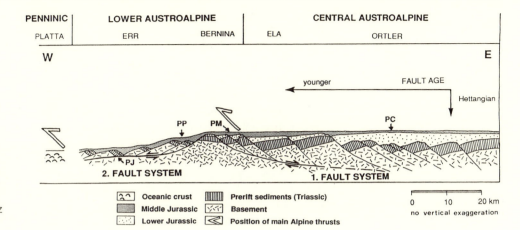

Fig. 12.8. Reconstruction of the late Jurassic Austroalpine passive margin (after Frotzheim and Eberli 1990). PJ: Piz Jenatsch; PP: Piz Padella; PM: Piz Mezzaun; PC: Piz Chaschauna.

Origin of Passive Margins

Fig. 12.9. The positions of Europe and Africa relative to that of America (after Dewey et al. 1973). The present positions are shown in black. The numbers (in Ma) indicate the ages of previous positions.

The extensional regime fragmenting the supercontinent Pangea continued from the Triassic into the Jurassic, when Africa "drifted" away. This displacement of the African plate led to the seafloor spreading of the Tethys and of the North Central Atlantic Ocean (Hsü 1971b; Smith 1971; Dewey *et al.* 1973). Whereas the former was eventually consumed through lithospheric subduction during the Cretaceous and Tertiary, the latter continued to grow, so that the history of the earlier movement between Africa and North America is preserved in the record of the magnetic lineations under the Atlantic seafloor (Figure 12.9). Since Europe remained attached to North America until the Late Cretaceous (81 Ma), the Atlantic record can be used to calculate the relative displacement between Africa and Europe. The record shows that Africa first moved east or south of east, relative to Europe, and that the Tethys Ocean was created by this movement.

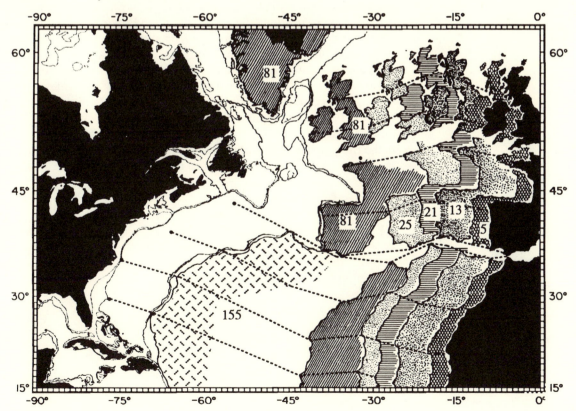

The Jurassic Tethys was a young and narrow ocean, similar to the Red Sea or the Gulf of California. The former has a central axis of seafloor spreading, whereas the spreading axis of the latter is offset by the presence of numerous transform faults. The Gulf is commonly held up as the actualistic analogue of the Jurassic Tethys between the Helvetic and the Austroalpine margins (Kelts 1978). The Alpine Tethys seafloor had been fragmented to form several basins, which were separated by submarine banks or islands (Figure 6.5). The continental basement under these was eventually to become the core nappes of Bernard, Tambo-Surretta and Adula. The ocean basins have been named the Valais Trough and the Piedmont Trough.

The Jurassic passive margins are not symmetrical. Lower and/or Middle Jurassic breccias are common on the Austroalpine, the southern Brianconnais, and the northern Tambo and Falknis margins. The European continental margin was characterized by less active faulting. The sediments on the outer Helvetic margin are thicker and more pelitic (Figure 4.7), but the common absence of coarse detritus in the Jurassic indicates a featureless seafloor in the Helvetic realm, as the basement relief had been smoothed out by sedimentary accumulation.

The nature of the Upper Jurassic sediments on the two sides of the Tethys is also different. The sediments on the southern ocean margin are mainly of radiolarite and pelagic limestone (Austroalpine), whereas the coeval sediments on the northern margin are mainly shallow marine limestones (Helvetic), changing southward to a thick monotonous sequence of dark gray shales (Ultrahelvetic). The relief on Jurassic continents was minimal: the southern continent was largely inundated by marine waters, so that little terrigenous detritus was supplied to the Austroalpine margin; the northern continent was fringed by a carbonate shelf, although fine detritus from land did reach the outer continental margin and accumulated there as hemipelagic deposits.

The Jurassic sediments in the oceanic realms of the Tethys (Penninic) were pelagic and hemipelagic, with rare distal turbidite deposits. Radiolarite and Aptychus Limestone were laid down where the input of the terrigenous detritus was minimal. Elsewhere the oceanic sediments are fine silts and clays, with varying amounts of carbonate detritus, which constitute the great bulk of the Bündnerschiefer or *schistes lustrés*. The turbidite deposits are mostly fine-grained siltstones or argillaceous limestones.

Whereas subsidence was most active on continental margins, the Jurassic continents were also inundated by marine waters. Jurassic marine strata constitute the bulk of the sedimentary rocks of the Jura Mountains, and they are widespread in southern Germany. Northern Italy, the promontory of Africa south of the Tethys, was deeply submerged. Pelagic limestones (*amonitico rosso*) and radiolarites are the most common Middle and Upper Jurassic deposits of the Southern Alps. An excellent section in the Breggia Gorge of Tessin is a Mecca for paleoceanographers who attempt to read in those rocks the record of the oceanic circulations of the Jurassic Tethys (Figure 12.10).

Fig. 12.10. A profile of the Middle Jurassic in the Breggia Gorge (after Bernoulli 1964). See Bernoulli's text for explanation.

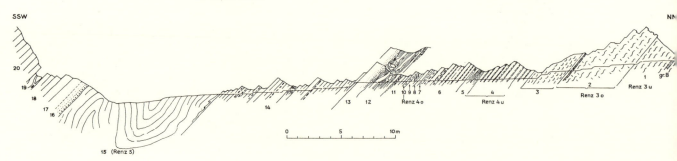

Cretaceous Eo-Alpine and Meso-Alpine Deformations

The first attempts to relate the Alpine deformation to the relative displacement between the European and African plates can be categorized as two-plate models (e.g., Smith 1971; Hsü 1971b; Dewey *et al.* 1973; Biju-Duval *et al.* 1977). These models predict a change from an extensional to a compressional Tethyan regime at 81 Ma (Santonian), when Europe first separated from North America and was displaced eastward at a faster rate than Africa. This Late Cretaceous timing is, however, contradicted by the evidence of Alpine geology. Eo-Alpine high-pressure metamorphism first took place at 120 or 130 Ma, very early in the Cretaceous (Hurford *et al.* 1989). The subduction of the southern Tethyan margin should have started even earlier, so that crustal material could be brought down to a depth of 80–100 km for the genesis of eclogite. The evidence thus indicates that the compression must have set in at the beginning of the Cretaceous, some 50 Ma earlier than postulated on the basis of the two-plate model. The numerous dates of Eo- and Meso-Alpine metamorphism indicate continuing deformation during the Cretaceous, although local extensional tectonics may also have been active—in back-arc basins, for example.

To harmonize the geological and geophysical evidence, Hsü (1989) proposed a three-plate model to interpret the geological evolution of the Western Alps. Italy moved with Africa during the Jurassic, but the two parted company after the beginning of the Cretaceous: while the Cretaceous movement relative to Europe of the African plate was still sinistral until 81 Ma, the movement of the middle plate, or the M-plate including Italy, was dextral starting at 130 Ma. This northwestward movement of Italy during the Cretaceous caused the Eo- and Meso-Alpine deformations in the Tethyan realm.

In such a three-plate model, the M-plate moved in unison with Africa during the time intervals when the Alpine Tethys was subjected to regional extension. Italy, also known as the Adriatic block, was the northern promontory of the African plate until 130 Ma. After that time, the M-plate moved independently of Africa. The direction of the movement is determined with reference to the geology of the Alps, and its magnitude is determined by using the principle of minimum overlap or offset of continental crust among the three plates (Hsü 1989). Maps for plate displacement during seven time intervals since the Jurassic, namely 160–140, 140–120, 120–100, 100–80, 80–60, 60–20, and 20–0 Ma, have been constructed (Figure 12.2).

The reconstructions indicate that the Alpine Tethys reached its maximum extent during the earliest Cretaceous (between 140 and 120 Ma), when the Tethys Ocean should have been more than 750 km wide. This figure is in general agreement with some geologists' estimate of a 700-km shortening in the Alps (Trümpy 1960). The paleogeography of the Early Cretaceous Alpine Tethys should be comparable to the geography of the western Mediterranean of today. Whereas the southern (African) margin had been turned into a convergent plate-margin, the northern continent was still bordered by its passive margin. Islands and banks stood high above abyssal plain; some of them may even have been mountainous like Corsica or Sardinia. The oceanic depressions between the islands and continents were floored mainly by hemipelagic sediments.

The reconstructions also indicate that the northern margin of the M-plate approached the southern European margin steadily starting in the Early Cretaceous (Figure 12.11). The two margins first came into contact, somewhere in the east, prior to 60 Ma. In other words, the Neo-Alpine collision took place during the Late Cretaceous in the area of western Austria and eastern Switzerland.

We have discussed the geological evidence for the consumption of the eastern segment of the Valais Trough prior to the Late Cretaceous collision of the Adula and Tambo "microcontinents." The Avers Trough between Tambo and Suretta was eliminated even earlier. Prior to the final overthrusting of the Austroalpine nappe onto the Helvetic in the Eocene, only a remnant of an originally narrow segment of the Piedmont Trough remained in existence.

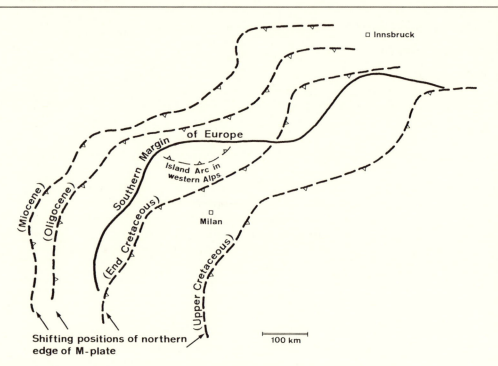

□ Innsbruck

of Europe

Southern Margin

Island Arc in
Western Alps

(Miocene)

(Oligocene)

(End Cretaceous)

□ Milan

(Upper Cretaceous)

Fig. 12.11. The advance of the
northern margin of the M-plate
(after Hsü 1989).

**Shifting positions of northern
edge of M-plate**

100 km

The position of the southern edge of Europe, shown by the solid line in Figure 12.11, is intermediate between the positions of the northern edge of the M-plate at 60 and at 20 Ma. This indicates that the collision in central and western Switzerland took place some 40 Ma ago during the Late Eocene, in good agreement with the geological evidence.

In the Paleocene, prior to the collision, the Alpine Tethys (underlain by ocean crust) should still have had a width of 250 km, which is the distance from Crete to Egypt. This Paleo-Mediterranean was fringed by a marginal sea some 150 km wide, which is the size of the Aegean. Our palinspastic reconstruction of the Schlieren/Gurnigel basins (Figure 5.7) is thus verified by this plate-tectonic kinematic analysis.

Cretaceous Subduction and Exhumation of High-Pressure Metamorphics

Eclogites and coesite-bearing rocks are found in the Western Alps, in the Dora Maira, Gran Paradiso, and Monte Rosa, which have been considered Upper Penninic units. High-pressure minerals are found in rocks of granitic composition, so that there cannot be any doubt that continental crust was taken down to a depth greater than 100 km (e.g., Smith 1984; Schreyer 1988; Wang *et al.* 1988). But how did these high-pressure minerals get back up again?

The uplift and exhumation of high-pressure metamorphic rocks is a tectonic problem. The two commonly suggested mechanisms are (1) they were driven up a subduction zone by the buoyancy force of individual slabs (Ernst 1975; Cloos 1982), and (2) they were uplifted by buoyant diapirism (England and Holland 1979). Platt (1987) pointed out the shortcomings of both schools of thought, and suggested instead an extensional-denudation model: the high-pressure metamorphics came up because the overlying lithosphere had been stretched thin under extension, and the underlying accretional wedge rose isostatically when the overburden was removed by erosion. The eclogites of the Dora Maira, Gran Paradiso, and Monte Rosa all supposedly came up during an episode of Alpine extension.

Platt's model is contradicted by kinematic analysis of plate movements. The M-plate moved toward Europe starting in the Early Cretaceous, and both the M-plate and the African plate moved toward Europe starting in the Late Cretaceous. Extension in

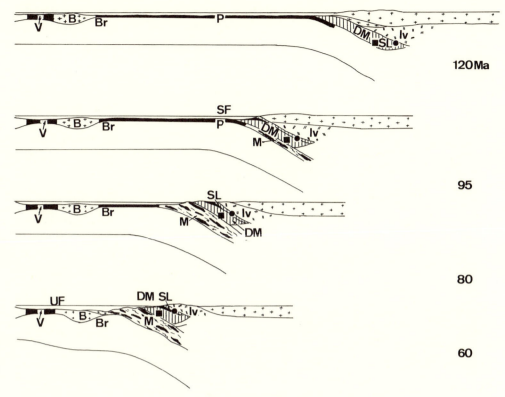

120 Ma

95

80

60

Fig. 12.12. Eo-Alpine metamorphism of high-pressure metamorphic rocks and their upward journey during the Cretaceous subduction of the Piedmont Ocean—a kinematic analysis by Hsü (1989). B: Brianconnais; Br: Breccia nappe; DM: Dora Maira; M: mélange; Iv: Ivrea; P: Piedmont Ocean; SF: Simmen Flysch; SL: Sesia-Lanzo; UF: Ultrapenninic Flysch; V: Valais Trough.

extra-Alpine regions did take place locally, but there is no evidence of large-scale Tertiary extensional structures to indicate an episode of extensional tectonics in the Southern Alps, as postulated by Platt.

The difficulty of finding a satisfactory model for the exhumation of Alpine eclogites, in my opinion, lies in the Kober-Stillean *Leitbild* of orogenesis. While the term "geosyncline" has been appearing less and less often in the geological literature since 1970, Elie de Beaumont's (1830) postulate of episodic orogeny is still deeply rooted in the thinking of many geologists who profess to adhere to the new *Leitbild* of plate tectonics. The term "orogenic phase," a relic from the days of Stillean dogma, is still the word used to describe the Alpine deformation. There was the "Meso-Alpine phase" in the late Eocene, which lasted some 5 Ma, represents the "paroxysmal" deformation (Trümpy 1973). In addition to that, there were "pulses," or short episodes, of Eo-Alpine and Neo-Alpine phases. Straitjacketed by episodic timing, no model of orogenic deformation has been able to accommodate a mechanism in which coesite-bearing eclogite is lifted up from a very deep to a very shallow crustal level.

I have repeatedly emphasized, in this book, that there are no episodes, no phases, only stages of Alpine deformation. An acceptance of the postulate of steady-state plate displacement could give a satisfactory explanation of the exhumation of high-pressure metamorphics in the Western Alps (Figure 12.12).

The Ivrea zone is underlain by mantle rocks of the Adriatic block, and the Sesia-Lanzo is a part of the Adriatic margin that has been subducted and exhumed. Accepting the postulate that the Monte Rosa basement was originally located on the southern Tethyan margin (Milnes *et al.* 1981; Laubscher and Bernoulli 1982), the eclogites of the Monte Rosa and the Sesia-Lanzo zone were formed when slices of continental basement were brought down to a depth of 100 km at the start of the Eo-Alpine stage of deformation. Assuming movement down a 30-degree ramp at a plate-displacement rate of 1.25 cm/yr, the eclogites could have been formed 20 Ma after the start of subduction at 130 Ma.

The subduction zone shifted oceanward during the Late Cretaceous (110–40 Ma). Displacement for 70–90 Ma, at a moderate rate of about 0.7 cm/yr, should have been sufficient to cause the consumption of the Piedmont Ocean and continental collision in the Eocene (Hsü 1989). The northward movement of the Adriatic block caused the Monte Rosa/Dora Maira and Sesia zones, now accreted onto the southern plate, to be displaced up the ramp of the Benioff zone (Figure 12.12). The high-pressure metamorphic rocks were thus brought back to upper crustal levels, as documented by the dating of regressive metamorphism. The vertical component of the movement up the ramp is an uplift rate of 1.5 mm per year—a rate comparable to that of Alpine uplift in recent years. This results in an ascent of 100 km, corresponding to a temperature drop of 400 °C, in 70 Ma (Figure 12.13).

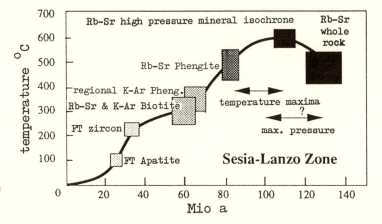

Fig. 12.13. Time–temperature path for metamorphic rocks in the Sesia-Lanzo zone (modified after Hurford *et al.* 1989). The temperature decrease verifies the postulated uplift of the high-pressure metamorphic rocks during Eo-Alpine deformation when they rode up the ramp of the Benioff zone.

As the overburden was eroded, the debris was deposited as Cretaceous deep-sea sediments in the Piedmont Ocean (now Flysch and *schistes lustrés*). The Simmen Flysch contains clasts of both continental basement and ophiolites (Hsü 1989). The Southern Alpine Flysch consists not only of quartz detritus derived from continental crust, but also of heavy minerals indicative of the exposure of high-pressure metamorphic rocks in a Cretaceous Coast Range on the southern margin of the Piedmont Ocean (Winkler and Bernoulli 1986). With the progression of ocean-plate subduction, those sediments subducted and were mixed with fragmented ocean lithosphere to form ophiolite mélanges in the Benioff zone (see Figure 12.12).

Meso-Alpine Collision and Post-Collision Deformations

The Brianconnais Swell is underlain by continental crust and its sedimentary cover. The sediments were detached from their basement to form cover thrusts after the collision of the M-plate (Italy) and the Brianconnais microcontinent, and they were brought northward under the "plunger action" of the Austroalpine nappes. The Prealpine nappes of central and western Switzerland were eventually pushed on top of the Helvetic nappes after the collision of the M-plate and Europe.

The rocks of the Schams nappes were also the sedimentary cover of the Brianconnais Swell, and they were detached from their basement to form cover thrusts. These thrusts have, however, a southerly vergence. Various lines of evidence indicate that this décollement resulted from a Late Cretaceous Meso-Alpine collision of the European margin (Adula) and the Brianconnais. The now united Europe-Brianconnais and the M-plate collided during the early Tertiary. Post-collisional Neo-Alpine deformation, with a northerly vergence, brought the Austroalpine nappes, the ophiolite mélanges, the Upper Penninic Flysch and the Bündnerschiefer above the Schams, and eventually above the Lower Penninic and Helvetic nappes.

The envelopment of the Helvetic nappes by the "Ultrahelvetic" flysch is evidence that the Helvetic margin was underthrust shortly after the plate collision. The continuous northward movement of the Austroalpine *traineau écraseur* on top of the Helvetic/ "Ultrahelvetic" causes the detachment of those sedimentary strata from their basement to form the spectacular recumbent folds that are the Helvetic nappes of the High Limestone Alps (see Chapter 3).

The Austroalpine nappes arrived at the Helvetic margin during the late Eocene, and continued to march northward during the Oligocene, causing the formation of the Molasse foreland basin. While the sedimentary cover on the outer European margin was stripped off and overthrust northward, the basement and part of the cover were underthrust (Hsü 1979; 1989). These rocks were metamorphosed under elevated temperature and pressure conditions to form the gneisses and schists of the Penninic core nappes. The overthrusting and underthrusting were continuous, not episodic. The Neo-Alpine deformation (A-subduction) and metamorphism continued in the Tertiary after the collision. The deformation of the Molasse and that of the Jura, as described in Chapters 1 and 2, are the most recent manifestations of Neo-Alpine orogenesis.

13.

The Tectonic Facies Concept

The use of comparative anatomy to understand mountain-building is not a new approach. The expression *vergleichende Tektonik* is essentially the title of Stille's 1924 classic monograph, and The *Anatomy of Mountain Ranges* is a book edited by Schaer and Rodgers in 1987 to narrate the life history of several mountain belts.

Anatomy is more than morphology. Jacob Scheuchzer, an eighteenth-century Swiss naturalist, misidentified, on the basis of morphological similarity, the skeleton of a giant salamander as *Homo diluviatis*. This mistake was corrected half a century later by Georges Cuvier, after Linnaeus introduced systematic taxonomy to biology, and after the functional utility of the various parts of living organisms was appreciated (see Hsü 1986). Cuvier emphasized "the importance of the connections that must exist between the various organs in order to make of them a living unit. He understood that certain functions have such a determining influence on the organism that they can easily be used as a general guide to classification; thus, for him, the nervous system gave the phyla, the respiratory and circulatory organs gave the classes" (Schaer 1987, p. 4). It was thus no problem at all for Cuvier to discriminate an amphibian from a hominid.

Animals and plants all have a blueprint (*Bauplan*), and each organ has to carry out its particular function, which is essential for the survival of the whole. A tree has, for example, a root in order to acquire water and nutrients, and the root further serves, in most instances, the auxiliary function of anchoring the plant. The stem contains vessels for water and nutrient transport, and to raise the crown to a level where it can spread out to collect carbon dioxide and photons for photosynthesis. The existence of one organ implies the existence of the others, so that palynologists can identify a tree on the basis of its pollen, and physical anthropologists can name new hominid species on the basis of two teeth or a jawbone. No paleontologist needs a complete skeleton or a whole plant to reconstruct the anatomy of a fossil organism; the existence of missing parts that are indispensable for vital functions can be assumed. The science of comparative tectonics is founded on the same principle: vital organs of mountains are to be identified to reconstruct their life history (Schaer and Rodgers 1987).

What are the "vital organs" of a mountain, and what is their functional utility? These problems have puzzled geologists since the time of James Hutton. For more than two centuries, the various parts of mountains have been described. The science of describing the geometry of the parts is structural geology and the science of depicting their kinematic evolution is tectonics, but there is a sad lack of understanding of the mechanics of mountain-building. The tendency for specialization in geology, especially in the first half of this century, is unfortunate: Structural geologists were engrossed in the detailed geometry of geological structures formed by strain and displacement, while stratigraphers were debating the timing of orogenic events. Structural geologists knew their structures very well, but many of them were not much interested in relating structural geometry to geological evolution; their orogenic chronology was coded with reference to "phases of deformation" producing structures S_1, S_2, etc. Stratigraphers, working on another set of data, looked for unconformities or other evidence to establish the chronology of "orogenic events." Once an "orogenic phase" was postulated, a name was given to it and little attention was paid to the kinds of structures produced during the various "phases." This type of effort can be compared to that of archeologists studying pyramids: some measure the dimensions of pyramids, and explore their hidden chambers, while others try to determine when they were built; but few concern themselves with the question of how they were constructed.

Geologists who were interested in the mechanism of mountain-building, i.e., the growth and the functional anatomy of a mountain range, were general geologists. Like general practitioners in various professions, generalists commanded less glamour than specialists so that the expression "tectonicians" had to be invented to categorize those "structural geologists" who strived to see the "connections that must exist" in order to make mountains comparable to living organisms. Tectonicians of the Alps have a glorious tradition in continental Europe, while structural geologists are gathering medals and prizes in the Anglo-Saxon world. At the same time geophysicists use deep-sounding seismic waves to investigate the geometry of subsurface structures, while rock physicists carry out deformation experiments. All speak their own languages in describing their random observations, and a paradigm on mountain-building processes is long overdue. Yet syntheses have been discouraged by establishment geologists, who cannot see the difference between scientific deductions and idle speculation. Theoretical treatises are seldom publishable in geological journals.

Random Observations and Controversies

I first learned geology in China, and we had to take courses in mineralogy, petrology, stratigraphy, paleontology, and structural geology. I did not see "the connections that must exist," and did not know where the knowledge was leading to. Then, in my junior year, I checked out an essay by A. W. Grabau (1924) on the migration of geosynclines. It was a revelation. I learned from Grabau that the precursor of a mountain had only two parts, or two "vital organs": the "old land" and the "geosyncline." The Appalachian mountains of eastern North America were his model. The old-land Appalachia supplied detritus to the Appalachian Geosyncline during the peaceful reign of geosynclinal sedimentation. Then came the orogenic revolution, when the geosynclinal sediments were squeezed and folded between Appalachia and the stable interior of North America. Grabau then proceeded to review the geological history of many other mountain ranges, and he found the same pattern everywhere.

Grabau conveyed the impression that geosynclinal deformation was induced by the lateral pressure exerted by the "old land," and he was expressing the orthodox opinion of his time that "bodies of crystalline rock materials near the earth's surface act as 'rigid bodies' in the course of the superficial (marginal) folding of weak sediments" (Bucher 1933, p. 244). Even then, however, Alpine geologists were aware that the "old land" was not all that "old," nor that "rigid." The metamorphic rocks of the Penninic core nappes constituted the basement under the sedimentary cover of the Brianconnais Swell; they did not deliver detritus to the Helvetic "geosyncline," nor were they rigid during the Alpine deformation. Nevertheless, the rigid and ductile deformations of the basement, be they of an "old land" or not, did seem to provide the compressive pressure for the folding of sedimentary rocks.

Where did the lateral pressure come from? People used to think that the compression resulted from a contraction of the earth: the crust was crumpled like the skin of a dried-up apple. After the discovery of radioactivity around the turn of the century, the contraction theory was no longer the ruling theory, and we were not even sure whether the earth has been contracting or expanding. "Fixists" saw in geosynclines and orogeny evidence of alternate global expansion and contraction. "Drifters," on the other hand, found in geosynclines and orogeny the manifestation of the rifting apart and coming together of wandering continents.

Wegener's theory of continental drift appealed strongly to Alpine geologists, because it gave an unlimited possibility for the amount of crustal shortening in mountain belts. A shortening that seemed to be closer to 1000 km than 100 km in the Alps (e.g., Kober 1921) could hardly be caused by episodic shrinkage of the earth. "Fixists" had to argue that crustal shortening in orogenic belts was an expression of tectonic thinning of layering during ductile deformation, and that "the actual amount of crustal shortening involved in orogenic folding is much smaller than the resulting structures lead one to expect" (Bucher 1933, p. 240).

Simple rock-folding is the most impressive and obvious geological structure, and the Jura and the Appalachian folds are good examples. Computing the depth of folds on the basis of geometric extrapolations, Chamberlin (1910) obtained figures up to 35 km (Moho depth!) for the Appalachian Valley and Ridge folding near Harrisburg, Pennsylvania. In contrast to this "thick-skinned model," Suess suggested as early as 1875 that the genesis of spectacular folds is merely a superficial phenomenon. This interpretation of thin-skinned deformation of sedimentary cover was verified by geologists studying the Jura (e.g., Buxtorf 1907), and the conclusion was eventually also accepted by students of Appalachian geology (e.g., Bucher 1933).

Physicists pointed out that thin piles of sedimentary rocks are not able to transmit stresses over great distances; the geological postulate of overthrusting was considered physically impossible (Smoluchowski 1909). Two alternative hypotheses out of this dilemma were proposed: either the force that leads to the folding may reside within the folded sedimentary series itself, or a tangential force is applied from above or from below. Advocates of the former hypothesis postulated gravitational sliding: the sediments were the cover above an "old land," and the cover that slid off became folded when the old land rose, for one reason or another (Haarmann 1930). Supporters of the latter speculated that there was distributive shearing stress by the "plunger action" of overthrusting from above (Bucher 1933), or they postulated convection currents in the earth's mantle below (Griggs 1939). Those who advocated gravity sliding were all "fixists." Some of those who invoked mantle convection were "drifters" (Holmes 1929), but the inventors of the tectogene concept did not particularly want to move continents and they postulated horizontal displacements of as little as 100 km (Vening-Meinesz 1930; Griggs 1939).

The occurrence of ophiolites in mountains was another troubling aspect for tectonicians in the first half of the century. The suite of ultramafic and mafic rocks, including peridotite, pyroxenites, gabbros, diabase, spilite, and pillow basalt, is found in the Alps in close proximity to radiolarite, pelagic limestone, and other deep-sea sediments. Steinmann (1905) found an actualistic analogue of ophiolites in the mafic and ultramafic rocks on ocean islands or in those dredged up from the bottom of the Atlantic. He and Suess (1909) both considered Alpine ophiolites relics of raised seafloor. The dominance of the "fixistic" doctrine in the first half of this century led many, however, to favor an alternative interpretation: pillow basalts were considered submarine lava flows and plutonic ophiolites were intrusives into a subsiding continental crust. Denying the evidence of an oceanic origin for ophiolites, the "fixists" were able to crow triumphantly on the permanence of continents and oceans.

Granites are found everywhere in the mountains. Obviously, "granitic magmas of the orogenic zones have come into existence as the byproduct of orogenic deformation. They rose to the surface differentially, as the most mobile of the material undergoing essentially 'plastic' deformation. . . . Most later intrusives reached the uppermost portion of the crust only as the orogenic pressure was dying down, reaching their final positions by changing places with solid crustal materials . . . (discordant type)" (Bucher 1933, p. 296). The bland phrase *changing places*, casually inserted, was at the root of the "granite controversy." The debate raged during the middle decades of this century, when I began my graduate studies. Granites did not intrude into caves, so there was the room problem: How did granitic magmas manage to change place with the solid crustal materials that had been there before? The magmatic origin of granites was questioned, while the advocates of a "replacement" origin of granites were divided into the camps of "wet" and "dry" granitizationists (see Gilluly 1948).

Equally puzzling to tectonicians was the time and place of metamorphism in orogeny. There seemed to have been three different trends: facies of regional metamorphism represent changes at increasing depth down a normal geothermal gradient; facies of contact metamorphism represent changes in regions of magmatic activity where the geothermal gradient is unusually steep; and the third category of metamorphic rocks seems to represent changes in regions with an abnormally low geothermal gradient. What is the significance of this type of metamorphic change? Glaucophane schists and eclogites are found here and there in mountains, and they commonly occur in association with ophiolites. Experimentalists vouched for their high-pressure genesis. Field geologists found those rocks, however, in close proximity to sedimentary rocks that show little sign of ever being dragged down to a very great depth. Professors told students that glaucophane schists were the product of "sodium-metasomatism" near the "igneous contact of ophiolite intrusions."

I can go to even greater lengths to describe the unsatisfactory state of our knowledge about the origin of mountains on the eve of the earth-science revolution. There were plenty of random observations, and each set of observations could be explained by two, three, or more *ad hoc* explanations, diverse and contradictory. Then came the theory of seafloor spreading and all the puzzles could be resolved; all the contradictions concerning mountain-building could find a solution in the model of plate tectonics.

Model Concept

What is a model?

Models are laws describing natural phenomena.

Physical laws are digital models. The path of ballistic motion as described by the formula $s = a t^2/2$ is a digital model. The model relates the distance traveled, s, to the gravitational acceleration a and the travel time t. The distance is deterministic, and one can predict where a cannonball would land. Not to be forgotten is the fact that each model is made with a certain set of assumptions, and this digital model assumes, for example, that the air resistance to the cannonball is negligible.

Geological laws are analog models.

My father-in-law, Hermann Eugster, was a geography teacher in a Swiss middle school, and he used to take his class to the mountains every year to perform an experiment on ground-water motion. They injected dye into ground water near the shore of a lake in the Alpstein Mountains. A few days later, they would go to a spring at the mountain and observe the outflow of ground water colored by the dye. The distance traveled along the tortuous path of the movement through limestones of various permeabilities cannot be related to the travel time by a digital model like that for the cannonball, but the kinematics of the movement is as deterministic as the ballistic motion and can be described by an analog model. A colleague was working on one such model of ground-water motion when I was a young research geologist with the Shell Company. In his model, the variable cross-sectional area and the permeability of the conduit were represented by the thickness and electrical conductivity of some very special cardboard, and the hydrodynamic potential by the electrical voltage difference, while the topography of the terrane and the geological structures were scaled geometrically. Such an analog model will yield reproducible deterministic results to describe the type of experiments carried out by Eugster's geography class. Analog models, like digitally expressed formulas, are statements of physical laws.

Cuvier's comparative anatomy is an intuitive reliance on an analog model. Each individual living organism is an analog model of its species: an embryo will develop and the young will grow according to the genetic codes stored in DNA molecules. Applying this concept of comparative anatomy to the study of mountains, we could conclude that the geology of the Swiss Alps is an analog model stating the physical laws of mountain-building processes. If the model is correct, then the tectonic evolution of every mountain range, young or ancient, could be predicted on the basis of a few outcrops sticking out of alluvium, or a few drill cores encountered in deep boreholes, just as the morphology and physiology of a birch tree can be reconstructed on the basis of birch pollen or a piece of birch bark.

Mountains and Mountains

A portrait painter sees in each person an individual, different from all others; Napoleon bears little resemblance to Emperor Kangxi of China. Geologists are likewise impressed by the difference of one mountain chain from another. Some of my colleagues in Switzerland are convinced that their Alps are unique, not to be compared with other, more mundane mountain chains; it has been said that the Helvetic Alps are fundamentally different from the foreland thrust belt of the Canadian Rockies, or that the giant basement nappes of the Austroalpine realm are patently Alpine. American geologists also emphasized that "different belts and terranes in mountain chains have had quite different relations in the past; especially in the North American Cordillera, such ideas have led to the recognition of many 'suspect' and 'exotic' terranes" (Rodgers 1987). The overemphasis on individual differences tends to encourage *ad hoc* hypotheses for the origin of the various mountains, while overlooking the unifying principles underlying all mountain-building processes.

The approach of comparative anatomy in biology is not to emphasize differences, but to identify the few basic blueprints (*Bauplan*) for all different classes of living organisms. The big difference between a redwood tree and a mushroom is obvious to all and need not be emphasized. The taxonomic difference is real, and a new scheme of classification excludes mushrooms from the plant kingdom. Yet despite all the differences, real or apparent, there is an intrinsic similarity. Both the redwood and the mushroom have the same blueprint, inherited perhaps from a common ancestor—the crown, the stem, and the root. Their blueprint is, however, different from that of animals, and this difference is a consequence of their different modes of life.

Mountain chains are different, and each has its individuality. There are also similarities, and similar chains could be considered to constitute a certain category of mountains. Mountains can thus be classified like living organisms, each class of mountains sharing a common blueprint. The collision of two continental plates produces mountains that have the same blueprint, and they belong to a category different from that of mountains formed by the subduction of ocean lithosphere under continent. It is therefore not surprising that Eduard Suess recognized two major kinds of mountains on the surface of the earth: the Tethyan and the Circum-Pacific.

Are the two really different, or do they represent two different stages of growth of one class of mountains? Caterpillars look quite different from butterflies, but the former are destined to undergo metamorphosis and be changed into the latter. Circum-Pacific mountains have structures distinctly different from those of Tethyan mountains, but the geology of Switzerland has indicated to us that the Cretaceous Alps of the Circum-Pacific stage evolved into the mountains of the Tertiary Tethyan stage (Chapter 12). Both types of mountains are formed by horizontal compression when two lithospheric plates approach each other. Circum-Pacific mountains made by ocean-continent interactions are the precursors of Tethyan mountains, which owe their origin to continent-continent interactions after the consumption of the intervening ocean.

Not all mountains in the world are formed by horizontal compression; some are formed by horizontal extension. If mountains are to be classified based on the stress orientations, what are the different categories in a classification that is all-inclusive and mutually exclusive?

Fortunately, the types of mountain chains in such a system can be reduced to three, because the earth's crust can be deformed by only three different systems of principal stresses. Designating the maximum, intermediate, and minimum principal stresses as σ_1, σ_2, and σ_3 respectively, the three stress systems and the three kinds of faults produced by them are as follows:

(1) horizontal compression with σ_1 and σ_2 horizontal and σ_3 vertical: overthrusts
(2) horizontal compression with σ_1 and σ_3 horizontal and σ_2 vertical: wrench faults
(3) horizontal extension with σ_2 and σ_3 horizontal and σ_1 vertical: normal faults

The Alpine type of mountain is characterized by the dominance of overthrusting (the *Alpinotyp* of Stille), and its evolution is marked by a Circum-Pacific and a Tethyan stage of development. Mountains with dominant strike-slip faulting, such as the Tertiary Coast Ranges of California, can be categorized as the California type (faulted-fold type of Stille; see also Hsü 1991b). Mountains formed by block-faulting are the Germanic type (*Germanotyp* of Stille), exemplified by the mountains of the Black Forest and the Vosges Mountains on both sides of the Rhein-Graben.

Circum-Pacific and Tethyan Mountains

We tend to refer to the anatomy of animals when we discuss the anatomy of mountains. This tendency may have been at the heart of the geosynclinal theory of mountains. Geosynclines were thought to be precursors, or embryos, of mountains, and mountain-building was supposedly episodic. Mountains, according to that theory, were made in all their splendor during the paroxysmal phase of orogeny. The Alps were supposedly formed by mountain-making during the late Eocene and the Hercynides during the mid-Carboniferous.

The plate-tectonic theory has supposedly ruled supreme since the geosynclinal theory of orogenesis was pronounced defunct during the Penrose Conference on Global Tectonics in Monterey, California, in 1969 (Dickinson 1970). Yet the practice of numerous tectonicians reminds me of the vintner who pastes a new label on a bottle of old wine that has turned into vinegar: the name has changed, but the contents are still vinegar. Two decades after the earth-science revolution, editors of geological journals do not think twice about publishing papers describing geological structures produced during the Variscanian *Orogeny*, Eo-Alpine *Phase*, etc., apparently not realizing that such a concept of episodic orogeny is contradictory to the very premise of the plate-tectonic theory (see Sengör 1991, p. 406).

An appreciation of the continuous growth of mountains led me to search for an analog model in the anatomy of plants. Animals and plants are both living organisms. Aside from the ability of locomotion, there is one important difference: "higher" animals such as hominids, domestic and wild animals, are born as they will be. When children are born, they have practically all their vital organs: head, body, limbs, sensory organs, internal organs, etc. The physiological processes of a child, aside from a few notable exceptions, are not basically different from that of an adult. Children see with their eyes, their blood circulates, they eat and the food is digested, they breathe and the oxygen is supplied to the bloodstream, they think with their brains and they walk with their legs. The development of a child into an adult does not involve fundamental changes. All children have a day of birth, on which they are presented to us with all the manifestations of a human life. The geosynclinal theory compared the anatomy of mountains and that of animals to propagate the preconceived idea that the structures in a mountain chain were formed mainly during the "birthday" (i.e., the paroxysmal episode) of orogeny. Plants differ from animals in that they undergo fundamental changes during their growth. Not only is the seed of a redwood greatly different in size from a fully grown tree; the process of seed germination is quite different from the process of photosynthesis.

I first began to appreciate this difference in the growth of living organisms when I was preparing my Fermour Lecture in 1987. Rome was not built in a day, and mountains were not made in a few million years during an "orogenic phase." We cannot assume that the Caledonides were born of Caledonian Orogeny in an episode that lasted only a few million years during the early or middle Paleozoic. Orogeny cannot be dated like progeny, or a child's birthday. Likewise, we cannot assume that the Alps grew in spurts during the Eo-, Meso-, and Neo-Alpine phases, interrupting stages of geosynclinal development. The evolution of the Alps cannot be compared to the life cycle of an insect, which changes only during the few days of its metamorphosis. The

making of mountains—an ultimate response to seafloor spreading—is more comparable to the steady growth of a plant, with all its gradual changes from seed germination
to full blossom.

Alpine *Bauplan* and Tectonic Facies

The *Bauplan* concept of comparative anatomy is derived by analogy with architecture.
If we overlook the decorative elements that distinguish one building from another, the
essential elements are the same: buildings have a foundation, floor, walls, ceiling, and
roof. Most have windows and doors; some have chimneys. Buildings do not come
wholly prefabricated, and there is also an orderly process for making a structure: the
superstructure, i.e., the roof and ceiling, of a house cannot be built without support, and
the support is seldom erected before a foundation is laid. Knowing the indispensability
of the vital parts, we could conclude, on the basis of a few broken pieces of tile in a
ruin, that a house had once stood there. Likewise for a plant—the finding of traces of
a crown implies the former presence of stem and root.

Mountains have their blueprint. Tethyan mountains must have a suture zone. Its
presence has to be assumed, even if it is not found, having been buried, eroded, or
faulted away. A knowledge of the blueprint permits us to conclude what must have
existed, even if no such relics are now extant. The methodology of geology is thus
not only induction on the basis of random observations, but also deduction on the basis
of a theory, an analog model, a paradigm. All the loose talk, now fashionable in China,
about "intracontinental collision" is an invention of those who have made incomplete
observations. Such speculation is not plate tectonics and has no theoretical basis.
Mountains formed by "intracontinental collision" are the same as "geosynclinal mountains"—the idea is old wine, now changed into vinegar, packaged in a new bottle
labeled "plate tectonics."

Grabau's blueprint identified two vital organs of mountains—the "old land" and the
"geosyncline." These have long been replaced by a model of threefold subdivision
favored by Alpine geologists, namely (1) rigid-basement nappes, (2) mobilized-
basement nappes, and (3) cover nappes (see Trümpy 1960). These "vital organs" have
been recognized not so much for their morphology, but for their functional utility in the
growth of a mountain system.

F. E. Suess (1937), son of the famous Eduard, recognized the dynamic behavior of
the three vital organs during mountain-building: (1) the rigid-basement nappes are the
bulldozer (*erzeugende Scholle*), (2) the mobilized-basement nappes are the overriden
parts (*überfahrene Zone*), and (3) the cover nappes are the escaped parts (*unbelastete
Zone*) (see Sengör and Okurogullari 1991). Phrased in terms of the plate-tectonic theory, the bulldozer is the overriding plate in a continental collision. The overriden plates
are the subducted elements, having been metamorphosed and deformed while they
were underthrust to deeper crustal levels. The escaped parts escaped because they had
been pushed to the foreland by the bulldozer. Suess's three vital organs are manifested
by the Austroalpine, Penninic, and Helvetic nappes of the Swiss Alps.

Alpine geologists have long recognized that the advance of the Austroalpine rigid-
basement nappes is the bulldozer of mountain-building; they used the term *traineau
écraseur* to describe the forward-marching rigid plate. The displacement of the plate
was driven by the seafloor spreading in the Atlantic and Mediterranean regions. The
Penninic elements include (1) ophiolite mélanges or tectonic mixtures of fragmented
lithospheres formed as an accretionary wedge at an active plate-margin, (2) core
nappes made of basement of the subducted continental plate, and (3) Bündnerschiefer
and *schistes lustrés*. Caught under the plunger action of the Austroalpine nappes, the
underthrust slabs were subducted down to a depth of 100 km or more, where they were
subjected to very-high-pressure metamorphism before they were welded onto the accretionary wedge; moving in unison with the overriding plate, they started their return
journey to the surface (see Chapter 11). Their rate of uplift is the vertical component

of a forward movement up an inclined ramp (see Chapter 12). The close proximity of rocks that have been subjected to different kinds and degrees of metamorphism in an ophiolite mélange is explained by this mechanism of tectonic mixing in a subduction zone. This mixing took place during the subduction and may have been intensified further during the collision. Mélanges are thus mainly subduction mélanges, although they are invariably found in suture zones of collision-type mountains. The Penninic core nappes and their enveloping sediments were the basement and cover of the under-thrust European continental margin, and those rocks were metamorphosed under a normal geothermal gradient (see Chapter 7). Those underthrust slabs were eventually also accreted onto the overriding plate and thrust on top of the lower Penninic, Helvetic, and autochthonous elements. Partial melting (anatexis) of mobilized base-ment led to the genesis of discordant granites such as those in Bergell. There is no room problem, because granite magma was derived from the solid crustal materials that were where the granite is now.

The escaped elements in Switzerland include not only the Helvetic nappes, but also the nappes of the Median Prealps, the Flysch nappes, the autochthonous and parau-tochthonous massifs, the foreland basin, and the Jura Mountains. Thin-skinned tecton-ics involving the folding and thrusting of the sedimentary cover is the dominant mode of deformation. Stress transmittal presents no problem, because of the weakness of the detachment horizon under the cover thrusts, and the folding is induced by the plunger action of the overriding Austroalpine and Penninic nappes (Hsü 1979). Basement is involved in the deformation of the massifs, of the underthrusting of the Molasse Basin, and of the crust under the Jura Mountains. The crustal underthrusting is related di-rectly or indirectly to motions of thermal convection in the earth's mantle.

Accepting the plate-tectonic theory, all the problems that troubled tectonicians dur-ing the early half of the century can be neatly resolved: the paradigm has provided explanations for all random observations, so that *ad hoc* explanations are no longer necessary. In this scenario, one finds a similarity between the life history of mountains and the growth of living organisms. Recalling Cuvier's lesson that "certain functions have such a determining influence on the organism that they can easily be used as a general guide to classification," the identification of fragmentary relics of vital organs could enable us to perceive the true identity of a mountain, despite its deceptive exter-nal morphology. The *Bauplan* of each Alpine-type mountain includes the *bulldozer*, the *overriden*, and the *escaped* of Suess Junior; these three tectonic facies are the three vital organs of orogenic belts, like the root, stem, and crown of plants. What have been recognized in the Alps should be found in all mountains of this type.

A Name Is a Name Is More Than a Name

The concept of tectonic facies evolved while I was working with Chinese colleagues on the geology of China. Conditioned by "fixistic doctrines," Chinese geologists con-sidered China to have been a consolidated craton since the middle Paleozoic or late Precambrian. Terms such as North China Platform, South China Platform, Tarim Intracratonic Basin, etc., have been given to describe an illusionary tectonic stability. Evidence for Mesozoic deformation and granitic intrusions are, in fact, everywhere apparent, so the term "unstable platform" had to be introduced to rescue orthodox interpretations. After the advent of the plate-tectonic theory, China was recognized as a collage of micro-plates, sutured along zones where mélanges cropped out (e.g., Wang 1986). This apparent conversion to the "new faith" was, however, not much more than old vinegar in new wine bottles. The classic terms "eugeosyncline," "mio-geosyncline" and "block" were replaced by *suture, passive-margin sequence* and *micro-continent* respectively. The concept of "paroxysmal deformation of a geosyn-cline" was replaced by continental collisions during recurring orogenic episodes.

For several decades prior to 1980, J. S. Lee's "geomechanical system" was the para-digm, and Chinese mountains were classified on the basis of morphology, i.e., ε-type, λ-type, etc. After the end of the Great Proletarian Cultural Revolution in China, T. K. Huang's theory of polycyclic orogeny was exhumed, and tectonic units were recognized on the basis of postulated chronology, i.e., the Caledonides, Hercynides, etc. When I first started in 1983 to work with my Chinese colleagues on a tectonic-map project, I found that they had little appreciation of the use of the comparative anatomy approach to analyze mountain-building. They made no attempt to compare the "folding of sedimentary cover on unstable platform" to the thin-skinned deformation in foreland thrust belts. Instead of recognizing that twigs and leaves are but parts of the crown of a tree, they considered those the whole of a new species of a living organism, variously called Yangtze Unstable Platform, Neo-Cathaysian System, Huanan Diwa (i.e., Geodepression of South China), etc. Likewise, ophiolite mélanges were not identifed as part of a vital organ of an orogenic belt in South China, but considered the whole of a Precambrian "organism" called Jingningianide (Wang 1986, p. 250). Finally, Chinese geologists considered klippes of rigid-basement nappes in the foreland thrust belt to be basement that had been uplifted several kilometers from below by block-faulting, not knowing that this was the mechanism wrongly invoked by Swiss geologists 150 years ago to interpret their Mythen klippes (see Chapter 5).

The tectonic facies concept evolved among us because of the need to communicate with my Chinese colleagues during discussions in the field. They told me that the folds south of the Yangtze were the structure of an "unstable platform" or of a "diwa." This kind of logic reminded me of the practice of calling the molars of *Sinanthropus pekinensis* "dragon's teeth." But the molars are not dragon's teeth, because dragons never existed; they lived only in people's fantasies. The molars of *S. pekinensis* are molars of hominids. My Chinese friends then argued that a name is just a name; it does not matter whether those particular objects are called molars or dragon's teeth. "But a name," I replied, "is more than a name, because naming is a conclusion that has implications. Dragon's teeth were pulverized and sold in drugstores as aphrodisiacs, but the molars were the basis for the establishment of a new hominid species." Unstable platforms and diwas are dragon's teeth, and dragons are mythical animals created by fantasy. Likewise, there is no place today where we can find an actualistic model for "unstable platforms" or "diwas"; they are the fantasy of ignorant and incompetent geologists. On the other hand, to call the same Yangtze structure a foreland thrust belt is not just a matter of using a different name, because the name implies that the Yangtze structure is a vital organ of an orogen: the foreland fold-and-thrust belt belongs to the *escaped* facies of F. E. Suess; it must have escaped from something, and the something was the bulldozing rigid-basement nappes pushing from above and behind.

My explanation of the need to find the right name made sense to my Chinese friends. We began to talk about the "décollement facies" or "cover-thrust facies" and started to search for the "motoring Silvretta." Calling a spade a spade led my Chinese friends to recognize that a spade is a spade. A fold-and-thrust belt should have overthrusts. Indeed, local geologists began to tell me that they had known all along about the overthrusting tectonics in South China. Drilling for coal in Carboniferous, they found Permian limestone beneath thin coal seams. Drilling for oil in porous Paleozoic carbonates, they encountered Cretaceous red beds below the Paleozoic. Thin-skinned deformation had been well documented by surface and subsurface geological data, and overthrusts are well known to local scientists, but Zhu Xia, a student of Rudolf Staub's at ETH, had to wait until J. S. Lee, the pope of Chinese geology, died before he could publish his article on the overthrusting tectonics in South China (Zhu 1983). His report was dismissed as too speculative. Dragon's teeth were "proletarian facts," but hominid molars were "capitalistic speculations." This was the logic of Chinese science under communism.

My Chinese friends and I did not have to search long for the "bulldozer" or the "Chinese Austroalpine": Precambrian granites crop out on isolated mountain peaks.

With steep cliff faces on all sides, these mountains rise some 2000 m above the hills or alluvial plains of the Yangtze region, like the Mythen of central Switzerland. It was not difficult to convince my Chinese friends who had been acquainted with Swiss geology that Fanjingshan, Jiulingshan, and Lushan are klippes of a "Chinese Silvretta" that has been thrust on top of a foreland thrust belt. This rigid-basement nappe was the bulldozer for the foreland deformation. Sandwiched between the bulldozer and the escaped is the Banxi Mélange—the accretionary wedge or the "Chinese Penninikum." Furthermore, Paleozoic rocks, those analogous to the Helvetic, crop out in windows near Lantian in Anhui Province and at Xishan in Jiangxi Province under the Chinese Penninikum (Hsü et al. 1988).

The 1-2-3 arrangement of the bulldozer, the subducted, and the escaped facies permits immediate recognition of the tectonic vergence of an orogenic belt and rapid clarification of its deformational history. We were cautious and tried not to coin new names, and we talked about (1) overriding crustal wedges, (2) mobilized basement and mélanges, and (3) cover thrusts (e.g., Hsü et al. 1987). It was more convenient, however, to use terms such as Helvetic, Penninic, and Austroalpine after my Chinese friends had visited the Swiss Alps, and those terms crept into the literature as names for the three tectonic facies (Hsü et al. 1991). They are, however, misnomers at best. Even in the broadest sense of the term, nobody could imagine that the tectonic facies "Helvetic," sensu lato, could include structures of autochthonous massifs or of foreland basins. I reverted, therefore, to the more cumbersome names of rigid-basement facies (Austroalpine), mobilized-basement facies (Penninic), cover-thrust facies (Jura, Helvetic, Flysch, klippe), and foreland-basin facies (Molasse) in my 1991 article on tectonic facies. That nomenclature was, in a way, even more unsatisfactory: instead of the simple 1-2-3 of motor/overrun/escaped, I now had four facies plus an unwieldy number of subfacies. I also had to name the Nordkalkalpen (a pile of Austroalpine cover thrusts) a subfacies of the rigid-basement facies! Fortunately this article was published in a Festschrift volume of limited circulation (Hsü 1991b), and my proposal has been overlooked or ignored.

Nomenclature of Tectonic Facies

I wrote to Pentti Eskola in 1953, when I was a graduate student at UCLA, criticizing his nomenclature of metamorphic facies. He said that all rocks of the granulite facies are granulites, but he could not refer to all rocks of the amphibolite facies as amphibolites: marbles, schists, and other rocks metamorphosed under amphibolite-facies conditions are not amphibolites. Rocks of the greenschist facies may be neither green nor schistose. "Pyroxene-hornfels" might contain no pyroxene nor be hornfelsic in texture. Furthermore, lacking a noun to denote the rocks of a certain metamorphic facies, we lack an adjective to describe "the characteristic textures or the characteristic mineral associations of a certain facies." I suggested to Eskola in my letter that we might adopt geographical names, especially when much of our understanding of the facies classification resulted from a few classical regional studies, such as his work in the Orijarvi region, Goldschmidt's work in the Kristiania district, Vogt's work in Sulitelma, Turner's work in Otago, etc.

The father of the metamorphic-facies concept was not entirely annoyed, and he did consider the advantage of having names such as Oslo facies (Kristiania is now Oslo!), Orijarvi facies, Sulitelma facies, Otago facies, etc. "The present moment," he added, "might be as favorable for a reform as any, because I am just preparing a new edition of Kristalle und Gesteine and also because I must soon write something about the facies . . . in the Bowen Volume. And you may start the praxis in your thesis. So I propose we will think over and you try to find out further good names and write me again." (Eskola, letter dated March 8, 1953)

I turned in my thesis in June 1953, and I did not start the practice of using new names for metamorphic facies. I cannot remember why I failed to do this; perhaps I was

discouraged by the all too conservative professors of mine. I remain convinced, however, that "Orijarvi facies" is a better name than "amphibiolite facies." In addition to
the other obvious reasons, naming a facies after a type locality has the advantage of
always having a standard reference: when one is no longer sure what Orijarvitic facies
means, one can always go back to study the rocks in Orijarvi.

This digression on the naming of metamorphic facies presents my conviction that
geographical terms are more suitable than genertic terms. The place will always be the
same: Helvetic, pertaining to the land of Helvetia, will always be Helvetic, but a foreland deformed belt could one day be turned into a zone of back-thrusting. But I cannot
use the name Helvetic for a tectonic facies, because the name has been preempted, as
it is used for a tectonic unit in the Swiss Alps; the name has too specific a meaning as
a tectonic designation. My colleague Trümpy, for example, was incensed by a less than
discriminating comparison of the Canadian Foothills to the Helvetic Alps. There is
thin-skinned deformation in both places, but the Foothill Thrust Belt is as different
from the Helvetic nappes as Emperor Kangxi was from Emperor Napoleon. Kangxi
was not Napoleonic, nor are the Foothills Helvetic. Likewise, the difference between
Franciscan mélanges and Antigorio nappes is too remarkable for both structures to be
grouped under one name; Franciscan is Franciscan and Antigorio is Penninic. Finally,
Austroalpine nappes, as mentioned previously, include not only Silvretta, but also
Nordkalkalpen; one is a rigid-basement nappe, while the others are cover thrusts.

Perplexed for almost a decade without an adequate solution, I was inspired one
Sunday morning while reading the entry on Switzerland in the *Encyclopaedia Britan-
nica*: "Rhaetians (Rhätier in German), Celts (Kelten), and Alemanni or Alamanni (Ala-
mannen) all have left their mark on Switzerland in the course of its historical evolution." The Rhaetians lived in the east since the Neolithic, and the Celts in the west in
Roman times, before the Great Migrations brought the Alemanni people to northern
Switzerland in the fifth century. The descendants of those three peoples are the majority inhabitants of the countries where the dominant structures are the Austroalpine,
Penninic, and Helvetic nappes, respectively. The new names, except perhaps Rhaetic,
have not yet been appropriated as technical terms in geology, and their adoption as
names for tectonic facies would lead to a minimum of confusion. I propose, therefore,
the terms raetides (Rhaetiden), celtides (Keltiden), and alemanides (Alamanniden) for
the tectonics facies.

Raetides are structures dominated by rigid-basement overthrusting. They include
the Lower and Central Austroalpine rigid-basement thrusts (see Chapter 11). Raetides
are commonly the "bulldozing structures" in the scheme of F. E. Suess, and they are
present in the overriding plate above an accretionary wedge on an active plate-margin.

Celtides consist mainly of metamorphic rocks, characterized by ductile deformation
of basement and of their sedimentary cover. In the Swiss Alps, the celtides are almost
exclusively Penninic structures, such as the Margna/Monte Rosa nappes, the ophiolite
mélanges (Zermatt-Saas, Furgg, Antrona, "Platta," Arosa, Misox, etc.), the Penninic
core nappes and mélanges (Bernard, Monte Leone, Lebendun, Antigorio, Tambo,
Suretta, Adula, Tessin, etc.), as well as the Bündnerschiefer and *schistes lustrés* (see
Chapters 7–10). Celtides in Tethyan mountains are the "overrun structures" in the
scheme of F. E. Suess, and the subducted structures on an active plate-margin or in the
suture zone between two plates are, as a rule, bounded above by raetides and below by
alemanides.

Alemanides are characterized by thin-skinned deformation of sedimentary cover
that has been detached from its underlying basement. The Swiss alemanides include
the Folded Jura, the Subalpine Molasse, the Helvetic nappes, the Prealpine nappes and
the Flysch nappes. They are the "escaped structures" in the scheme of F. E. Suess, also
called *foreland folds and thrusts* or *foreland deformed belts* because of their tectonic
position.

Although most Austroalpine nappes are raetides, the two terms are not synonymous. Neither are all Penninic nappes celtides, nor all Helvetic nappes alemanides. One set of terms consists of names of tectonic facies, whereas the other designates tectonic positions in an orogenic system. Reverse faulting of autochthonous and parautochthonous massifs is rigid-basement overthrusting. Those structures are raetides even though they occupy a position in a foreland deformed belt, whereas their sedimentary covers have been largely sheared off and overthrust as alemanides. The Schams nappes are Penninic structures, but they were cover thrusts, or thin-skinned alemanides formed after the Cretaceous Meso-Alpine collision of Europe and Brianconnais. The Sesia zone is correlative to the Austroalpine nappes in its tectonic superposition, but the basement and sedimentary cover of this zone have been overridden by higher Austroalpine elements, and they are thus celtides similar to those in underlying Penninic nappes of the Valais. The Northern Limestone Alps are upper Austroalpine, but the sedimentary strata have escaped to the foreland, having been pushed from behind by higher Austroalpine raetides. Their style of deformation is similar to that of the Helvetic nappes, and both are alemanides.

In an idealized model of two-plate collision, the Helvetic structures are the escaped alemanides, the Penninic are the overridden celtides, and the Austroalpine are the bull-dozing raetides. In actuality, each of the three tectonic units is cut by thrust faults, forming the numerous Helvetic, Penninic, and Austroalpine nappes. Tectonic facies alemanides, celtides, and raetides thus do not necessarily correspond to units of tectonic superposition.

14.

Global Tectonic Facies

A Swiss friend in the medical profession read my geology of Switzerland and was very pleased to understand as a layman the anatomy of the Alps, which he climbed. He thought that I should extend my efforts beyond the national border and find my alemanides, celtides, and raetides in the geology of Germany, France, Italy, etc. Eduard Suess did make a global synthesis with his *Das Antlitz der Erde*. I also thought about that when I first discussed this opus with Princeton University Press: we considered the possibility that this Swiss-based introduction to tectonic facies be followed by a volume illustrating the worldwide applications of the new concept. I eventually decided against such a venture. I shall retire from geology in a few years, and there are younger colleagues who can do a better job. Instead of a new edition of *Das Antlitz*, I would rather make some informal comments summarizing my understanding of the anatomy of other classic mountain ranges, and give a preliminary report of my work during the last decade in applying the facies concept to the geology of China. The purpose is not to describe the "root," "stem," and "crown" of all mountains, but to point out that such vital organs do exist in every orogenic belt.

The subtitle of the German edition of this book is *A Textbook for Beginners and a Discourse with Experts*. I have, in the first thirteen chapters, deferred to the beginners, and I hope they will excuse these final chapters which are tailored to suit the experts. For the sake of brevity, I have to presume a certain familiarity with the subject matter. I shall mention names and use technical terms with a minimum of explanation. The sections are written in the style of John Rodger's (1983) essay *The Life History of a Mountain Range—the Appalachians*, except that I shall have several maps instead of just one geographical sketch. My mini-essays are best viewed as autobiographical narratives of my search for answers to problems that have puzzled me during the last four decades. I hope beginners will not feel completely left out. In adopting this approach of "name-dropping," I am following the lead of my teacher George Tunnel. He came to inspire, not to inform. By talking over our heads, he motivated those of us who were ignorant but curious to dig into the literature and find out what he was talking about.

Appalachian Mountains

I gave my inaugural lecture at the ETH 25 years ago on comparative tectonics, and I chose to compare the Appalachians to the Alps. The tectonic units of the Appalachian Mountains include those of thin-skinned deformation, namely the Allegheny/Cumberland Plateau, the Valley and Ridge, and the Great Valley (Figure 14.1). They are readily comparable to the Plateau Jura, Folded Jura, Molasse foreland basin, the Helvetic nappes, and the Flysch nappes of the Swiss Alps. The plutonic rocks of the Inner Piedmont Province are comparable to those in the Penninic Alps. Core nappes of the Piedmont have not yet been mapped out individually, whereas the mélange nature of the ophiolite-bearing rocks of the Kings Mountain Belt is obvious to all (Misra and Keller 1978).

I did not know where should I place the Appalachian Blue Ridge Province in the Alpine scheme. Some of the Precambrian rocks there were little metamorphosed during the Appalachian deformation; they have retained their 1000 to 1100 Ma radiometric ages. From this point of view, the Blue Ridge is comparable to the rigid-basement nappes of the Alps. On the other hand, "the Blue Ridge belt is literally dotted with small bodies of ultramafic rocks" (Misra and Keller 1978, p. 391), and they are present both in the basement complex and in the metasedimentary-metavolcanic sequence. The age of the ultramafic rocks is unknown; they could represent exotic slabs in a late

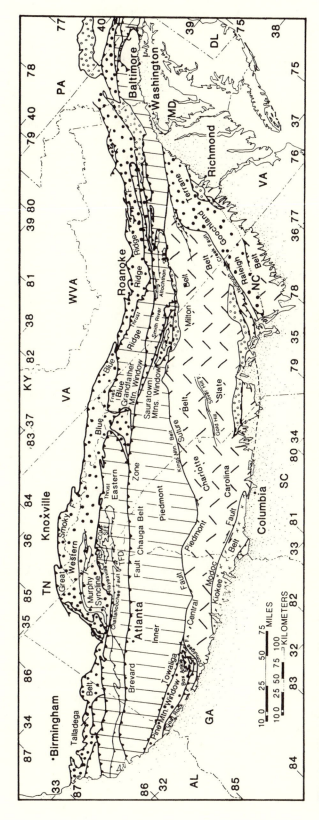

Fig. 14.1. Major structural features and belts of the Southern and Central Appalachians (after Hatcher 1987).

Precambrian accretionary complex, and/or slabs in the mélange of the suture of a Paleozoic arc-continent collision.

What is on the east side of the Appalachian Mountains? Flat-lying lower Paleozoic carbonate strata on continental basement are encountered in wells in central Florida, and this is a fragment of Africa left behind on the west side of the Atlantic (Opdyke *et al.* 1987). The proto-Atlantic Ocean came into existence when Africa was rifted apart from North America (Wilson 1966), but this ocean was not always a simple ocean bounded on both sides by passive margins. The eastern North American margin of the Paleozoic may have been comparable to that of East Asia today—the passive margin of the continent was fringed by island arcs and/or microcontinents (Rodgers 1983).

Paleogeographic reconstructions left little doubt that the late Precambrian/Paleozoic North American margin extended then for many kilometers eastward beyond the Valley and Ridge Province of today. A passive-margin sequence is present under the Blue Ridge and the Piedmont, as evidenced by the exotic carbonate slabs of Cambro-Ordovician age in the Brevard zone (Hatcher 1978) and by the COCORP reflection studies (Cook *et al.* 1979). A strip of continental crust, having been split off from North America, formed an early Paleozoic island arc called the Piedmont Swell ("Piedmont Terrane" of Hatcher 1987). This intra-oceanic elevation finds its analogue in the Brianconnais Swell of the Tethys. The marginal basin between North America and Piedmont was a site of detrital sedimentation during the late Precambrian and early Paleozoic. The crust under the basin was consumed along an easterly dipping Benioff zone under Piedmont (Figure 14.2), and the basin was eliminated by an arc (Piedmont)–continent (North America) collision during the early Paleozoic (Hatcher 1987).

The structures of the Blue Ridge zone, according to the simplistic scenario, are celtides, and the ultramfic rocks of the zone should be the exotic blocks of the ocean lithosphere under a back-arc basin between North America and Piedmont. The Precambrian radiometric ages of basement rocks in the Blue Ridge and the mild metamorphism of their sedimentary cover indicate, however, that the Blue Ridge structures were raetides, not celtides, during the Paleozoic deformation. The presence of a tectonic unit similar to the Austroalpine between units comparable to the Helvetic (Valley and Ridge) and the Penninic (Piedmont) was the puzzle that was troubling me when I gave my inaugural lecture at the ETH. The basic difficulty lay in our inability to jump out of the straitjacket of geosynclinal theory, which had already been falsified in more than one aspect. We were thinking of one ocean (alias geosyncline) and one intercontinental collision (alias orogenic phase), so that there could be only one Helvetic alemanide, one Penninic celtide, and one Austroalpine raetide. Of course, the geological evidence for Brianconnais is undeniable, but the elevation between the Valais and Piedmont basins was considered a *Zwischengebirge*, a cordillera, or an intra-oceanic swell.

The awakening came only recently, after I had already sent the last draft of this manuscript to the publisher. Most mountains were not formed by plate collisions; they were formed by back-arc basin collapse, involving arc-continent and arc-arc collisions. Adopting a West Pacific model, which will be presented in Chapter 17 of this opus, I finally realized that Brianconnais is not a *Zwischengebirge*, not a cordillera, not an intra-oceanic swell. Brianconnais is a remnant arc like the Palau-Kyushu Ridge between the West Philippine and the Parece-Vela basins (Karig, Ingle *et al.* 1975). Furthermore, remnant arcs do not always occur singly; there are at least two remnant arcs in the region of the Philippine Sea today: the Palau-Kyushu and the Mariana ridges behind the Mariana Volcanic Arc. Applying this model to interpret the paleogeography of the Paleozoic North American margin, one can recognize two remnant arcs—the Blue Ridge and the Piedmont. The Blue Ridge raetides have been emplaced directly on top of the Southern Appalachian alemanides; the celtides, representing the Paleozoic back-arc basin between North American and the Blue Ridge, are now buried

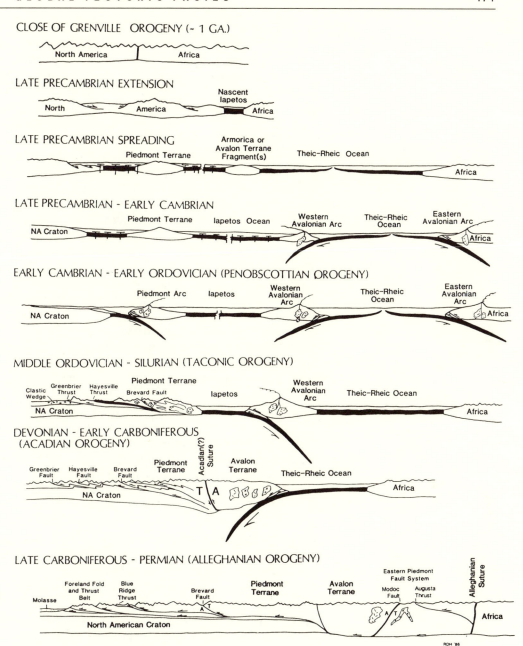

Fig. 14.2. Geological evolution of the Southern Appalachians (after Hatcher 1987).

beneath the 200-km-plus overthrust identified by the COCORP study. The metamorphic rocks of the Piedmont are the celtides formed by the collision of the Blue Ridge and Piedmont arcs.

The Piedmont is separated from the Charlotte and the Carolina Slate belts by the Kings Mountain Belt (Figure 14.2). The Carolina Slate Belt is a distinctive succession of upper Precambrian and lower Paleozoic rocks, which are found several places along the eastern North American margin (Skekan *et al.* 1978; Rodgers 1983). The granite batholiths yields radiometric ages in the range 550–650 Ma, and they are intrusive into volcanic and volcaniclastic rocks. These rocks are overlain by Cambrian and Lower Ordovician sediments characterized by the presence of Acado-Baltic faunas, and the stratified sequence is little metamorphosed and relatively undeformed (Skekan *et al.* 1978). It has been proposed that the slate-belt rocks, well exposed on the Avalon Peninsula in Newfoundland, were the sedimentary cover of Avalonia, the outer volcanic arc of Paleozoic North America (Williams 1964; Rast *et al.* 1976; Hatcher 1987).

The presence of ultramafic rocks as exotic blocks in the Kings Mountain Belt indicates the existence of a Paleozoic ocean, called Iapetus, between Piedmont and Avalonia (Figure 14.2). Avalonia was an east-facing island arc, and the Iapetus a large back-arc basin. The consumption of the ocean lithosphere of this basin along an east-dipping Benioff zone caused the ensuing arc (Avalonia)–continent (North America) collision in the Devonian (Figure 14.2). The final collision of Africa and North America, eliminating the Theic-Rheic Ocean between Africa and North America, did not take place until later in the Paleozoic (Figure 14.2).

Three collisional events have traditionally been recognized in the Appalachians, and they are referred to in the literature as the Taconic, Acadian, and Allegheny orogeny, respectively (see the review by Hatcher [1987]). Radiometric dates of metamorphic rocks indicate, however, continuous compressive deformation from the Middle Ordovician to the late Paleozoic. I would like to suggest to my American colleagues that they consider adopting an Eo-, Meso-, Neo- nomenclature for the Appalachian deformation, as I proposed for the Alpine orogenesis: those stages are not defined by any particular time or time intervals, but they refer to the modes of deformation, namely the Eo-(Alpine/Appalachian) subduction stage, the Meso-(Alpine/Appalachian) arc-continent or arc-arc collisions, and the Neo-(Alpine/Appalachian) post-collisional deformation.

Within the framework of the tectonic facies model for the Appalachians, the Cumberland and the Valley and Ridge structures are the Southern Appalachian alemanides. The Precambrian basement gneiss northwest of the Hayesville-Fries Fault in the Blue Ridge is overlain by shallow marine sedimentary cover (Hatcher 1987). This Blue Ridge massif, with its more than 200 km thrust displacement, could hardly be called parautochthonous, but the rigid basement was once the outer margin of North America, comparable to the Gotthard massif as the "marginal high" of the European margin. According to our recognition of the Blue Ridge structures as raetides, the Blue Ridge rocks were the basement and sedimentary cover of a remnant arc east of the Paleozoic North American continent. Ultramafic rocks do occur in the Blue Ridge. They could be a part of a Precambrian accretionary-wedge complex, formed when the proto-Atlantic Ocean was subducted under the North American continent.

Referring to circum-Pacific analogue, eastern North America may always have been an Andean type of active-plate margin since the close of the Grenville deformation at about 1 GA. I suggest that the Piedmont and Avalon terrances were not formed by extension during late Precambrian seafloor spreading, as portrayed by Hatcher (1987) in Figure 14.2. Instead, these terranes, together with the Blue Ridge, were all split away from North America as an island arc on the eastern margin of the North American continent. Continued subduction of the Theic-Rheic Ocean (Figure 14.2) led to the genesis of three rows of arcs separating three back-arc basins: one between North America and the Blue Ridge, another between the Blue Ridge and the Piedmont remnant arcs, and the outermost Iapetus basin between Piedmont and Avalonia, which was the outer volcanic arc on the active margin of the Paleozoic North American plate.

According to this scenario, the Inner Piedmont metamorphics, the mélanges of the Kings Mountain Belt, and the igneous and metamorphic rocks of the Charlotte and Carolina Slate belts were all celtides, although they include both overrun and overriding structures. The basement of Avalonia should have been a raetide during Meso-Appalachian collision, but it was mobilized during the late Paleozoic (Allegheny) orogenesis, to form the core nappes of the Kiokee belt. The granitic basement was also mixed with slabs detached from the Theic-Rheic Ocean to form the ophiolite mélange of the Raleigh Belt (Hatcher 1987). The Meso-Appalachian raetides were thus changed into Neo-Appalachian celtides after the collision of North America (with Avalonia) and Africa.

Corresponding tectonic facies are recognizable in the Central and Northern Appalachians. Stanley and Ratcliffe (1985) reviewed the Taconic problem and came up with

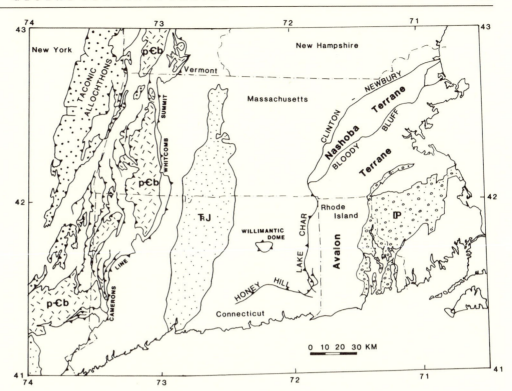

Fig. 14.3. Map of southern New England, showing the distribution of Taconic allochthon, Grenville basement (pCb), Pennsylvanian sediments of the Narragansett Basin (P), overlying Avalonia, and Triassic Jurassic sediments (TrJ) (after Hatcher 1984).

a synthesis on the basis of the plate-tectonic theory. The Paleozoic carbonate sequence of New York, the Taconic allochthon and the Grenville basement (with its autochthonous carbonate cover) are the foreland, the flysch nappes, and a parautochthonous massif of the Appalachians (Figure 14.3); the first two are alemanides and the last a raetide. Farther east in New England, the deformation of metamorphic rocks has been considered evidence of an arc-continent collision (Armstrong *et al.* 1992): the West and East Acadian metamorphics, including schists, gneisses, and ophiolitic mélanges, are the Meso-Appalachian celtides. The raetides of the Northern Appalachians are the Avalonian rocks that extend from southeastern New England and maritime Canada to several places in Europe (Skekan *et al.* 1978; Hatcher 1984).

Scandinavian Caledonides

I am not a student of the Caledonides, and I have never had a chance to see the geology of the Scandinavian Caledonides. I was selected, however, by John Rodgers to review a monograph, *The Caledonide Orogen* (Gee and Sturt 1985), probably because all the experts had become authors of that magnificent opus. I thumbed through the 1266-page volume, while mentally translating the geosynclinal nomenclature into that of tectonic facies. This is what I managed to understand:

The structure of the Caledonides in Scandinavia is dominated by thrust faults, or nappes, emplaced from west to east, and the piles of nappes reached the Baltic Platform during the mid-Paleozoic. The nappes have been grouped into four complexes: the Lower, Middle, Upper, and Uppermost Allochthons (Roberts and Gee 1985; see also Figure 14.4).

The Lower Allochthon consists mainly of sedimentary successions. They were deposited on a passive margin of latest Proterozoic and early Paleozoic age, and their structure is typical of mountain-front décollement zones. Precambrian crystalline rocks and thick unfossiliferous metasandstones dominate the Middle Allochthon, and the Caledonian metamorphism of those rocks generally belongs to the greenschist facies but reaches amphibolite facies in westernmost areas. Dolerite dykes of Cambro-

SCANDINAVIAN CALEDONIDES

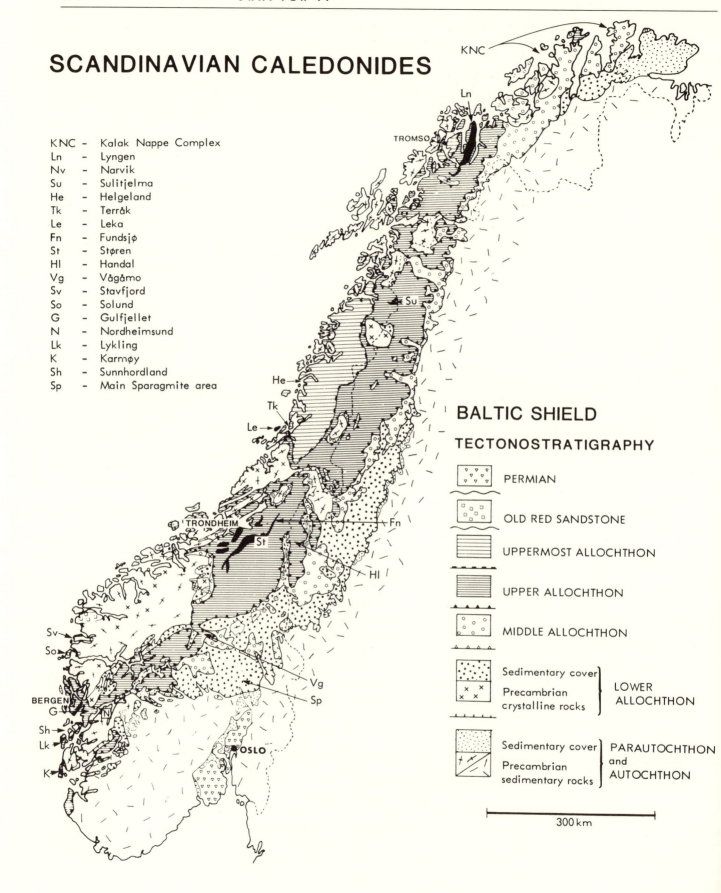

KNC – Kalak Nappe Complex
Ln – Lyngen
Nv – Narvik
Su – Sulitjelma
He – Helgeland
Tk – Terråk
Le – Leka
Fn – Fundsjø
St – Støren
Hl – Handal
Vg – Vågåmo
Sv – Stavfjord
So – Solund
G – Gulfjellet
N – Nordheimsund
Lk – Lykling
K – Karmøy
Sh – Sunnhordland
Sp – Main Sparagmite area

BALTIC SHIELD

TECTONOSTRATIGRAPHY

PERMIAN

OLD RED SANDSTONE

UPPERMOST ALLOCHTHON

UPPER ALLOCHTHON

MIDDLE ALLOCHTHON

Sedimentary cover
Precambrian crystalline rocks } LOWER ALLOCHTHON

Sedimentary cover
Precambrian sedimentary rocks } PARAUTOCHTHON and AUTOCHTHON

300 km

Ordovician age are evidence of igneous activity during the extensional phase of the life history of the Scandinavian Caledonides.

The Upper Allochthon is the most heterogeneous and complex tectonic unit. The lower unit, called the Seve nappes, comprises high-grade schists, gneisses, amphibolites, migmatites, ultramafic rocks and eclogites (Andreasson et al. 1985). This mélange of different rock types and the deformed island-arc complexes of the Köli nappes have been considered parts of the "eugeosynclinal terranes" of the Upper Allochthon (Roberts and Gee 1985).

The Uppermost Allochthon consists of the continental rocks east of a late-Precambrian and early Paleozoic ocean. The terrain is underlain by Sveconorwegian basement (about 1000 Ma) and its Riphean/Vendian sedimentary cover. Granitoid intrusions are dated mid-Ordovician to Silurian; they apparently represent the root of a volcanic island-arc facing the ocean (Stephens and Gee 1985). The Uppermost Allochthon has been thrust over the Köli nappes and ophiolite mélanges.

Thus viewed, the Scandinavian Caledonides are not basically different from the Alps or from other collision-type orogenic belts. The Uppermost Allochthon rocks are the raetides. The Seve mélange of the Upper Allochthon is definitely a celtide. The Köli nappes with all their ophiolite mélanges and metamorphic rocks are apparently also celtidic. Comparisons with other orogenic systems point, however, to the possibility of a more complicated scenario. The tectonic evolution of the Köli nappes could be comparable to that of the Appalachian Piedmont. Some Köli elements, such as the intrusive granites, may have been the root of an island arc, and those rocks should thus have been a raetide during a Meso-Caledonian collision. But they were eventually *overrun* by the *bulldozer* that was the Uppermost Allochthon. A Meso-Caledonian raetide thus became Neo-Caledonian celtides, like the rocks of the Piedmont Arc in the history of Appalachian orogenesis. Lower down in the tectonic superposition are the nappes of the Middle Allochthon, and they are also celtides, comparable to the Penninic in the Alps. The foreland thrust belt of the Lower Allochthon consists of alemanides. The Scandinavian Caledonides have been sufficiently well preserved that the whole pattern of the alemanide/celtide/raetide trinity is easily recognizable.

The ocean west of the Uppermost Allochthon terrane is commonly considered the ancestral Atlantic and has been called Iapetus (Gee and Sturt 1985). The term has, however, also been used to designate a North American marginal sea west of the Avalonian Arc, and the Iapetus of students of the Caledonides corresponds thus to the Theic-Rheic Ocean of Hatcher (1987), as portrayed in Figure 14.2 of this opus.

British Caledonides

When I was a Chinese undergraduate learning the geology of the British Isles, we had to memorize a whole string of names: Lewisian, Torridonian, Moine thrust, Great Glen Fault, Grampian Highlands, Dalradian, Midland Valley, Highland Boundary Fault, Southern Uplands, Lake District, Anglesey Mélange, Graywackes of the Welsh Basin, shelly facies on Midland Platform, etc. (Figure 14.5), but we were not told of their interrelations or that they were vital organs of an orogenic system.

The one feature that impressed us all was the Moine thrust. We were told of the controversy throughout the nineteenth century, when geologists were split between "fixists" and "transporters." The great achievement of Peach et al. (1888) was extolled: they had recognized the foreland with its Lewisian basement and its Torridonian and lower Paleozoic sedimentary cover, and they had delineated the Moine thrust! As we were told, the Moine Schist is thrust northwestward onto the foreland, so the tectonic vergence of the British Caledonides must be northwestward. The direction of tectonic transport was thus directly opposite to that of the Scandinavian Caledonides. Yet in the Southern Highland, the dominant structure is the southward-facing Tay complex. This reversed vergence has been the cause of much controversy, and has been considered by some a type of backthrusting (see review by Fettes et al. 1984).

The Paleozoic rocks in England and Wales have been divided into a graptolite facies

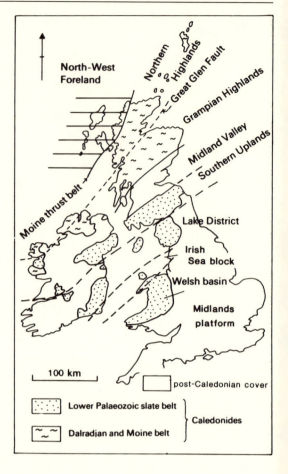

Fig. 14.5. Location of the main tectonic zones of the British Caledonides (after Park 1988).

in "geosyncline" and a shelly facies on platform (Jones 1938). The sedimentary strata and interstratified volcanic rocks are folded. On the Island of Anglesey, where Greenly (1919) innovated the idea of mélange, Precambrian basement, late Precambrian and lower Paleozoic sedimentary rocks, as well as ultramafic rocks, have been fragmented and mixed. The Ordovician and Silurian beds of the Southern Upland are mainly gray-wackes, whereas the Ballantrae Complex, consisting of shales, radiolarites, and ophio-lites, has been variously referred to as "olistostrome deposits" or tectonic mélanges (see the review by Kelling *et al.* 1984). All the Paleozoic deformations south of the Highland Boundary Fault took place prior to the deposition of the Old Red Sandstone, but their relation to the Highland Caledonides seems obscure to a novice.

Now that we have the Scandinavian Caledonides as a model, the pieces of the puzzle of the British Caledonides can easily be put back together. The Dalradian, certainly the Tay nappe complex, is analogous to the Penninic of the Alps; it is the metamorphic celtide of a southerly vergent Caledonide. The Ballantrae Complex and the Anglesey Mélange were the accretionary-wedge complex on the northern margin of the ances-tral Atlantic during the Ordovician and Silurian (see Mitchell and McKerrow 1975; Leggett *et al.* 1979; Bluck *et al.* 1980); they are thus comparable tectonic facies equiv-alent to the Saas-Zermatt and Arosa Mélanges of the Alps. The graywackes of the Southern Uplands and Wales are the flysch nappes and the "shelly facies" carbonates constitute the foreland folded belt, or the alemanides of the British Caledonides.

In making this comparison, I noted that the core nappes of the Dalradian constitute a higher tectonic element than the Highland Border mélanges (Figure 14.6). The Tay nappe complex is thus not to be correlated to the Bernard nappe of the Alps, but to the Margna/Monte Rosa, being a subducted fragment of the continental crust from the overriding plate.

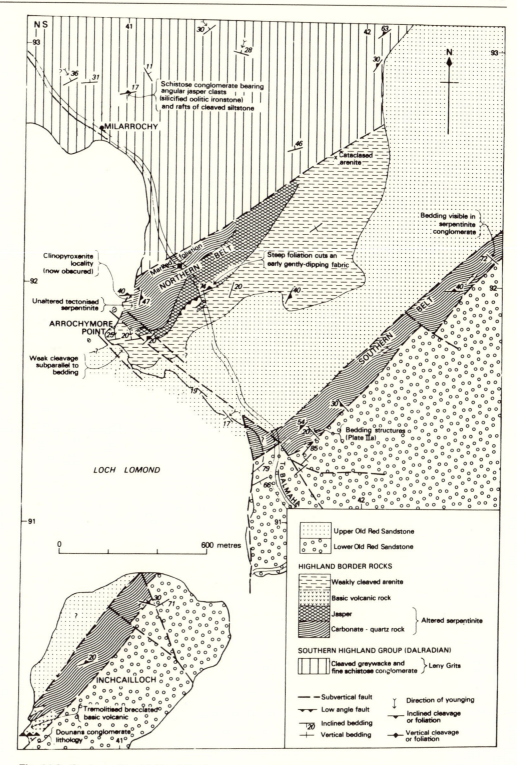

Fig. 14.6. Geology of the Highland Border near Loch Lomond (after Harris and Fettes 1984).

What was the role of the Moine thrust in Caledonian deformation?

The Moine rocks of the so-called North Highland Terrane were carried over 50 km to foreland rocks on a foreland folded belt during a long time span (430–415 Ma) (Coward 1984). The movement took place after the suturing of the British Caledonides. Near the Moine thrust the Proterozoic rocks have been thrust as rigid basement above the lower Paleozoic rocks of the foreland. Farther to the west, the metamorphic rocks of Proterozoic age in the zone west of the Loch Quoich Line were redeformed under amphibolite-facies conditions, with crustal shortening "in excess of 140 km" (Hutton 1989, p. 49). The deep-seismic survey MOIST revealed a series of crustal thrusts decoupling at a depth of about 18–20 km (Smythe *et al.* 1982) (Figure 14.7). The thrusting of this wedge-shaped continental crust onto the foreland has

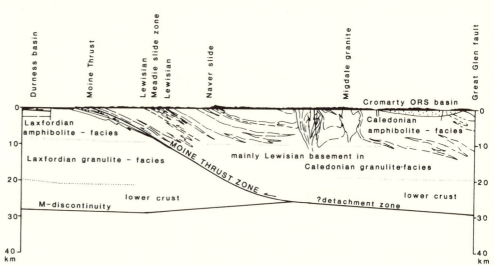

Fig. 14.7. Crustal section through the Moine thrust zone (after Soper and Barber 1982; used by permission of the Geological Society of London).

been compared with the thrusting of the Blue Ridge/Piedmont in the Appalachians and with the Neo-Alpine overthrust of the Aar massif in the Alps (Hatcher 1984). The deformation history suggests that the North Highland rocks were first thrust under the Moine rocks like the Helvetic nappes under the Penninic and/or Austroalpine; the sedimentary sequence of the foreland has been deformed as alemanides of a northwesterly vergent Caledonides. The Moine rocks are celtides or raetides, depending on whether they were subjected to metamorphism during the Paleozoic deformation. They were brought to Scotland by strike-slip movement along the Great Glen Fault, and they are thus, strictly speaking, not British Caledonides. Coward (1984, p. 275) correctly perceived this fact, and he wrote:

> There is no obvious plate collision zone in Scotland coeval with this [Moine] deformation. . . . In the Silurian the Grampian deformation had finished and the area now southeast of the Midland Valley was the site of an ocean trench. . . . The source of the compression, which produced the Moine thrust crustal ramp and then carried the Moine over 50 km to the WNW, is unknown and may well have been removed by late to post-Caledonian movement on some major faults, such as the Great Glen fault, juxtaposing regions affected by different ages of Caledonian crustal compression.

Coward recognized the fact that the bulldozer which once pushed the Moine thrust from behind is to be found somewhere on the other side of the Great Glen Fault, perhaps under the North Sea, unless a small relic is still to be found in the area between the Loch Quoich Line and the Great Glen Fault. The northern extension of the alemanides/celtides of the Northern Highland is the orogenic belt of eastern Greenland, where the Laurentian/Greenland Shield is fringed on the east by westerly vergent Caledonides (Figure 14.8).

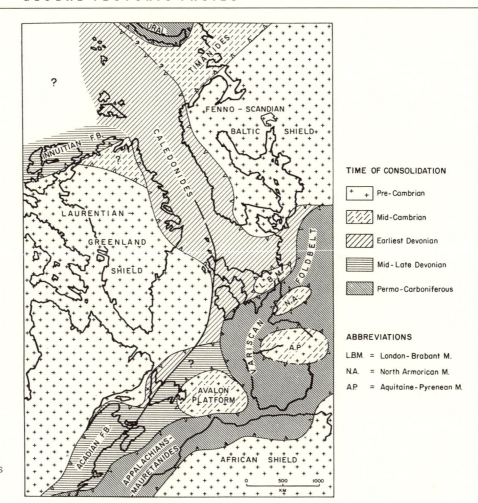

Fig. 14.8. Spatial relationship between Paleozoic orogenic belts in the Arctic–North Atlantic realm (after Ziegler 1985).

North American Cordilleras

In 1950 I started my dissertation work on experimental rock deformation with David Griggs. A year later he took a sabbatical leave for a temporary assignment in Washington, D.C., and I had to find another thesis advisor. Ken Watson suggested that I work on the granulites and mylonites of Cucamonga Canyon so that I could make field observations on rock-deformation processes. Not having any better ideas, I became a "hardrock man" and worked for three years in the San Gabriel and San Bernardino Mountains (Figure 14.9). In Cucamonga Canyon of the southeastern San Gabriel Mountains, I found that the basement rocks were first metamorphosed under granulite-facies and subsequently under amphibolite-facies conditions; they were called Aurela Ridge Granulite and Cucamonga Canyon Mylonite, respectively. The sedimentary cover consists of rocks of the carbonate-orthoquartzite association, which were also subjected to Cretaceous deformation under amphibolite-facies conditions; they may have been the Paleozoic sedimentary cover on a passive continental margin, because similar rocks in the adjacent San Bernardino Mountains and elsewhere in Southern California are dated by Paleozoic fossils. Intruded into the metamorphic rocks are Cretaceous quartz-diorite and Miocene quartz-monzonite (Hsü 1955).

Perry Ehlig, who also decided to work on "hardrocks," came to talk to me in 1952 about a thesis area. I suggested Mount Baldy, which had just been developed as a ski resort. The Pelona Schist was a well-known group of rocks, but nobody seemed to know what it was doing up there on Baldy. Ehlig liked the area and eventually became

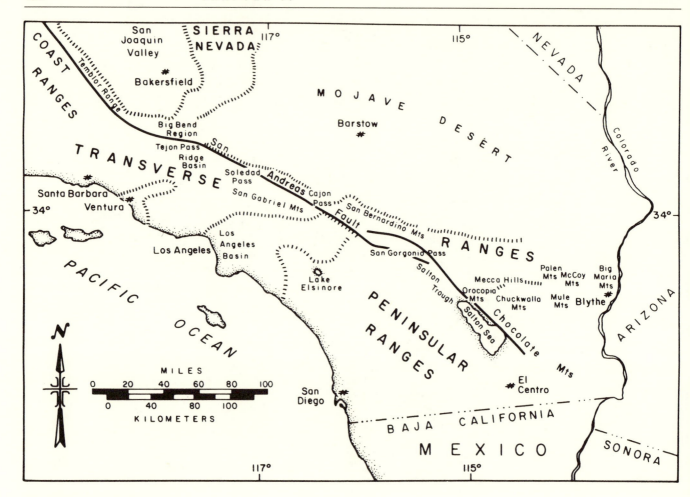

Fig. 14.9. Geographical location map (after Crowell 1981). Note that Pelona Schist in the San Gabriel Mountains is offset by the San Andreas Fault from Orocopia Schist of the Chocolate Mountains.

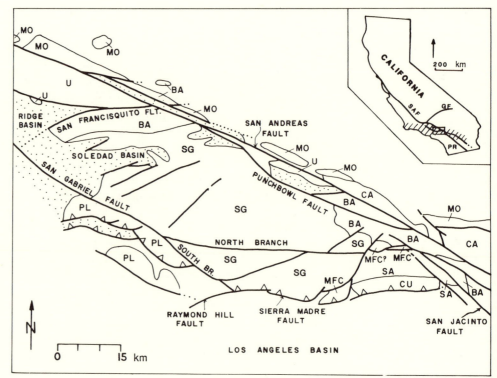

Fig. 14.10. Basement terranes of the San Gabriel Mountains (after May and Walker 1989). Terranes relevant to discussions in the text are BA (Baldy), CU (Cucamonga), SA (San Antonio), and SG (San Gabriel). The Baldy and San Gabriel terranes are separated by the Vincent thrust.

an expert on the Pelona Schist and other greenschists of Southern California. One of his major contributions was the discovery of a northeasterly vergent thrust under the Pelona Schist. The basement rocks, including Precambrian anorthosite and Cretaceous quartz diorite, have been thrust on top of the schist (Figure 14.10), and a mylonite zone is present to mark the base of the Vincent thrust (Ehlig 1958).

Ehlig and I worked in adjacent areas, but we seemed to be working on entirely independent problems. We were working in basement terranes bounded by faults, and we did not worry about the relation of our San Gabriel units to one another, and even less about their affinity to those in the San Bernardinos, the Mohave Desert, the Sierra Nevada, or the Peninsula Range. We did not care then. We students of Jim Gilluly at UCLA were iconoclasts, and we did not think very much of theoreticians and their "dogmas." California geology did not seem to fit any model; we could neither see geosyncline nor find episodicity in orogeny. In hindsight, our iconoclasm was not completely unjustified: the Cenozoic tectonics of California is dominated by strike-slip faulting, and mountains of the California type are neither Tethyan nor Circum-Pacific. This "Wild West" tradition of sticking to the "facts" has found its ultimate development in the "terrane concept" in tectonics. The rule is to describe the "terrane" in detail, but to avoid "theoretical deductions" or "speculations." The consequence is, of course, that we know only that pieces of a jigsaw puzzle have been scattered about; we forbid ourselves to make any effort to recognize the picture that was there before the puzzle was broken up.

Yet we all knew that the Upper Cretaceous and Cenozoic strata of Southern California constituted only the "Superjacent Series." Older rocks, or the "Subjacent," were called "basement," and two types of "basement" were recognized: the granite basement and the Franciscan basement (Reed 1933). The basement blocks have been dissected and displaced by strike-slip faults. But what did the jigsaw puzzle look like before the pieces were scattered about?

The "basement" geology of California is a complex jigsaw puzzle. It is like Humpty Dumpty, Phil King (1977, p. 179) lamented—"all the king's horses and all the king's men cannot put the Coast Range back together again." The tectonic facies concept is designed to do what all the king's horses and all the king's men could not do. If one could describe a new hominid species on the basis of two molars, or a new plant species on the basis of a few pollen grains, why could we not reconstruct the life history of a mountain system that has the Aurela Ridge Granulite, Cucamonga Mylonite, and the Pelona Schist as its vital organs?

The Pelona Schist is a metamorphosed sequence of graywackes with some pelitic rocks, mafic and ultramafic rocks, as well as limestone and radiolarite. Similar units in the Orocopia and Rand mountains are called Orocopia Schist and Rand Schist, respectively (Figures 14.9 and 14.11). These mélanges of oceanic rocks are certainly celtides. Two alternative hypotheses have been suggested:

(1) The Pelona-Orocopia Schist and the Rand Schist are correlative to the Franciscan rocks of the California Coast Range. They are Late Cretaceous and early Tertiary celtides formed in the easterly dipping Benioff zone between the North American and the Pacific plates (Yeats 1968; Burchfiel and Davis 1981; Crowell 1981, p. 589).

(2) The Pelona-Orocopia Schist was the celtide formed in a southwesterly or southerly dipping Benioff zone on the inner margin of an island arc (Ehlig 1981; Haxel and Dillon 1978; Crowell 1981, p. 588). The collapse of the back-arc basin caused an arc-continent collision (Figure 14.12).

Burchfiel's postulate that the Pelona Schist is a celtide of a Circum-Pacific type mountain requires the scenario that the schist, like the Franciscan, is overthrust on top of an ocean plate. This is, however, not the case for the Pelona Schist. In fact, the Pelona Schist is underthrust toward the southwest below the basement complex of the San Gabriel Mountains (Figure 14.10).

The idea of Ehlig and others that the Pelona Schist was the accretionary complex

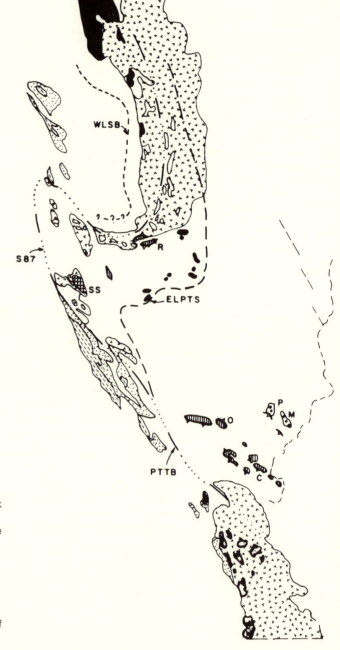

Fig. 14.11. Distribution of the Mesozoic magmatic arc and other rock units in California, plotted on a reconstructed basis with Cenozoic displacements on faults removed (after Burchfiel and Davis 1981). The Pelona Schist as well as the greenschists of the Rand Mountains (R), Orocopia Mountains (O), and Chocolate Mountains (C), shown as vertically hatched areas, are metamorphosed ocean sediments and ocean lithosphere. Note their position east of the Peninsula, Coast Range, and Sierra Nevada batholiths, suggesting their deposition in back-arc basins between the North American continent and an island arc chain on the outer margin of the continent. WLSB: western limit of Sierran basement.

formed during the process of back-arc collapse could be likened to the "molars that permit the reconstruction of Peking Man." I looked up the literature and found that a young generation of geologists has made great contributions to our understanding of the geology of the San Gabriel Mountains. Daniel May and Nicolas Walker (1989) recognized four major units (Figure 14.10): the granulites and mylonites in the Cucamonga Canyon constitute the Cucamonga Terrane; the largely Precambrian amphibolite-facies gneisses and anorthosites make up the San Antonio Terrane; the San Gabriel Terrane consists of Upper Cretaceous quartz diorite with roof-pendants of metasedimentary rocks of probably Paleozoic age; and the Mount Baldy Terrane consists of Pelona Schist. Are these four components of an orogenic belt? None of

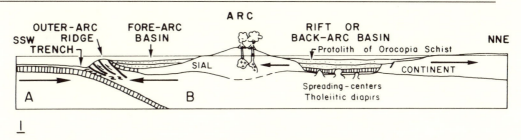

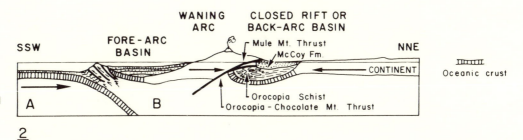

Fig. 14.12. Ehlig's and Haxel's model for the origin of Pelona-Orocopia Schist (after Crowell 1981). (1) Late Cretaceous back-arc spreading, and deposition of deep-sea sediments in the basin; (2) earliest Tertiary collapse of the back-arc basin, and the greenschist metamorphism of the ophiolite mélanges.

them could be considered alemanides; the metamorphic rocks are thus celtides and/or raetides.

The mineralogy of the rocks in the Cucamonga Terrane indicates granulite facies metamorphism at 8 kb and 700–800 °C at about 110 Ma. Later, at about 80 Ma, the rocks in the shear zones were subjected to metamorphism at 5 kb and 550–650 °C (May and Walker 1989; Barth and May 1992, Barth *et al.* 1992). The layered granulites have a steeply dipping N-S trend, resulting from Early Cretaceous deformation. That they should have been dragged down to a depth of 30 km can be interpreted as evidence of Early Cretaceous subduction, probably of the Pacific Ocean lithosphere under western North America. Similar granulites are present in the Salinian Block of the Coast Range and in the Sierra Nevada (Barth and May 1992), marking the active plate-margin, where the Farallon plate was thrust under the North American plate (Hamilton 1969; Dickinson 1970; Cross and Pilger 1978; see Figure 14.13).

Such a tectonic framework has a place for a back-arc basin, in which the deep-sea sediments now constituting the Pelona Schist were deposited. The granulites of the Cucamonga Terrane and the Precambrian rocks of the San Gabriel Terrane were the basement, and the Paleozoic (?) sequence the sedimentary cover of this island arc, which was located between the Pelona Basin and the open Pacific. The back-basin collapse process started in the Late Cretaceous, when the island arc was underplated by an accretionary wedge. The Benioff zone had a more nearly E-W trend and dipped to the south. The granulites of Cucamonga started in the middle Cretaceous to move up the Benioff zone at a rate of about 1 mm/year, corresponding to a 2 mm/year convergence rate up a 30-degree ramp. The mylonites of Cucamonga were formed under amphibolite-facies conditions in the Late Cretaceous; they marked the base of the overthrusting granulites. The Cucamonga Terrane was an Early Cretaceous celtide, but it became, together with the San Gabriel and San Antonio terranes, a rigid-basement nappe complex; they all were the raetides of the Late Cretaceous back-arc collapsing deformation.

According to this scenario, the quartz diorite of Mount Baldy was the root of andesitic volcanism of the San Gabriel segment of a magmatic arc. The magmatic arc, 150–80 Ma in age, extended from the Peninsula Range to the Sierra Nevada (Figure 14.13). In Southern California, the arc was separated from the passive margin of North America by the Pelona basin. Its elimination caused an arc-continent collision, and the Pelona Schist marks a segment of the suture zone.

Farther north, in the "greenstone belt" in the northwestern Sierra Nevada, Burke *et al.* (1976, p. 121) noted: "The ultramafic and gabbroic rocks in this area are almost

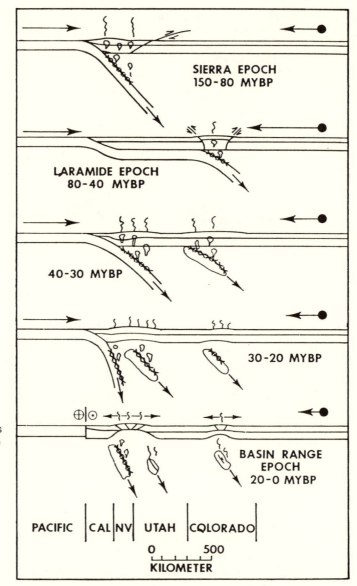

Fig. 14.13. Cross and Pilger's model of the tectonic evolution of western North America (after Cross and Pilger 1978). The Benioff zone between the North American and Farallon plates has always dipped eastward. Note, however, the postulate here of easterly vergent overthrusting during the "Sierra Epoch." This structure corresponds to those produced by the collapse of the Pelona-Orocopia back-arc basin in the Ehlig/Haxel model.

all or perhaps entirely variably dismembered ophiolite complexes including . . . well preserved sheeted dykes. . . . The ophiolite complexes and the mafic volcanics and volcaniclastics are probably at least partly of marginal basin origin."

The continuation of this suture zone has been interrupted by the emplacement of batholithic intrusions of the Mohave Desert and the Sierra Nevada. East or northeast of the suture zone, the mobilized continent crust of the underthrust North American margin has been described by Hamilton (1987, p. 765): "Broad tracts of the middle crust of southeastern California northeast of the last exposures of subcrustal oceanic schist underwent pervasive regional metamorphism and severe deformation during latest Cretaceous time."

The mobilized basement with its passive-margin cover finds its Alpine analogue in the Penninic core nappes. I am hoping to visit the region with Hamilton in the not too distant future. If the Ehlig/Haxel model of a collapsing back-arc basin is correct, the vergence of the ductile deformation should be eastward or to the northeast.

Having identified the raetides and celtides, we could start to look for the alemanides of the Mesozoic Cordilleran Orogen. There is no lack of easterly vergent thrust faults in western North America; they are the foreland deformation belts of an arc-continent

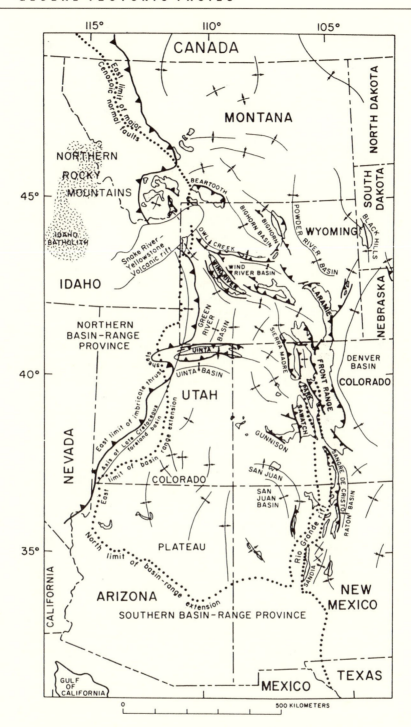

Fig. 14.14. Selected structural elements of the Rocky Mountain region (after Hamilton 1987). Note the easterly vergence of the foreland thrust belt. Most of the uplifts and basins in the region are also of "Laramide age," very late Cretaceous and early Paleogene. Precambrian-basement outcrops east of the thrust belt are shaded.

collision. In the Foreland Thrust Belt of the Western Interior and in the Rocky Mountain region (Figure 14.14), the craton "was distorted by crustal shortening during very late Cretaceous, Paleocene, and early Eocene time" (Hamilton 1987, p. 765). These are the Cordilleran alemanides.

Southern Andes

I devoted the best part of my professional life to the Deep Sea Drilling Project of the Joint Oceanographical Institutions Deep Earth Sampling Program (JOIDES). It was started in 1968 as an American project, and I had been working toward its internationalization ever since 1970. I was pleased to see the establishment in 1975 of the International Program for Ocean Drilling, but Switzerland was left out, because the membership fee was too high for a country with a population of only 6 million. The smaller and financially weak countries of Europe had to get together. I worked for a decade to promote such an effort, until twelve European countries banded together to become, under the auspices of the European Science Foundation, a consortium member of the Ocean Drilling Program (see Hsü 1992). I then resigned my duty as the Swiss delegate to the ESF Scientific Committee, which I had chaired, and I was rewarded with an appointment as the ESF representative to the ODP's Tectonics Panel.

I was pleased to be back among scientists after all those years of science politics. My work with DSDP had been mainly in the field of paleoceanography, but I chose the Tectonics Panel because I wanted to learn something new. I was not disappointed. One of my fellow panel members was Ian Dalziel, who had worked mainly in Antarctica and the Southern Andes.

My perception of the Andean geology was naive. I adopted the stereotype plate-tectonic model: The Andean Margin is an active margin, along which ocean lithosphere is subducted. The distinguishing feature is andesitic volcanism. The elevation of the Andes is related to the subduction. The Andean structures are mainly block-faulting, forming intermontane basins in which terrestrial sediments and volcanics are laid down. Ophiolites are present in the Northern and Southern Andes, but not in the Central Andes. In my ignorance, I proposed two types of Circum-Pacific mountains (Hsü 1973): the Franciscan type, typified by the California Coast Range, is underlain mainly by ophiolite mélanges; the Andean type, typified by the Central Andes, is characterized by a dominance of batholithic intrusions.

Chatting with Dalziel during the cocktail hour between panel sessions, he told me that there was more to Andean geology than the simple subduction-model. The eastern margin of the Pacific was once fringed by island arcs like that of the western Pacific, and the elimination of the basins behind the arcs is the model for Andean orogenesis. Dalziel used the expression "back-arc basin collapse" to designate this process.

I used to think of arc-continent collisions in terms of frontal confrontation between the apex of an arc and a wandering continent. This is the case for the collision of the Indonesian Arc and Australia. Along a north-dipping Benioff zone the arc is thrust on top of the continent and the tectonic vergence of the arc is frontal. In Dalziel's model, an island arc is thrust backward onto the passive margin of a continent. An actualistic analogue for such a process is to be found on the eastern margin of the South China Sea.

The small ocean basin under the South China Sea was formed by seafloor-spreading behind an island arc that extended from offshore Sarawak to Palawan to the Philippines. The back-arc spreading started during the Oligocene and ceased in the Middle Miocene, when subduction started. The ocean lithosphere under the South China Sea has been thrust under the Philippine plate ever since that time, and the Manila Trench marks this active margin (Figure 14.15). The plate-margin extends northward into the Backbone Range of Taiwan, where an ophiolite mélange marks the suture of the passive margin of Eurasia and a west-facing volcanic arc of the Philippine plate (Biq 1977; Lu and Hsü 1993). The suturing is a result of the collapse of the northern extremity of the back-arc basin under the South China Sea (Figure 14.15).

With reference to such a model of back-arc basin collapse, the geology of the Southern Andes is easily understood. The latest Wilsonian cycle of the Southern Andes began in the Jurassic, when extensive normal faulting and volcanism took place above an easterly dipping subduction in the Andes. The root of this Jurassic arc is now exposed as the Patagonia Batholith of the Outer Islands (160–140 Ma). A back-arc basin, floored by ocean crust, came into existence behind the west-facing arc (Katz 1973; Dalziel 1981). The elimination of the basin caused an arc-continent collision, and the suture zone is marked by the ophiolite mélange of the Inner Islands and South Georgia. Farther east is the "Early Andean Orogenic Belt" of the Cordillera Darwin (Figure 14.16): In this zone a pre-Mesozoic basement was deformed and metamorphosed during Andean deformation. A foreland thrust belt with an easterly vergence is present in

Fig. 14.15. An actualistic model of back-arc basin collapse. The back-arc basin under the South China Sea is collapsing. Its ocean floor has been thrust down an eastward-dipping Benioff zone under the Manila Trench since the Miocene (see inset map). The suture of an arc (Philippine Arc)–continent (Eurasia) collision on Taiwan, where the ocean floor is completely consumed, is marked by the Lishan Fault (Li.F.) and Laonungchi Fault (La.F.). An ophiolite mélange (Kenting Mélange) is found near the southern termination of the latter.

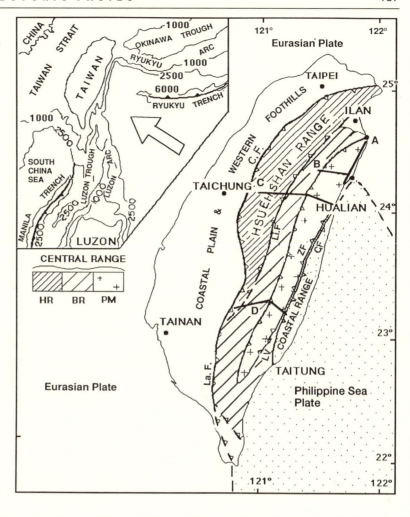

Fig. 14.16. Composite cross section through Tierra del Fuego (after Milnes 1987).

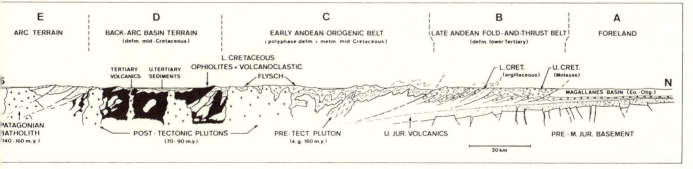

sub-Andean foothills, whereas the foreland basin is underlain by flat-lying sediments (Milnes 1987).

Patagonian Batholith is the raetide, the ophiolite mélange and the mobilized basement are the celtide, and the "Late Andean Fold-and-Thrust Belt" is the alemanide of this late-Mesozoic arc-continent collision.

Central Andean Cordilleras

After talking to Dalziel about the back-arc basin collapse in the Southern Andes, I began to wonder if the same process might also have been operative in the Central Andes. I asked several American and European friends who had worked there if they had ever seen any ophiolites. No, they had not; they had only mapped sediments and volcanics in faulted basins, and the basement is everywhere granitic or metamorphic.

When I was invited by the Argentine Academy of Science to be a consultant for their lake-research project, I accepted on the condition that I be given a tour of the Andes. Arturo Amos, who was my host, very happily agreed; he had mapped for many years in the Cordilleras near Mendoza and he was always happy to show "his rocks" to a foreign visitor.

The Cordilleras of the Central Andes have been divided into three ranges: the *Cordillera Principal*, the *Cordillera Frontal*, and the *Precordillera* (Figure 14.17). We started our excursion from Bariloche and traveled north through the Pampas grassland to Neuquén. My prior perception of the Andes was confirmed. We drove across desert landscape and saw only volcanics and continental sediments in faulted basins; the geology reminded me of the Tertiary geology of the Basin and Range Province of western North America.

We had lunch at Piedra l'Aquilla (Eagle Rock). Amos told me that the red sandstones around the village are Triassic continental deposits.

"What are they doing here?" I asked.

"Oh, they are the 'Andean basement.'"

I was puzzled, because the red beds are not metamorphosed; they are the foreland-basin deposit of a pre-Andean orogenic belt.

We flew from Neuquén to Mendoza, and rented a car to drive toward Aconcagua, the highest peak of the Andes in the Cordillera Principal. We did no climbing, of course, but we did drive beyond Puente del Inca to see a shallow-marine Jurassic and Cretaceous sequence. The passive-margin deposits were laid down in a basin behind a west-facing arc, and they now constitute an easterly-vergent foreland-thrust-belt in an orogen of arc-continent collision. The Cordillera Principal is underlain by an Andean alemanide like that of the sub-Andean foothills of the south. I now began to understand what Amos meant by the "Andean basement": the Triassic red beds at Piedra l'Aquilla were considered "basement" because they were deposited prior to the Jurassic-Cretaceous-Cenozoic Wilsonian cycle of the Andean tectonics.

On the way back to the village of Uspallata, at the southern end of the Valle de Uspallata-Calingasta (Figure 14.17), Amos showed me some more of the pre-Jurassic basement. In addition to the Triassic, the basement rocks are metamorphosed Permo-Carboniferous detrital, volcaniclastic, and volcanic rocks, including the Gondwana tillites. Those rocks were intruded by Mesozoic granites, 350–250 Ma in age, at a time when South America was bounded on the Pacific border by a magmatic arc. It was hot and dry in the desert, but Amos could not suppress his enthusiasm to show me some "strange" rocks in the vicinity of Uspallata. They did not seem strange to me at all, because the pillow lavas and flysch looked just like the Franciscan Mélange of the California Coast Range; they are the pre-Andean celtide.

The next day, we drove north up the valley to the mining town of Calingasta, and I was shown one outcrop after another of ophiolite mélange. Silurian graptolite and Devonian fossils have been collected from the flysch in the mélange. Pillow lavas, gabbros, and serpentine slabs are common in a sheared matrix; they have been variously mapped as "sedimentos marinos Devonico" and "Paleozoico inferior" or "Sedimentitas carbonicas" (see Figures 14.17 and 14.18). Their resemblance to the Franciscan Mélange of the California Coast Range is, however, so obvious that Arturo Amos spontaneously renamed the ophiolite mélange the *franciscan calingasta* of the Cordilleras. Now neither he nor I had any hesitation in considering this mélange a celtide formed during a Paleozoic arc-continent collision. We then proceeded from Calingasta down the River San Juan Gorge eastward to the city of San Juan, and we drove across a foreland-thrust-belt of Paleozoic carbonate strata in the Precordillera.

The pattern is clear: the raetides of the Cordillera Principal are thrust above the celtides of the Cordillera Frontal and the alemanides of the Precordillera. The age of collision is later than Silurian, because Ordovician and Silurian fossils have been found in the flysch sediments of the mélange. The presence of late Devonian fossils in

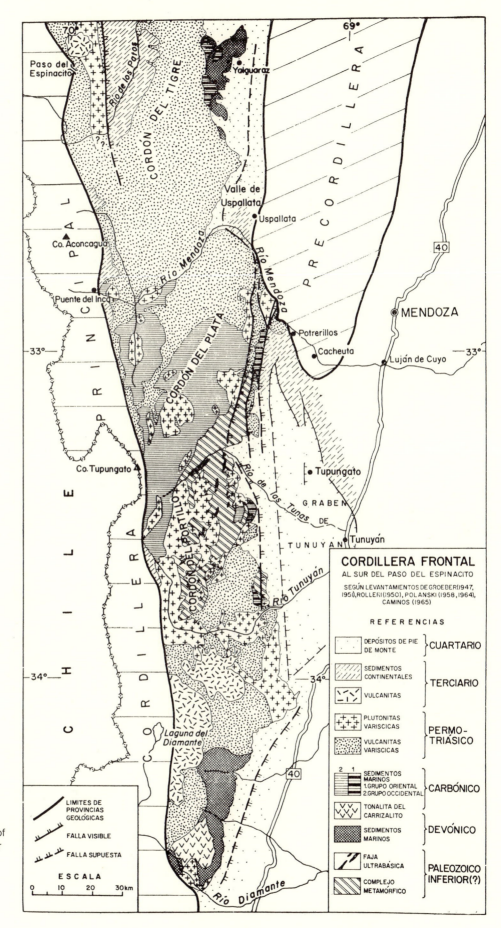

Fig. 14.17. Geological sketch map of the Frontal Cordillera, Argentina (after Caminos 1979). What has been described as pre-Carboniferous basement is, at least in part, an ophiolite mélange.

CORDILLERA FRONTAL
AL SUR DEL PASO DEL ESPINACITO

SEGÚN LEVANTAMIENTOS DE GROEBER (1947, 1951), ROLLERI (1950), POLANSKI (1958, 1964), CAMINOS (1965)

REFERENCIAS

DEPÓSITOS DE PIE DE MONTE	CUARTARIO
SEDIMENTOS CONTINENTALES	TERCIARIO
VULCANITAS	
PLUTONITAS VARISCICAS	PERMO-TRIÁSICO
VULCANITAS VARISCICAS	
SEDIMENTOS MARINOS 1. GRUPO ORIENTAL 2. GRUPO OCCIDENTAL	CARBÓNICO
TONALITA DEL CARRIZALITO	DEVÓNICO
SEDIMENTOS MARINOS	
FAJA ULTRABÁSICA	PALEOZOICO INFERIOR (?)
COMPLEJO METAMÓRFICO	

LIMITES DE PROVINCIAS GEOLÓGICAS

FALLA VISIBLE

FALLA SUPUESTA

ESCALA

0   10   20   30km

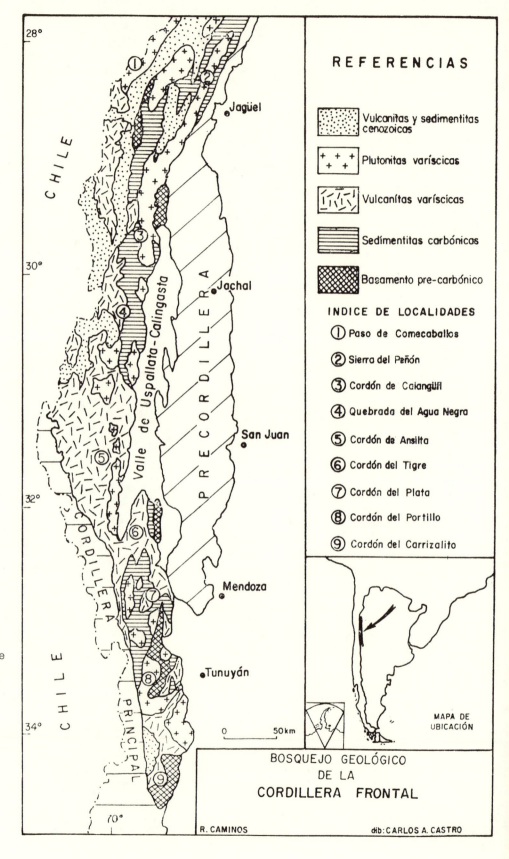

Fig. 14.18. Geological map of the Espinacita Pass district of the Frontal Cordillera, Argentina (after Caminos 1979). The lower Paleozoic ultramafic rocks and metamorphic complex constitute an ophiolite mélange, which is a celtide between a raetide (represented by the pre-Jurassic basement of the Cordillera Principal) and an alemanide (represented by the foreland fold-and-thrust belt of the Precordillera).

REFERENCIAS

Vulcanitas y sedimentitas cenozoicas

+ + + Plutonitas varíscicas

Vulcanitas varíscicas

Sedimentitas carbónicas

Basamento pre-carbónico

INDICE DE LOCALIDADES

① Paso de Comecaballos
② Sierra del Peñón
③ Cordón de Colangüil
④ Quebrada del Agua Negra
⑤ Cordón de Ansilta
⑥ Cordón del Tigre
⑦ Cordón del Plata
⑧ Cordón del Portillo
⑨ Cordón del Carrizalito

MAPA DE UBICACIÓN

BOSQUEJO GEOLÓGICO
DE LA
CORDILLERA FRONTAL

R. CAMINOS   dib: CARLOS A. CASTRO

conglomerates in the Precordillera suggests a collision in the Devonian (Amos, personal communication), when the first molasse deposits, now cropping out in the San Juan River Valley, were laid down in a foreland basin. The compressive deformation in the Cordilleras continued until the Permian, which includes the youngest stratum of the foreland folded belt. Plutonic rocks were intruded into the raetides of the Cordillera Principal. The Wilsonian cycle of this older Cordilleran deformation ended in the Triassic, when normal faulting and volcanism took place on a rifted South American margin. The extensional deformation of the next Wilsonian cycle continued until the Jurassic, when a back-arc basin floored by ocean lithosphere came into existence, and when the passive-margin sediments of the Andean orogen were laid down unconformably on the raetides of the Cordilleran orogenic belt.

15.

A Tectonic Facies Map of China

A neighbor noticed that I had been away to China almost every year since 1976, and asked what was I doing there. Thinking of comparative anatomy, I made an allusion to early European explorers traveling to faraway places. Trees in China were given Chinese names, but trees in the botanical literature were called by their Latin names. One might get more or less an impression, on the basis of Chinese descriptions, what kinds of trees grow in China, but one could not be sure, because there was no technical dictionary to correlate the Chinese and Latin nomenclatures; one had to go there in person to find out. The problem in botany was solved after the adoption of the Linnean nomenclature by Chinese botanists, but the problem in geology is still there. Chinese interpretations of their geology are not easily comprehensible to a foreigner, because different parts of mountains in China were described in terms (such as unstable platform or diwa, Indosinian, etc.) that are incomprehensible to most of us. My job in traveling to China is thus to introduce a set of nomenclature, that of tectonic facies, to designate the various geological units in Chinese mountains in order to facilitate communication between geologists in China and those in other parts of the world.

Huanan Alps and Gunanhai

I left China in 1948 for graduate studies in America, and was not permitted to return until 1976. After the end of the Great Proletarian Cultural Revolution in China, the xenophobia of the Chinese government was cured, and I was asked by the Chinese Ministry of Geology to give a six-week course in 1979 on the theory of plate tectonics and sedimentology. After two weeks consulting in Canton, I received a *Geological Atlas of China* (1973) as a gift, and then proceeded to Chengtu to start the short course. Waiting for a connecting flight at the Changsha airport, I thumbed through the atlas and was immediately impressed by the similarity of the map pattern of Guizhou and Sichuan to that of the Appalachian Valley and Ridge Province.

I remembered those folds in the Yangtze Deformed Belt (Figure 15.1). As a sophomore in National Central University, I went on an excursion to the hills north of Chungking. I was impressed by the nearly vertically dipping strata of a Jurassic orthoquartzite. I asked my instructor about the depth of the fold. He was not sure, but thought that it might go straight down to the Moho. He had no knowledge of disharmonic folding in foreland fold belts.

The sediments of Sichuan, Guizhou, and other provinces of the Yangtze region are mainly carbonate rocks, typical of deposition on a carbonate platform or a passive margin. Recognizing, however, that the "platform" sediments are folded, various expressions such as "unstable platform" or "diwa" (geo-depression) were adopted to describe the thin-skinned deformation of the sedimentary strata (see Wang 1986; Chen 1989). A native, returning after 30 years, recognized the similarity in anatomy of the

ON OPPOSITE PAGE:

Fig. 15.1. Geological sketch map of South China. Yangtze and Huanan are two major tectonic units of South China. The Banxi Mélange and the rigid-basement klippes shown here were the basement of the early Paleozoic Banxi Arc. The rigid basement and the mobilized basement shown here were the basement of the early Paleozoic Cathay Arc. The Huanan microcontinent was formed after a mid-Paleozoic arc-arc collision, and the Huanan passive-margin sequence was the neoautochthonous cover. The Paleozoic rocks under the Nanpanjiang nappes constitute a mélange. The Paleozoic rocks in the Qinzhou Window area (see Figure 14.17) are mainly mobilized basement. The Yangtze Folded Belt is the alemanide of the Yangtze-Huanan collision. The Gunanhai Mélange is the suture between the Huanan and the Donanya microcontinents.

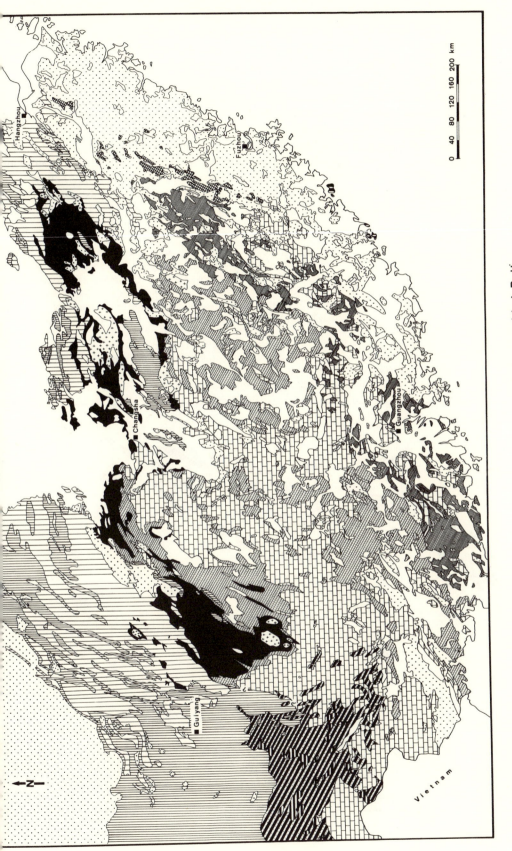

0 40 80 120 160 200 km

Yangtze Folded Belt

Foreland Basin Sequence (Upper Triassic to Lower Cretaceous)
Platform Cover Sequence (Devonian to Middle Triassic)
Passive Margin Sequence (Sinian to Silurian)

Nanpanjiang Nappes

Flysch Nappes (Triassic)
Mélange (Devonian to Triassic)

Granites and Cenozoic

Huanan Nappes

Foreland Basin Sequence (Upper Triassic to Lower Cretaceous)
Passive Margin Sequence (Devonian to Middle Triassic)
Flysch Nappes (Late Precambrian to Silurian)
Rigid Basement Klippes
Mobilized Basement
Rigid Basement
Banxi Mélange (Late Precambrian to Silurian)

Gunanhai Mélange

Mesozoic Metamorphism

Yangtze folded belt to the Appalachian thin-skinned deformation (Hsü 1981). The Yangtze is obviously an alemanide, but where are the celtide and raetide?

An assortment of rocks of diverse origins, including ophiolite, flysch, and volcanic and volcaniclastic rocks, is widely distributed in the Yangtze region, and the mélange is called Banxi (Figure 15.2). These rocks are mainly unfossiliferous, and radiometric dating of the ophiolites and eclogites has yielded ages of 850–900 Ma. The data thus indicate that Precambrian rocks are an important constituent of the Mélange, but the field evidence is unmistakable that the Banxi is superposed tectonically above the Yangtze Deformed Belt, so that the Banxi Precambrian is not the basement of the Sinian and Paleozoic strata of the Yangtze passive margin (Hsü *et al.* 1988). Despite heated debate, there seems to be a consensus that thin-skin overthrusting is a dominant tectonic style of the Yangtze deformation. Rodgers and Hsü (1990, p. 192) agreed:

"Whatever the Banxi is or the different Banxis are, they are *not* basement to the stratigraphic sequence of the Yangtze fold belt. Such basement does not seem to be

Fig. 15.2. Tectonic units of South China. The Banxi Mélange and the associated Precambrian granites (shown by crosses) form an arcuate belt extending from the area east of Guiyang eastward to the vicinity of Hangzhou. These are the basement rocks of an early Paleozoic island-arc, called the Banxi Arc, south of the Yangtze passive margin. The Huanan basement of the southeastern coast underlay another island arc, called the Cathay Arc. Two marginal basins were present, back of the Banxi and Cathay arcs, respectively. The Huanan flysch nappes were the accretionary wedge and the Hunan metamorphic complex was the subducted Cathay basement, deformed while the marginal basin between the Banxi and Cathay arcs was partly consumed during the early Paleozoic, when the two arcs collided in the east to form the Huanan microcontinent. The suture zone of this mid-Paleozoic collision is covered by the neoautochthonous cover. Later, the marginal basin between the Yangtze and Huanan was eliminated, when the two microcontinents collided in the early Mesozoic. The Paleozoic rocks of the Qinzhou and Tianyang areas cropped out in two windows under the Huanan Thrust.

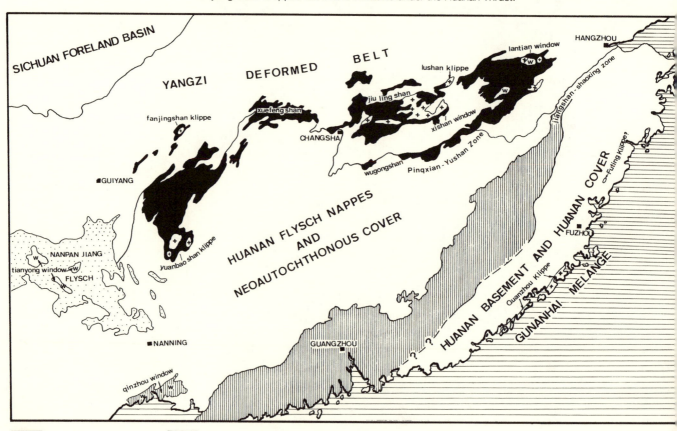

exposed anywhere within the fold belt; it appears farther north and northwest—for example, in the major dome northwest of Yichang where it is exposed in the 'Gorge of the Yangtze,' unconformably beneath the type Sinian strata. . . . The group of rocks . . . that is thrust over the Yangtze sequence . . . contains mélange, ultramafic rocks, and greenschist metamorphism."

That the Yangtze basement is not exposed in the fold belt is understandable. The Yangtze folds are alemanides, comparable to the décollement folds of the Jura Mountains. The Precambrian Banxi Mélange and the associated Precambrian granites are a rigid-basement that has been thrust above the décollement folds. The presence of ophiolites, flysch, and high-pressure metamorphic rocks such as eclogites in the Banxi Mélange has been a source of confusion. Radiometric ages indicate that subduction and high-pressure metamorphism took place during late Precambrian time. In fact, active-margin rocks similar to the Banxi Mélange extended from the Huanan and Yangtze and westward to Qinling and Tianshan. Considering that Serindia (south of Tianshan), Yangtze, and Huanan are island arcs split off from a Siberian craton, the Precambrian components of the Banxi may have been derived from the celtide of a Andean-type continental margin of late Precambrian age (see Wang 1986). The Precambrian rocks were mixed with Paleozoic flysch and carbonate rocks during the Mesozoic deformation after the collision of the Yangtze and Huanan microcontinents (Hsü *et al.* 1990).

Huanan was a late Paleozoic microcontinent, which owed its origin to an arc-arc collision during the mid-Paleozoic. The Precambrian Banxi Mélange and the Precambrian granites crop out in an arcuate belt of mountains (Yuanbaoshan, Fanjingshan, Xuefengshan, Jiulingshan, Lushan) extending from the area east of Guiyang to the vicinity of Hangzhou (Figures 15.1 and 15.2). Those basement rocks are overlain unconformably by a shallow marine sequence, ranging from Devonian to Triassic in age. Also present under this sedimentary cover are Late Precambrian and lower Paleozoic broken formations of the Huanan flysch nappes (Figure 15.2). Farther southeast are the metamorphic rocks of the mobilized and rigid basements, which yielded Paleozoic and Precambrian radiometric ages, respectively.

Assuming that the lower Paleozoic flysch was deposited in a back-arc basin, the Precambrian rocks north and south of the marginal sea should have constituted the basement of two island arcs—the Banxi Arc to the north of the flysch basin, and the Cathay Arc to the south. The subduction of the flysch to form an accretionary complex and the elimination of the flysch basin led to an arc-arc collision. The transgressive unconformity under the neoautochthonous sequence in southeast China suggests a pre-Devonian date for this collision, when the Huanan microcontinent was formed as a result of the suturing of the two arcs.

The Yangtze and Huanan microcontinents were separated by the back-arc basin north of the Banxi Arc. Late Paleozoic pillow lavas and deep-sea sediments have been found in the Tianyan Window. Also, the deep marine sedimentation of lower Paleozoic flysch and upper Paleozoic radiolarite continued until Early Permian time in the Qinzhou Window area. The Nanpanjiang Flysch is mainly Triassic, deposited in a foredeep just to the north of the advancing Huanan nappes. The suturing of the Huanan and Yangtze took place in the Late Triassic or Early Jurassic (Dobson 1993; Gu Tongming, personal communication, 1992). The Banxi rocks had, however, become a rigid basement and had assumed the role of a "bulldozer" during the deformation of the Yangtze fold belt; this Precambrian celtide was a raetide during the Mesozoic collision.

The sedimentary cover of the Huanan microcontinent forms a foreland thrust belt, whereas the Jurassic and Cretaceous volcanic and volcaniclastic rocks of coastal southeastern China are the deposits of a foreland basin (Figure 15.2). They are the alemanides of a Mesozoic collision between Huanan and a microcontinent farther to the southeast (Hsü *et al.* 1990). Where are the celtide and raetide of this orogenic belt?

The Upper Paleozoic flysch in coastal Fujian was subjected to greenschist metamorphism from 180 to 90 Ma (Fujian Bureau of Geology 1985, p. 487). Serpentinites have been found perched on top of Jurassic volcanics south of Fuzhou (*Geological Atlas of China*, 1973; Hsü *et al*. 1990). A metamorphic complex, including quartz-biotite schist, quartz-feldspar gneiss, and amphibolite and yielding radiometric ages ranging from 90 to 120 Ma, is present near Quanzhou in coastal Fujian. These are the celtides of a late Mesozoic collision of the Huanan microcontinent with another continental fragment, and the collision took place in the Early Cretaceous after the elimination of the Gunanhai Ocean between the two microcontinents (Hsü *et al*. 1990).

The raetide of this orogenic belt is found on southern Hainan Island. Studies of the lower Paleozoic faunas of southern Hainan indicated their close affinity to those of Australia and Antarctica (Lu 1981). The discovery of Permian glacial marine sediments on Hainan Island implies that it was still part of Gondwanaland then (Yu *et al*. 1988). Hsü *et al*. (1990) suggested the name Dongnanya, meaning Southeast Asia in Chinese, for this Mesozoic microcontinent, which may have been an eastern extension of what has been called the Sibumasu microcontinent (Siam-Burma-Malaysia-Sumatra). Dongnanya was separated from western Australia by seafloor-spreading during the Triassic and Jurassic. The northward march of Dongnanya led to the consumption of the Gunanhai Ocean and its eventual collision with Huanan. The presence of an ophiolite in central Hainan Island marks this suture, which dives under the South China Sea but reappears on the coast of Fujian as discussed previously.

The Early Cretaceous Huanan/Dongnanya collision led to the deposition of molasse sediments in foreland basins on Hainan Island and in coastal Fujian. The sudden appearance of the Gondwana faunas and floras in the Early Cretaceous continental strata of southeastern China (Chen 1991; Yu Ziye, written communication, Feb. 1992) is confirming evidence for the Cretaceous suturing of a Gondwana fragment (Dongnanya) onto Huanan.

The Qinling Mountains traverse the Gansu, Shaanxi, and Henan Province of China and are the central segment of an E-W trending mountain chain, which extends westward to the Kunlun and eastward to the Dabi mountains (Figure 15.3). Within the Qinling orogenic belt, the following units have been recognized (Figure 15.4):

(1) Precambrian basement and its sedimentary cover
(2) Qinling metamorphic complex
(3) Flysch nappes
(4) Foreland folded belt

Qinling has been formed by the collision of the North and South China blocks. The Archaean basement and its Proterozoic and lower Paleozoic sediments are the raetide, whereas the flysch nappes and foreland thrust belt are the alemanides. The role of the metamorphic complexes in Qinling is, however, not yet ascertained. The northern complex that extends from Tianshui to Shangnan is the celtide of the Paleozoic/Triassic Qinling deformation. The southern complex in the Hanzhong and the Wudang Mountain regions is underlain by mélanges of Precambrian age. I once thought that they were ophiolite nappes thrust on top of the Qinling foreland thrust belt (Hsü *et al*. 1987). Its late Precambrian age suggests, however, correlation with ophiolites and metamorphic rocks formed in late Precambrian time along an Andean-type margin that extended from Tianshan to Huanan. Accordingly, those rocks are foreland raetides during Qinling deformation; they are the parautochthonous massifs of the Yangtze microcontinent.

The age of Qinling deformation has been variously considered to be "Caledonian," "Hercynian," or "Indosinian," because the radiometrically determined dates of granitic and metamorphic rocks range from 500 Ma to less than 100 Ma (e.g., Chen 1975;

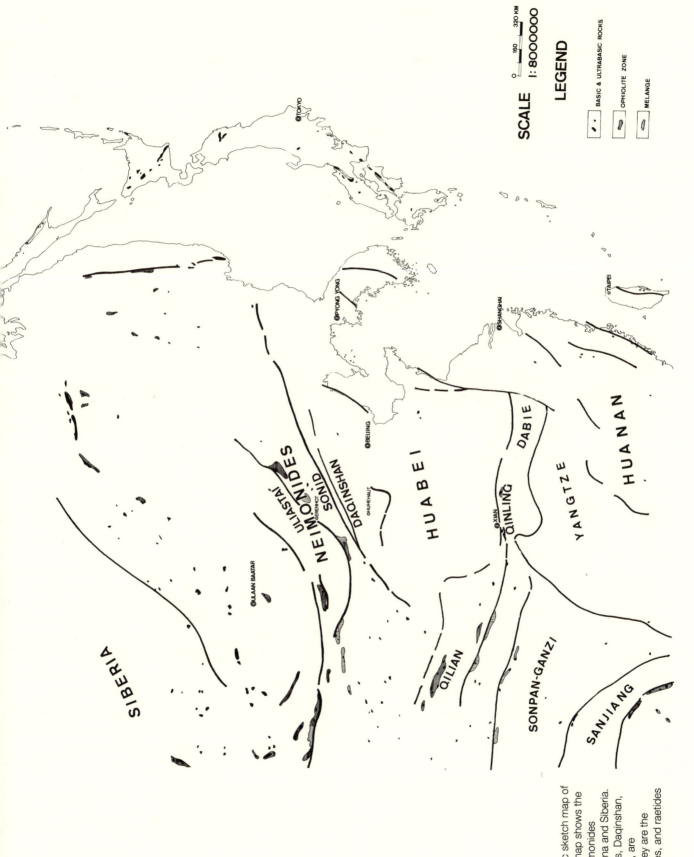

Fig. 15.3. Tectonic sketch map of North China. The map shows the position of the Neimonides between North China and Siberia. Three tectonic units, Daqinshan, Sonid, and Uliastai, are recognized, and they are the alemanides, celtides, and raetides of the Neimonides.

SCALE 1:8000000

0   160   320 KM

LEGEND

BASIC & ULTRABASIC ROCKS

OPHIOLITE ZONE

MELANGE

SIBERIA

ULAAN BAATAR

ULIASTAI

NEIMONIDES

SONID

ERENHOT

DAQINSHAN

HUHEHAUT

BEIJING

PYONG YONG

TOKYO

HUABEI

QILIAN

XIAN

QINLING

DABIE

YANGTZE

SONPAN-GANZI

SANJIANG

HUANAN

SHANGHAI

TAIPEI

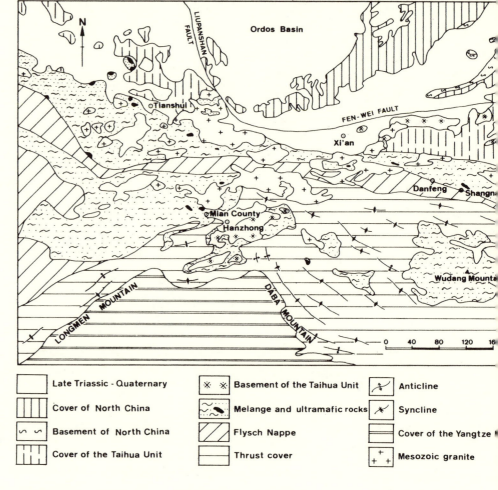

Fig. 15.4. Geological sketch map of Qinling. The basement and sedimentary cover (Taihua) of North China are the raetide, the northern belt of the mélange (Paleozoic) is the celtide, and the flysch nappes, the cover thrust, and the southern belt of the mélange (parautochthonous massifs of Precambrian basement) are the alemanides of the Qinling Mountains.

	Late Triassic - Quaternary		Basement of the Taihua Unit		Anticline
	Cover of North China		Melange and ultramafic rocks		Syncline
	Basement of North China		Flysch Nappe		Cover of the Yangtze
	Cover of the Taihua Unit		Thrust cover		Mesozoic granite

Zhang 1980; Mattauer *et al.* 1985; Li *et al.* 1989). The age of intercontinental collision is Triassic, as indicated by the youngest age of the passive-margin strata in the Qinling Fold Belt. The marine Triassic is overlain by red beds of Late Triassic, Jurassic, and Cretaceous age, deposited in a Qinling Foreland Basin. The current interpretations of Devonian, or polycyclic, collisions are incorrect, having been the result of misguided faith in the episodic nature of orogeny. Adopting the tectonic-facies model, the radiometric dates indicate continuous igneous activity and metamorphism from mid-Paleozoic Eo-Qinling subduction to Mesozoic Neo-Qinling post-collisional deformation (Hsü *et al.* 1987).

The Neimonides

The early successes in delineating the tectonic facies of South China and of the Qinling Mountains led Sun Shu, director of the Geological Institute, Academia Sinica, to launch a Sino-Swiss Cooperative Project to prepare a tectonic-facies map of China. My Chinese colleagues from the Academy and I have been driving across all 30 provinces and autochthonous regions of China to identify alemanides, celtides, and raetides. We have now almost completed the field work, and hope to send the first draft of the map to the printer in 1994, the year of my retirement.

Three tectonic units have been recognized in the Neimonides, the mountains of Inner Mongolia: the Daqinshan, the Sonid, and the Uliastai units.

The Daqinshan structures are foreland folds and thrusts; they belong to the alemanide facies. This east-west trending belt of thin-skinned deformation extends for several hundred kilometers near the southern border of Inner Mongolia (Figure 15.3). The shallow-marine Daqinshan strata were deposited on the northern margin of North China during the late Precambrian and early Paleozoic. They form southerly vergent recumbent folds, such as those of the West Hills near Beijing, and those nappes have been thrust southward onto folded strata of North China, which was a Paleozoic micro-continent.

The Sonid zone (Figure 15.3) consists mainly of igneous and metamorphic rocks. Those ophiolite mélanges are the celtides of the Neimonides, but they crop out in two belts north and south of a terrane underlain by continental crust. Precambrian basement and mid-Paleozoic granitoids crop out east of Erenhot (Figure 15.3). This Paleozoic "intra-oceanic swell" was a remnant arc, called the Sonidzuoqi Arc (Hsü *et al.* 1991), which had a paleogeographic position similar to that of the Brianconnais of the Alps.

The Uliastai unit is the raetide of the Neimonides, and this belt extends from west to east along the northern border of Inner Mongolia (Figure 15.3). Overlying Cambrian and lower Ordovician shallow-marine strata are volcanic and volcanoclastic strata ranging from Middle Ordovician to Permian in age. Also present are Devonian and Carboniferous granites. The onset of igneous activity indicates that the Uliastai margin changed during the Ordovician from a passive margin of carbonate-quartzite sedimentation to an active margin (Hsü *et al.* 1991). The continued deformation of the Neimonides led to the collision of Uliastai and North China. The collision, as indicated by the paleomagnetic studies now in progress (J. Dobson, personal communication), took place during the Permian Period. Post-collisional orogenic and igneous activity continued well into the late Mesozoic, to the time of the so-called "Yengshanian Orogeny" (see Cheng 1986).

Uliastai magmatism was active during the middle and late Paleozoic time. Uliastai is separated from North China by the mélanges of the Sonid zone, and is bounded on the north by the Altaides. The Mongolian Altaides include ophiolites (Ordovician to Carboniferous), radiolarian cherts (Silurian/Devonian), marine clastics (Ordovician to Carboniferous), volcanic and volcanoclastic clastics (Silurian to Permian), and reef limestones (mainly Devonian). Sengör *et al.* (1993) considered those broken formations or exotic slabs in tectonic mélanges that constituted the accretionary wedge on the southern periphery of the Paleozoic Angaran Craton. I have recognized, however, evidence for the existence of several late Precambrian and Paleozoic remnant arcs in Mongolia. Instead of one giant accretionary wedge comparable to the Franciscan Mélange of western North America, the Altaides of Mongolia have been formed by arc-arc collisions at or near the active southern margin of the Paleozoic Angara, before its collision with North China.

According to this scenario, Uliastai was a volcanic arc on the northern margin of North China, facing the Paleozoic Altaide Ocean. The mélanges of the Sonid zone are the suture zones of late Paleozoic arc (Uliastai)–arc (Sonidzuoqi) and early Paleozoic arc (Sonidzuoqi)–continent (North China) collisions.

Tianshan and Kunlun Mountains

Turning our attention to the Chinese Northwest after the completion of our studies in eastern China, I was impressed by the scarcity of fold belts west of Lanchow, the geographical center of the country. The only well-defined alemanides are the Kelpin Belt in the southern foothills of Tienshan, where the sedimentary cover of Serindia is folded by thin-skinned deformation. Elsewhere the mountains are underlain by mélanges, metamorphic complexes, and batholithic intrusions (Figure 15.5).

The geology of northwest China is best interpreted with reference to a West Pacific model (see postscript). Slices of continental crust were split off from a craton (Angara) during the late Precambrian to late Paleozoic to form numerous island arcs and remnant arcs. They were, from north to south, the Middle Tianshan Arc, the Serindia Arc, the Middle Tarim Arc, the North Kunlun Arc, and the Middle Kunlun Arc. Between the arcs were back-arc basins: the Junggar was a back-arc basin north of the

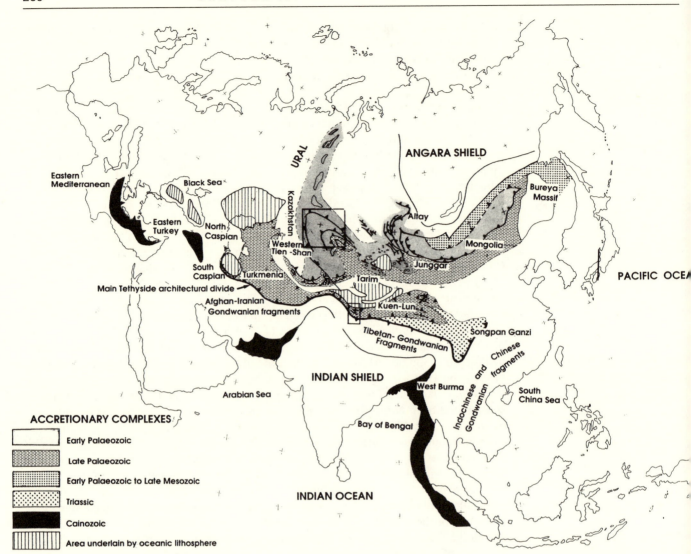

Fig. 15.5. Tectonic sketch map of Asia showing the distribution of accretionary complexes (after Sengör and Okurogullari 1991).

Middle Tianshan Arc, and the Tarim Basin between the Serindia and North Kunlun arcs was a composite of two back-arc depressions that were separated by the Middle Tarim Relic Arc. The arcs are identified by their Precambrian basement, by the batholithic intrusions that are interpreted as the root of arc volcanism, and by their shallow marine sedimentary cover. The back-arc basins, having been eliminated by the process of back-arc basin collapse, are evidenced by ophiolite mélanges of the Tianshan and Kunlun mountains. Only the Junggar and Tarim basins have survived the Paleozoic and Mesozoic compressional deformations; those basins have been filled up by Mesozoic and Cenozoic terrestrial sediments, and they are thus considered relic back-arc basins (Hsü 1988; Aplonov *et al.* 1992).

The postulate that the Junggar Basin is underlain by ocean crust leads to the prediction that the mountains surrounding the basin should resemble Circum-Pacific mountains, and I had a chance in 1988 to verify this prediction. The Karamay, the Altai, the Northern Tianshan, and the Beishan mountains are all underlain by mélanges formed during Paleozoic subductions (Hsü *et al.* 1990; Hsü *et al.*, 1992). The genesis of the mélange is evidence of the ocean-continent interactions: to the north, the Junggar ocean lithosphere was thrust under the Angaran Craton, and to the south, the ocean lithosphere was thrust under the Middle Tianshan Arc (Figures 15.6 and 15.7). The

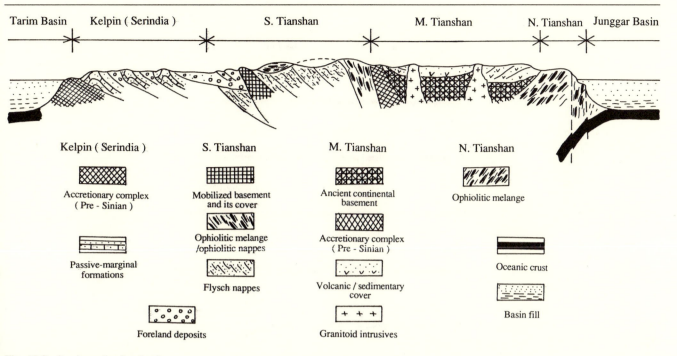

Tarim Basin   Kelpin (Serindia)   S. Tianshan   M. Tianshan   N. Tianshan Junggar Basin

Kelpin (Serindia)

Accretionary complex
(Pre - Sinian)

Passive-marginal
formations

S. Tianshan

Mobilized basement
and its cover

Ophiolitic melange
/ophiolitic nappes

Flysch nappes

Foreland deposits

M. Tianshan

Ancient continental
basement

Accretionary complex
(Pre - Sinian)

Volcanic / sedimentary
cover

Granitoid intrusives

N. Tianshan

Ophiolitic melange

Oceanic crust

Basin fill

Fig. 15.6. A schematic structural cross section of Tianshan. The Middle Tianshan is the raetide. The North Tianshan Mélange and South Tianshan Mélange are the celtides, whereas the Junggar Basin and the Kelpin Fold Belt/Tarim Basin are the alemanides of Tianshan.

present Junggar Basin, triangular in shape, is the remnant of this process of back-arc collapse.

I have made four traverses across Tianshan, in 1983, 1988, 1990, and 1992, and was able to identify two island arcs and an intervening back-arc basin in the mountains. The Middle Tianshan Arc was underlain by a Precambrian basement and its Sinian and lower Paleozoic sedimentary cover. The marginal sea, or the Tianshan Ocean, south of the Middle Tianshan Arc, was a site of radiolarite and flysch sedimentation during early Paleozoic time. Farther south was the Serindia Arc. The subduction of the lithosphere under the Tianshan Ocean started in the Late Ordovician or Silurian, and its consumption resulted in the collision of the Middle Tianshan and the Serindia arcs. The South Tianshan Mélange is now found in the suture zone of this collision, and the mélange consists of slabs of ultramafic and mafic plutonic rocks, Ordovician pillow lavas, lower Paleozoic radiolarites, and Silurian flysch, as well as shallow marine limestones and siliciclastic rocks detached from adjacent arcs. The mélange is intruded by mid-Paleozoic batholiths, which may represent the root of arc volcanism.

The alemanide of the Middle Tianshan and Serindia collision is the foreland thrust belt of the Kelpin region. A passive-margin sequence of shallow marine quartzites, limestones, and shales, ranging in age from late Sinian to Silurian, is deformed by décollement folding and thrusting. Devonian and Carboniferous foreland-basin deposits overlie unconformably the lower Paleozoic rocks of Tianshan.

The geology of the Kunlun Mountains was the dissertation theme of my late student Yao Yungyun, and he made a tectonic map of the Kunlun Mountains (Figure 15.8) shortly before he died. I was finally able, under very difficult circumstances, to make two geological traverses in 1992 across a western segment of this majestic range, and was thus able to verify in part Yao's interpretations.

Two island arcs were identified in the Kunlun Mountains by their Precambrian basement and Paleozoic sedimentary cover. The North Kunlun Arc is identified by its pre-Sinian basement and by its Sinian and Paleozoic sedimentary cover, called the Tekiliktag unit by Yao (Figure 15.8). Fossiliferous Devonian, Carboniferous, and Permian sediments were deposited in shallow marine environments and interbedded with arc-volcanics. While traveling south across Kunlun from Yecheng (Figure 15.8), I

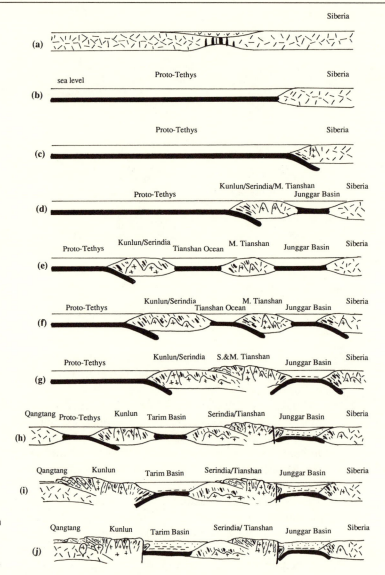

Fig. 15.7. Geological evolution of Xinjiang: (a) and (b) Proterozoic; (c) Late Proterozoic; (d) Cambrian; (e) Early Ordovician; (f) Middle Ordovician to Silurian; (g) Devonian to Early Permian; (h) Late Permian to Triassic; (i) Late Triassic to Jurassic; (g) Cretaceous/Cenozoic.

observed in river gorges north of Akarz spectacular disharmonic folds of those upper Paleozoic strata. The Middle Kunlun Arc is recognized by its Precambrian basement and by the numerous Paleozoic batholiths. The granites in the vicinity of Kudi were mapped by Yao as the Kongur Intrusive. A belt of ophiolite mélange, the Kudi Mélange, is present between the two arcs. Pillow lavas, with pillows 2 or 3 meters across, have been observed in the valley north of Kudi, whereas the mélange has been penetratively sheared to form the greenschists of the range between Kudi and Akarz. Farther west, along the China-Pakistan Highway, serpentinite slabs are common in the Kudi Mélange. Triassic fossils, according to geologists of the Tarim Office of the Chinese Petroleum Company, have been found in flysch sediments in the collision zone, suggesting a Late Triassic age for the arc-arc collision. The tectonic vergence is directed northward: the Middle Kunlun Arc is the raetide, the mélange the celtide, and the Tekiliktag unit of the North Kunlun Arc the alemanide.

South of the Middle Tianshan Arc is the Main Paleotethys Suture. The colored mélange is well exposed in the valley of Mazar (Figure 15.8). I was permitted to travel 80 km upstream and eastward from Mazar toward the 30-Li Barracks, but not farther. The mélange is very similar to that in the Main Paleotethys Suture of the Three Rivers

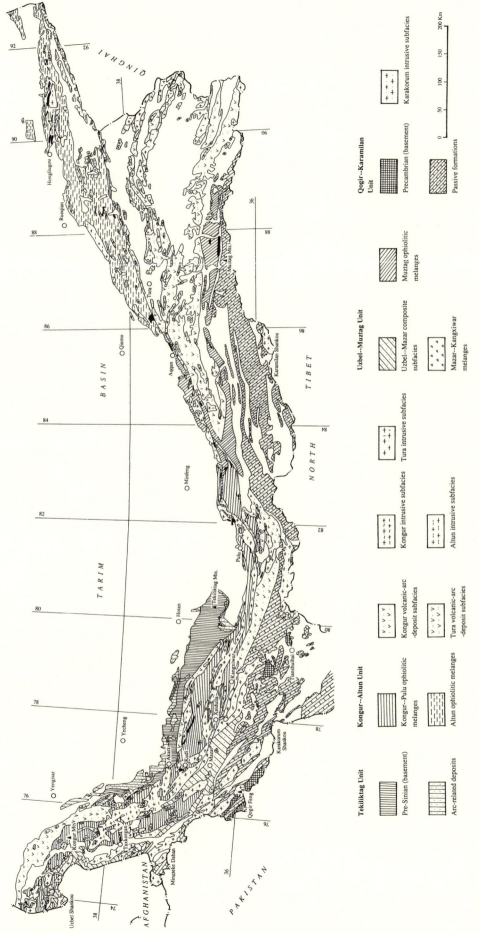

Fig. 15.8. Tectonic sketch map of the Kunlun Mountains.

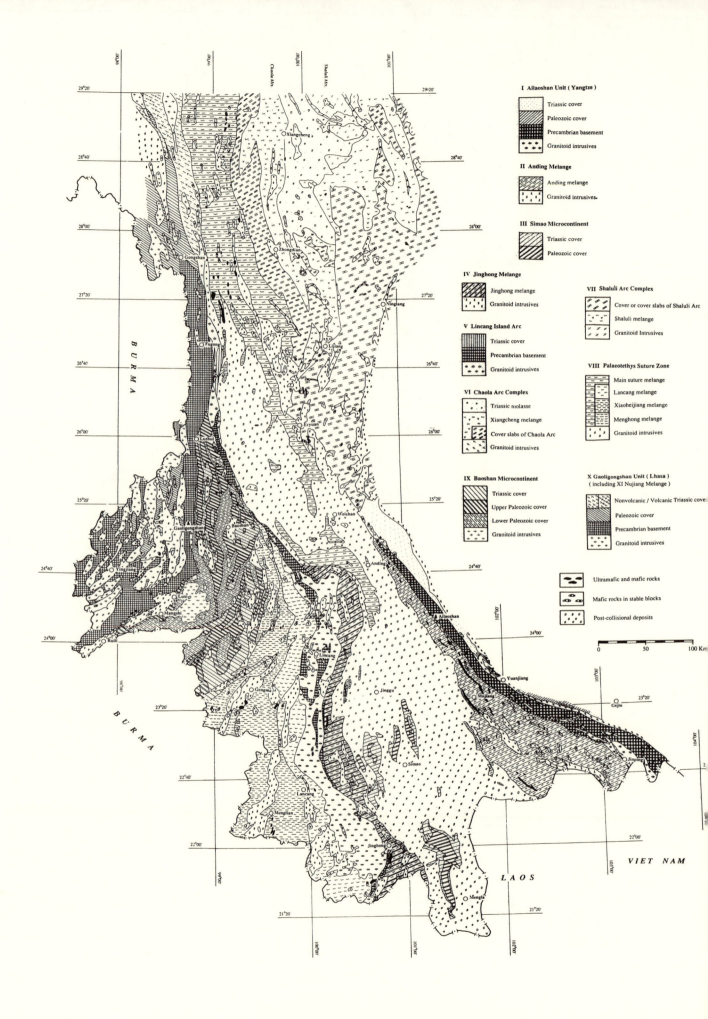

2 9 9 0 4 2 6 9 4 0 0 5 9 9 0 6 9 9 0 8 8 0 4 0 6 9 9 0 9

region. Slabs of ophiolites and radiolarite are enclosed in a schistose matrix. Silurian fossils have been found in the mélange, so that the Uzbel-Mazar Mélange of Figure 15.8 was mapped as a Silurian formation in the *Geological Atlas of China* (1973).

The Suturing of Eurasia and Gondwana

The geology of southwest China and Tibet is a record of the suturing of Gondwana fragments onto Cathaysia. Between the Yangtze Province in Sichuan and the Lhasa block in Tibet are several belts of arc-arc, arc-continent, and continent-continent collisions. We have made several geological traverses across the suture zone, and we are still trying to sort out the alemanides from the celtides and the raetides in the Paleotethys Zone of the Sanjiang (Three Rivers) region. A preliminary interpretation of the southern traverses in Yunnan and Sichuan is shown in Figure 15.9.

In southwestern Yunnan, the Mesozoic continental margin of Eurasia is represented by the Ailaoshan unit (I) of Precambrian basement and its sedimentary cover of the Yangtze facies. The Anding Mélange (II) marks the suture zone between Ailaoshan and Simao (III), and the Jinghong (IV) between Simao and Lincang (V). The Anding celtide is thrust eastward under the Ailaoshan raetide, and the Jinghong celtide is thrust westward under the Lincang raetide.

The Lincang is underlain mainly by Precambrian basement and by Paleozoic intrusive, and it was the outermost volcanic arc of the Mesozoic Eurasian plate. The existence of an east-facing outer arc of Gondwanaland is evidenced by the granitoids of Baoshan (IX), which are intrusive into lower Paleozoic strata of the microcontinent. Between the two outer arcs are the remnants of the great Paleotethys Ocean; they are present in the form of Menghong and Xiaoheijiang mélanges (VIII).

The sedimentary cover of Baoshan has been deformed to form an alemanide under the Gaolingongshan raetide (X), which was a volcanic arc on the outer fringe of the Lhasa block.

In western Sichuan the existence of island arcs and back-arc basins near the western margin of the Eurasian plate is evidenced by the rocks of the Chaola (VI) and Shaluli (VII) arc-complexes and by Paleotethyan mélanges (VIII). I visited western Sichuan again during the Second Sino-Swiss Expedition to Tibet in 1993 and recognized that the Chaola Arc was not the outermost volcanic arc of Eurasia. In the Jinshajian area north of 26°30'N (near Fugong) the Lhasa block (X) is faulted against a Paleotethys mélange (VIII); the Lincang Arc seems to be cut off by strike-slip faulting (Figure 15.9).

Our second expedition to Tibet verified the paleogeographical interpretation of a Permo-Triassic archipelago near the southwestern plate-margin of Eurasia. The Indosinides of Yunnan, Sichuan, eastern Tibet, and southern Qinghai were deformed by back-arc basin collapses long before the intercontinental collision of Cathaysia and Gondwanaland. We shall discuss the details of our 1993 observations in our explanatory notes to the Tectonic Facies Map of China (1:4,000,000) to be published in 1996.

On opposite page:

Fig. 15.9. Tectonic sketch map of the Three Rivers region, Yunnan Province.

16.

Theoretical Geology

Theoretical physics and theoretical chemistry (alias physical chemistry) are courses taken by science students at universities, but theoretical geology is not. I asked why when I was a young student, and my geology professor told me that there can be no theoretical geology. Theoretical physics and theoretical chemistry teach physical laws such as Newton's three laws of motion, or the first, second, and third laws of thermodynamics. The so-called "laws" in geology, the laws of superposition, of lateral continuity, and of paleontological dating, are merely statements of common sense; they are not generalizations deduced from observations of diverse natural phenomena. Not only can we not make laws in geology, we *should* not, he emphasized. Conclusions in geology must depend upon observations, field work, laboratory measurements, or experimental studies; all other endeavors are just idle speculation.

Codes of Stratigraphical Nomenclature

There may have been no physical laws in geology, but there are plenty of legal codes. Carried almost to judicial perfection is the International Code of Stratigraphical Nomenclature. There are many international commissions, subcommissions, and working groups to codify stratigraphical nomenclatures and to settle boundary disputes. I agree that codification is necessary—otherwise chaos prevails. In order that the words formation, group, series, stage, etc. do not mean different things to different geologists, one has to assign precise geological definitions to those words. Precision in language is a prerequisite in science, but precisely aimed shots can be inaccurate and thus miss the target altogether, as I have learned from personal experience.

The area where I did my dissertation research is underlain by metamorphic rocks. Granulite facies rocks are cut by low-angle shear zones, where they are mylonitized and converted into amphibolite-facies rocks (see Chapter 14). An effective way to portray this tectonic evolution is to delineate different kinds of shear zones on a map. I was able to divide the basement complex of the Cucamonga Canyon into three map units: the granulites in the Aurela Ridge Zone are cut by a few shear zones spaced far apart, the granulites in the Cucamonga Zone have been largely converted into mylonites in densely spaced shear zones, and the granulites in the West Ridge Zone are now sheared amphibolites. One day while I was in my office coloring my map, a professor of stratigraphy dropped by and noticed the legend of my map units.

"You cannot do that," he told me.

"Why not?"

"Zones are biostratigraphical units. Mappable units are lithostratigraphical; you are mapping formations!"

"I am not mapping formations. I cannot call them formations, because they are metamorphosed strata of different stratigraphy. The rocks are the same in all these zones, but they have been altered to different degrees in various shear zones. Perhaps I should call them series."

"No, you cannot do that either. Series is a term for time-stratigraphical units."

"Just the same, I will not call them formations, and I can decide. It is my thesis, after all."

"No, you cannot, not if you want to have your thesis approved by the faculty!"

The professor of stratigraphy was not to be deterred by the protests of a student; the code of stratigraphical nomenclature had to be upheld. He went to see Ken Watson, and I was called in. The three of us finally hammered out a compromise, after much acrimony. Those units cannot be zones, cannot be series, are not formations; Watson the mediator suggested the term *group*. This is the origin of the names Aurela Ridge

Group, Cucamonga Group, West Ridge Group, etc., which have been cataloged in the lexicon of North American stratigraphy. They may have been appropriate names, but they are useless or even misleading. None, as far as I know, have ever been recognized elsewhere. Nowadays, when the rocks of the Cucamonga Canyon are described, they are referred to by their lithology: granulites, mylonites, amphibolites, etc. (see May and Walker 1989).

Misuse of a Great Methodology

This summer I joined a geological excursion to Shandong Peninsula in China. Forty years have elapsed and a revolution in earth science has taken place since I was admonished by my professor of stratigraphy, and field geologists are still making geological maps as William Smith did two hundred years ago. But William Smith did not have to work in a metamorphic terrane. The Shandong Peninsula is underlain by metamorphic rocks. I was given a "stratigraphical columnar section" as we were heading out to the field, and I counted more than 20 formations of granulites, amphibolites, and schists, portrayed in a stratigraphical superposition on the basis of the law of superposition: gneisses that seem to "lie" above schists are considered younger, although all rocks are separated by steeply dipping surfaces of penetrative shear. The formations turned out to be useless names, because our field trip leader was not able to tell us, whenever we stopped at an outcrop, which formation we were looking at. His excuse was that those formations have been recognized only in a type section; they cannot be identified elsewhere.

So I was told. But if they are not mappable, they are, by definition, not formations. Yet hard-rock geologists in China, like the young student at UCLA, are being forced by their superiors to adopt a stratigraphical code that is designed for stratified rocks. Units of metamorphic rocks have to be called formations or groups, regardless of whether they are mappable or not. They have to be assigned an age of deposition, and they have to have a sedimentary thickness. A geological report that does not fulfill the requirements is not acceptable.

The application of Smithian stratigraphy to interpret the Shandong geology has led to endless confusion. After the recent discovery of fossils, there occurred the anomaly that Precambrian and Paleozoic formations are "interbedded." The radiometric dates of those rocks range from 2 Ga to 100 Ma. One and the same formation may include such strange bedfellows as greenschist and eclogite. This contradiction encourages proliferation of the geological literature, because everyone can have different ideas concerning the age of sedimentation, the age of metamorphism, and the deformational history, and they can invent different names to portray the geological history of the region. A visitor from Europe with a distant view could, however, plainly see that those 20 "formations" were one big mélange in the suture zone of collision between North China and the Yangtze.

If our understanding of the geological evolution of an orogenic belt depends upon the recognition of tectonic facies, why should the big bosses continue to dictate that the rocks have to be divided into "formations"? The answer is not difficult to find: incompetent geologists do not know what to do when they are placed in the midst of a mélange terrane; they have to do the wrong things because they do not know what else to do. Then, after being carried to the top of their incompetence, the younger generation is required to repeat their mistakes. The bureaucrats who formulated the guideline for field mapping forgot that the wonderful invention by William Smith cannot be used to map terranes of non-Smithian stratigraphy.

The Principle of Mélange

The word *mélange* was coined by Greenly in 1919, but few of those in my generation ever heard of the term, because our teachers never knew about it. Rock bodies of known mixed origin were recognized, but the American Commission on Stratigraphic

Nomenclature (1961) pontificated: "If a mass of rock is composed of diverse types of any class or classes or is characterized by highly complicated structure, the word 'complex' may be used as part of the formal name instead of a lithologic or rank term."

By recognizing "complex" as a lithostratigraphic unit, the Commission swept the dirt under the carpet. A "complex" is a lithostratigraphic unit like a formation, except it is composed of more diverse types or characterized by more highly complicated structures. Like a formation, the rocks in a complex have an age and a paleogeographic site of deposition. The rocks of the Franciscan Complex are thus a rock-stratigraphic unit of Upper Jurassic to Cretaceous age, deposited at a site west of the Great Valley Sequence. The tectonic contact between the two rock-stratigraphic units is the so-called Great Valley Thrust (Bailey *et al.* 1964).

I took a group of 90 prominent geologists to the California Coast Range in 1969, and they were surprised to find that the Great Valley Thrust is in fact the upper limit of penetrative shearing of a mélange called Franciscan. While we were there looking at those broken and sheared rocks, everyone agreed that the three laws of Smithian stratigraphy are irrelevant in a mélange terrane. Yet many of them quickly forgot, after they went back home, the distinction between a rock-stratigraphic unit called complex and a tectonically sheared unit called mélange. The Franciscan rocks are still called Franciscan Complex by many who want (or are requested by their editor) to adhere to the International Code of Stratigraphical Nomenclature.

When I first came to Switzerland in 1967, I had difficulty convincing my colleagues that the difference between mélanges and olistostromes is more than semantic. Olistostromes are sedimentary formations and thus rock-stratigraphic units, but mélanges are not. In treating the Penninic ophiolite mélange as an olistostrome deposited on a continental crust, Alpine geologists could hold on to the notion of a Platta nappe, and this invention served for more than a decade to deter the invasion of the theory of plate tectonics (see Chapter 9).

I tried to call to the attention of my colleagues the need to invoke the five principles of mélange, and not the three laws of stratigraphy, when mapping and interpreting the geology of mélange terranes, but the manuscript was accepted only after I agreed to publish the paper in the Geological Note section for trivia (Hsü 1968). It is trivial to point out that layered rocks in a mélange are separated by shear surfaces, whereas layered rocks in a formation are separated by sedimentary surfaces; we all know that. A stratum separated by a sedimentary surface from an underlying stratum is younger than the latter; we all know that too. However, when mélanges are mapped as formations, a reader of the map has no way of knowing that the strata within the "formation" are separated by shear surfaces, and that the law of superposition is not applicable. The reader has no way of knowing that such a "formation" cannot be given an age of sedimentation on the basis of fossil occurrences. Nobody argues that the unfossiliferous flysch or ophiolite may have an age quite distinct from that indicated by a fossil coral found in a block of limestone, but we do not worry about that if the whole mélange is mapped as a formation. Geologists in China do not think twice before they assign an age of sedimentation to a mélange on the basis of fossils in exotic blocks. This practice led to the anomaly that the Main Paleotethys Mélange is a Silurian formation in the Kunlun Mountains of Xinjiang, Permian in Qinghai, and Triassic in Sichuan and Yunnan. The unfossiliferous mélange terranes of China have become Middle Proterozoic formations, because those slightly metamorphosed rocks look younger than Archaean gneisses, but older than Late Proterozoic Sinian sediments. Once rocks have been designated Middle Proterozoic, there is a tendency for the error to be perpetrated. If fossils should now be found in a mélange, the fossiliferous slabs would be excluded and given the name of a Paleozoic or Mesozoic formation, so that the unfossiliferous part of the mélange could retain its false identity.

Young Chinese geologists working with me are well aware of the problem. Coming to Switzerland, they found that Swiss geologists are, for better or for worse, not straitjacketed by the American Commission on Stratigraphic Nomenclature. A textbook on the geology of Switzerland is full of "inappropriate" terms, such as *Einsiedler Schuppenzone* (Chapter 4), *Wildflysch* (Chapter 5), *Série de la Mocausa* (Chapter 6), *Zone du Combin, schistes lustrés* (Chapter 7), *Tomül-Lappen, Bündnerschiefer* (Chapter 8), *Gotschnagrat-Schuppe* (chapter 9), *Averserbündnerschiefer* (Chapter 10), *Champatsch Zone* (Chapter 11), and so on. What are those zones, scales, and slices? What are the *schistes lustrés* and Bündnerschiefer?

They are mélanges. I pleaded 20 years ago that "in mapping mélanges, tectonic units must be recognized to supplement rock-stratigraphic units" (Hsü 1968, p. 27). I emphasized that a geological map of a mélange terrane should include, in addition to formations, two types of tectonic units—slabs and mélange units. My pleadings went largely unheeded: Mélange units are not recognized as cartographic units on most quadrangle maps of recent years; the international codifiers and their editorial representatives have seen to that. Only the neutral and independent Swiss have been violating the code; they have been mapping units of mélanges ever since the turn of the century, even though they did not invent the name.

Law-abiding citizens adhering to the International Code of Stratigraphical Nomenclature could give us wonderful maps of the alemanides in Switzerland. The geological map of Glarus by Oberholzer will be an eternal classic. On the other hand, if we had not had the genius of the freethinking Emil Argand, we might still be mired in the mud of a hundred "formations" in the Alpine celtides. How did Argand do it?

Argand followed the simple postulate of his time that the gneisses were the basement and the *schistes lustrés* the sedimentary cover. He recognized six Penninic nappes, each with a gneiss core enveloped in the schist. Once the framework was established, it was time for the mopping up. It was obvious to all that the envelope is not a homogeneous body of mica schist. So we had the Zone du Combin between the Dent Blanche and the Bernard/Monte Rosa nappes. More detailed studies revealed even greater heterogeneity, so we now had a subdivision of the Combin Zone into the Tsaté Decke, zones of Saas-Zermatt and of Antrona, Frilihorn Series, Evolène Series, Metailler Series, Barrhorn Series, Pontis Decke, and *Zone Houillère interne* (see Chapter 7). Pioneers in Alpine geology recognized that those nappes, zones, or series are not formations. Therefore, they did not think in terms of the Smithian stratigraphy when they interpreted the geology of the Pennine Alps. None of them seemed, however, to have read Greenly. Otherwise, those tectonic units would be known to us as Tsaté Mélange, Saas-Zermatt Mélange, Antrona Mélange, Frilihorn Slab (or Broken Formation), Evolène Slab, Barrhorn Slab, Pontis Mélange, and Internal Houillère Mélange. In fact, even the rocks of the Monte Rosa and Dora Maira nappes are mélanges, except they have attained a certain degree of homogeneity, because the different components of the mélanges were metamorphosed under about the same high temperature and pressure conditions.

When I was in Shandong, objecting to the use of the term "formation" to designate map units in metamorphic terranes, my Chinese colleagues advised me that they had to put more than one color on their map; they could not map all the rocks of the whole peninsula as one celtide unit between the Yangtze raetide and the North China alemanide. They could not do that when mapping the region on a 1:50,000 scale; they could not even do that when mapping on a 1:200,000 scale. But they do not have to, if they can learn from Alpine geologists. During the two days of the excursion, we visited outcrops of eclogites near Wei-Hai, of metamorphosed Paleozoic passive-margin sequence near Qixia, and of a marble broken formation east of Yentai; all of them are exotic slabs in a greenschist matrix that has been called Penglai. The geology of Shandong has not been served by the invention of a mess of 20 "formations" that are not formations. An understanding of the tectonic evolution must wait for the recognition of tectono-stratigraphic units, such as Wei Hai Mélange, Qixia Slab, Yentai Slab, Penglai Mélange, etc. Those are the map units in the celtide country.

Tectonic Facies and Subfacies

Classification has to be based on criteria that lead to all-inclusiveness and mutual exclusiveness. The two criteria for classification are

(1) whether the basement is or is not involved in the deformation; and

(2) whether the basement deformation is predominantly ductile or brittle.

If only the sedimentary cover, not the basement, is involved in the deformation, the structures are alemanides.

If the basement deformation is predominantly ductile or "mobilized," the structures are celtides.

If the basement deformation is predominantly brittle or "rigid," the structures are raetides.

There can thus be three, and only three, tectonic facies, classified on the basis of the two criteria. As a rule, the various manifestations of the tectonic facies are related to the history of deformation. In a collision zone, rocks above the Benioff zone go up and those below go under. As a rule, raetides are the rocks that went up; celtides are those parts that went down first before they were carried back up again; alemanides are the parts that did not go under but were pushed out. In a two-plate collision, the rocks of the overriding plate (Austroalpine) are thus mostly raetides, the rocks that are overrun (Penninic) are mostly celtides, and the rocks that have "escaped" (Helvetic) are mostly alemanides. However, the correspondence of tectonic superposition and tectonic facies is not always exact. The examples of the Monte Rosa and Margna nappes in the Alps suggest that Austroalpine elements could have been thrust under first to form high-temperature and/or high-pressure metamorphic rocks, before they were exhumed at the end of their journey back up. Those metamorphic units are thus considered celtides. Less clear-cut is the question of those underthrust Lower Austroalpine rocks that are less intensely metamorphosed. In such a borderline case, we have to use judgment when doing our pigeonholing. We face the same problem in classifying autochthonous and parautochthonous massifs: they were pushed out, but they actually also went down relative to the décollement overthrust of their sedimentary cover. They are commonly raetides, but some may become celtides because they have plunged far enough down to become mobilized like the Penninic core nappes. Arbitrary decisions are unavoidable when natural phenomena are classified.

In my first attempt to propose tectonic facies (Hsü 1991b), I suggested numerous tectonic subfacies. I was overenthusiastic in my effort to portray the distinctive histories of the various alemanidic, celtidic, and raetidic elements. In naming a Jura subfacies, I was distinguishing the thin-skinned deformation in the Jura Mountains from that of the Helvetic Alps. Then I had to recognize a Flysch subfacies and a Klippen subfacies because of their different paleogeographic origins. In addition, the autochthonous massifs and foreland basins have to be recognized as subfacies of foreland deformation. For the celtides, I suggested Tessin subfacies to designate the Penninic core nappes, Monte Rosa and Sesia subfacies to designate subducted continental basement that had been detached from the overriding plate, Bündnerschiefer subfacies to designate metamorphosed ocean sediments, ophiolite mélange subfacies to designate the tectonic mixture of ocean sediments and ocean lithosphere, and Bergell subfacies to designate granitic intrusions formed by anatexis of mobilized basement. Finally, I had to propose two subfacies to distinguish the tectonic styles of the overriding plate; the Nordkalkalpen and the Silvretta represent the thin-skinned deformation of sedimentary cover and the rigid-basement thrusting, respectively.

After thinking over the matter of subfacies during the last two years, I decided to drop the nomenclature. First of all, the Alpine subfacies may be inappropriate designations for similar tectonic units elsewhere. The Appalachian Valley and Ridge is not exactly the same as a tectonic unit of the Jura or the Helvetic subfacies. The Rocky Mountain foreland thrust belt is certainly not to be compared with the Jura, and the thrusts are very different from the Helvetic nappes. Obviously, we would have a horrifying proliferation of terminology if we began to build up a superstructure of tectonic subfacies.

Tectonic units of different mountain belts share common traits, but they also have characteristics of their own. Comparisons could be made. Hemipelagic sediments in many orogenic belts, for example, have been deformed to resemble the Bündnerschiefer or *schistes lustrés*. Flysch nappes or ophiolite mélanges look pretty much the same in all mountains. My Chinese friends and I spoke immediately of Silvretta when we saw rigid-basement thrusts in China. We can use those names, think of those concepts, when we travel in faraway lands, but there is no need to construct an arbitrary nomenclature of tectonic subfacies. To invent a lot of inappropriate names is counterproductive.

The Art of Seeing and the Art of Classifying

The English word *science* first entered the Oxford English Dictionary in 1867; it is derived from the Latin word *scientia*, which means knowledge. The German word Wissenschaft is literally the art of knowing, and the word *wissen* came from the Latin word *vide*, signifying that knowing is seeing. This derivation reflects the Baconian scientific philosophy of the nineteenth century, which stressed that the basis of science is observation. The Chinese word for science is *Koxue*. *Ko* is the Chinese word for categories, classes, or divisions. In choosing *Koxue*, the Chinese seem to have emphasized science as the art of classifying.

Karl Popper considers the Copernican and Darwinian revolutions the two greatest intellectual achievements of mankind. That may well be the case, but the scientific achievements by Newton and Linnaeus are, in my opinion, the most extraordinary. Science is not a mere fragmentation of knowledge through observations; that would be "stamp collecting." The objective of science, it is said, is to reach a series of points from which the mind can return along its track to arrange its findings into a theory or generalization. Tycho Brahe was a stamp collector; he devoted his life to collecting astronomic observations. Kepler was a data processor; he found patterns in random observations. Newton was a scientist. From the phenomena of motions, he investigated the forces of nature. Knowing these, he formulated the three laws of motion and he invented the law of universal gravitation to explain not only Brahe's observations and Kepler's calculations, but also apples falling from a tree, or rivers flowing into the sea. Ever since, there has been this division of labor in physics: Theorists are not necessarily observers or experimentalists; they can draw conclusions on the basis of observations made by their predecessors or their colleagues.

Linnaeus provided the inspiration for the biological sciences. He may or may not have been the greatest observer of his time, but he perceived that the beginning of science is classification. Physics searches for the universal truth, the gravitation, the electricity and magnetism, the propagation of light, the constitution of matter. Physical laws are, however, not always applicable without limitations. Stoke's law for calculating the settling velocity of sedimentary particles is, for example, valid only if the fluid system has a viscous motion. There are other kinds of motions. The less fundamental regularities in nature can be considered logical consequences of conjunctions of more fundamental regularities. To recognize the distinctions of the various "less fundamental regularities," there is a need for classification. Each living organism is unique, an individual. We know our friends by their names; Swiss farmers know their cows by their names. There is, however, this "fundamental regularity," or universality, shared by all people, all cows, and all living beings: they all are alive or have been alive. Between individuality at one end and the fundamental regularity of being alive at the other end are the various "less fundamental regularities" of anatomical structures and physiological processes. Linnaeus invented a scheme in which all individuals could be grouped into species, genus, family, order, class, phylum, and kingdom, and members of different hierarchical groups share increasingly more fundamental regularities.

Science and
Stamp Collecting

When Lord Kelvin, the famous physicist, was asked around the turn of the century to enumerate the various disciplines of science, he gave a brusque answer:

"There is science, which is physics, and there is stamp collecting."

He was, of course, unfair, but he was expressing his irritation with his contemporaries, whose scientific endeavors consisted mainly of collecting and describing minerals, fossils, plants, or butterflies. We have come a long way from that. Biologists of this century have recognized that the establishment of an all-inclusive and mutually exclusive taxonomy was only the first step toward defining the "truth" shared in common by all members of a group. As biology advances, the morphological differences so apparent in distinguishing species have become less and less relevant, while the fundamental regularity of life shared in common by vastly different groups, such as genetic coding, enters center stage. No wonder my friend Werner Stumm made fun of his Harvard colleagues in biology, saying that they don't know the difference between a fox and a rabbit. Indeed, they might not be able to tell a rabbit from a fox, because it may not be a relevant problem in modern biology.

Earth science has not come that far. Field geologists are still afraid of being ridiculed if they cannot take accurate measurements of dips and strikes. Experimental geochemists are embarrassed if they mistake anorthosite for marble. We are proud of being "stamp collectors," probably because of our glorious victory in the interdisciplinary debate on the age of the earth. Lord Kelvin forgot that a scientific deduction from a premise is no longer valid when that premise is proven false. When he continued to argue for a young earth long after the discovery of radioactivity, his opinions were no longer respected; nothing he said could be meaningful.

Of course, some of us would like to think that geology is science, because our way of "stamp collecting" is science. We no longer collect minerals or fossils; those are left to "rockhounds" and to amateur paleontologists. We collect scientific data, carry out scientific experiments, and make scientific calculations. Geology articles have plenty of digitized information, obtained after laborious observations or measurements. Hidden behind this appearance of being scientific, the "stamp collector's" mentality could prevail.

I have learned to understand the stamp collector mentality of my son Peter. We collect Chinese airmail covers. My motivation was to gather documentary verification of my theoretical knowledge of the history of Chinese aviation. I spent much time, for example, searching for the reason why two postal flights (Peking–Tientsin on May 7, 1920, and Peking–Tsinan on July 1, 1921) should both claim to have carried the first official airmail of China. When Peter and I prepared for a philatelic exhibit, I wanted to speculate on this problem in our explanatory notes. Peter was horrified. Great philatelists are those who have assembled a great collection, i.e., a very costly collection. Great exhibits are judged by the "quality and importance of the collection." Collectors who give factual descriptions of their rare and expensive items will win Gold Medals or even Grand Gold. Speculations of the sort contemplated by me would cause the judges to downgrade our exhibit.

Having learned to write philatelic articles for stamp collectors, I got a best-paper award in a philately journal. However, when I write manuscripts on geology, I am aiming at scientists and I forget Lord Kelvin's observation that I should write for stamp collectors. It is perhaps no coincidence that I never did receive a best-paper award for my 300-plus articles in geology journals. My batting average is getting worse, as all three manuscripts I submitted in 1991 to refereed geology journals were rejected. Some editors of geology journals, in my prejudiced opinion, may have received their appointments because of their stamp collector's mentality; they would probably reject a theoretical article in the style of Newton on gravitation because the author had not documented his or her conclusions by personal observations. This attitude is the source of "regrettable errors of editorial judgment" such as those involved in an injustice done to L. W. Morley.

Morley wrote a manuscript in December 1962, after he read the 1961 papers by Mason and Raff about magnetic stripes and by Dietz about seafloor spreading. He invoked the idea—now known as the hypothesis of seafloor spreading—to explain the observational data reported by Mason and Raff. The manuscript was rejected by *Nature* in early 1963. He then submitted it to the *Journal of Geophysical Research*. It was rejected in late August or September, about the same time *Nature* published the paper by Vine and Matthews on magnetic anomalies over oceanic ridges. Many opinions have been expressed about those "regrettable errors of editorial judgment"; the most persuasive was offered by Menard (1986, p. 219):

> The letter (to Morley) from the editor indicated that a reviewer had stated that the ideas were suitable for discussions at a cocktail party but not for publication. . . . Reviewers are apt to reject wild ideas in which an author apparently has invested little time or effort.

Menard shared the opinion of many earth scientists that "ideas are cheap" (p. 217). He indicated that Morley apparently had not "done a reasonable amount of work before coming to his opinions" (p. 219). Menard did note that Vine had not collected any data either, but "Matthews had spent the months at sea" (p. 219), and Matthews was a coauthor. Menard seemed to think that Morley's idea would not have been cheap if he had offered coauthorship to Raff and Mason. Following this kind of logic, one might conclude that Newton's *Principia* could not be published today if Brahe and Kepler were not coauthors.

The prejudice that theorists can theorize only on their own data is not sufficient to explain many "regrettable errors of editorial judgment." The more fundamental cause is the "stamp collector's mentality" of journal editors who do not understand what constitutes science and who insult theorists and their ideas. It might seem petty to discuss what I consider poor editorial judgments in the rejection of my three 1991 manuscripts, but an analysis might serve to focus attention on the need to recognize priorities in making science.

I wrote a paper, coauthored with Chinese geologists and geochemists, on the potash brines of the Qaidam Basin. The origin of the brines is not just a theoretical question, because the brines are exploited commercially. My coauthors had collected, during a 30-year span, a wealth of data on the climate, hydrology, and hydrochemistry of the region. They had also conducted numerous experiments on the chemical evolution of ground water as a consequence of evaporation. Their work led me to the conclusion that the potash-rich brines are simply evaporated ground water, after calcium and sulfate ions have precipitated as gypsum or anhydrite, sodium and chloride ions as halite, and magnesium ions as bischofite. Our conclusion led my son Peter to invent a process by which commercial-grade potash brines could be produced from ordinary ground water. At about the same time, we submitted a manuscript to the *Geological Society of America Bulletin* to present the data and the theory. The editors and reviewers found our idea attractive, but they had to reject the manuscript because, in their opinion, the conclusion "is not well documented."

One cannot defend, in this case, the editorial judgment with the remark that the authors have not "done a reasonable amount of work before coming to their opinions." The reviewers and editors are not blind to the wealth of data, but they fail to see the connection between the data, which had been collected to advance some discredited ideas, and the new idea. They probably would have concluded that Newton's theory of universal gravitation was not well documented, because Tycho Brahe's data had been collected for a different scientific purpose. Making the case of the rejected Qaidam paper seem even more absurd is the fact that the published papers on the origin of Qaidam brines all postulated a derivation of potassium brine from an unknown source at an unknown depth. Documentation is no requirement for the publication of idle speculation.

The second manuscript of mine rejected in 1991 was coauthored with Daniel

Müller. The fossil mollusks *Inoceramus* spp. are found in hemipelagic or pelagic sediments, and they are present in laminated sediments of anoxic basins where benthic organisms could not have survived. A few well-preserved specimens were fossilized while attached to driftwood; they apparently sank with the wood after it became waterlogged. Some paleontologists suggested, therefore, that some species of *Inoceramus* may have had a pseudoplanktonic habitat. Müller and I carried out an experiment to test if the hypothesis could be falsified: If the *Inoceramus* shells studied by us were pseudoplanktonic, they should have a carbon-isotope composition similar to that of the skeletons of planktonic organisms that had lived in the same surface water. The test was positive, and we had to conclude that those species were planktonic or pseudoplanktonic, or that not all *Inoceramus* species were benthic. The paper was rejected twice, because (1) the authors had not cited all the monographic studies postulating a benthic habitat for *Inoceramus* spp., and (2) each of the arguments falsifying the theory of a benthic habitat could be explained by some special pleadings.

The deterministic philosophy from Newton to Linnaeus was succinctly stated by Karl Popper: truth cannot be proved, but falsehood can be recognized through the falsification of a hypothesis. My motivation in bringing up the case histories of the rejected manuscripts is to illustrate what I consider a fundamental fallacy of the scientific philosophy now prevailing in certain circles. The editors asking for more observational data "to prove a theory of origin of potash brines" failed to recognize that theories are not deduced from observations. Each theory is first put forward as a hypothesis, and each working hypothesis restricts the field of possible explanations. In the case of Qaidam brines, the field of possibility had been unlimited; previous workers thought that the brines could have come from anywhere. The purpose of our theorizing was to restrict the possibility to the one that postulates an origin by evaporative pumping. The hypothesis was formulated to explain all the information contained in the Qaidam files. Therefore, all pertinent data, accumulated over a period of three decades, are the documentary evidence. To claim that the hypothesis is not supported by sufficient data is a manifestation that the reviewers and editors did not understand the philosophical wisdom expressed by Celal Şengör: observations do not lead to the palace of truth; they are only the guards that keep imposters away from the palace of truth. The editors who rejected the *Inoceramus* manuscript were even more ignorant of scientific logic. They were evidently unaware of the famous dictum by Popper that seeing one black swan is enough to falsify the theory that all swans are white, no matter how many sightings of white swans were previously reported. They failed to see that the postulate that *Inoceramus* spp. could be pseudoplanktonic is a working hypothesis. The alternative that all *Inoceramus* spp. are benthic is obviously not working, because many special pleadings are required to explain away the "black swans."

Nature of the Mediterranean Ridge

Journal editors in geology were not always as intolerable as those today. Marcel Bertrand invented the nappe theory to explain the Glarus Overthrust without ever having set foot in Glarus (see Chapter 3). Bertrand was a theorist, and his duty was not data collection. It does not take brilliant intellect to recognize that Newton might never have had the time or inclination to write his *Principia* if he had had to collect astronomic data all over again from the start just as Tycho Brahe did.

A theorist has quite a different mentality from that of a good observer, and both types of scientists have their place. Yet research planners today do not always appreciate the fact that significant experiments are often those designed to test theories that have been formulated on the basis of very incomplete data; the greatest surprises come from the greatest lack of knowledge. The Ocean Drilling Project, which is spending some 50 million dollars annually for marine geology research, has repeatedly turned down research proposals because their proponents had not submitted sufficient multiple-channel seismic profiles. If such a research philosophy had been adopted, our knowledge about the Peking man would still be limited to the two molar teeth; bureaucrats blessed with a "catch-22" mentality would insist that no funding be granted to search for *Sinanthropus* skeletons until a nearly complete skeleton was found.

This last chapter on theoretical geology was written after I came back frustrated from a meeting at Trieste to plan Mediterranean drilling. I went there with the naive belief that the big money for ocean drilling was to be spent to investigate processes and to formulate general principles in geology. I went with a proposal that the Mediterranean Ridge holds a key to understanding mountain-building. No, I was "shot down in flames" by my peers; I was throwing a monkey wrench into well-oiled machinery. The multiple-channel seismics (MCS) have all been taken, and the sites chosen; a transect of shallow boreholes across the ridge will be drilled. My "wild ideas" would only delay the scheduling of the cruises.

It is useless to drill those shallow holes; we cannot learn Alpine tectonics by digging into glacial drift. But the shallow holes are to be drilled because their near-surface structures are well documented by MCS. A working hypothesis based upon the wealth of data on island geology and marine geophysics is considered a "wild idea"; it is "speculative," and no holes can be drilled without the "facts" supplied by MCS. At Trieste I came to the realization that journal editors in geology do not have a monopoly on the stamp collector's mentality. It seems petty to complain about rejected manuscripts, but the future of geology is at stake when that sort of mentality becomes the prevailing attitude in science planning. I felt sufficiently concerned to write this addendum for posterity.

The origin of the ridge is the history of the evolution from the Paleotethys to the Mediterranean, and there have been many papers written on the subject. There is even consensus on several salient "facts":

(1) Europe and Africa were sutured together as part of the Permian supercontinent Pangea (Figure 16.1).

(2) Permo-Triassic ophiolites are found in the Main Paleotethys Suture Zone of China and western Asia, testifying that the consumption of the Paleotethys led to the

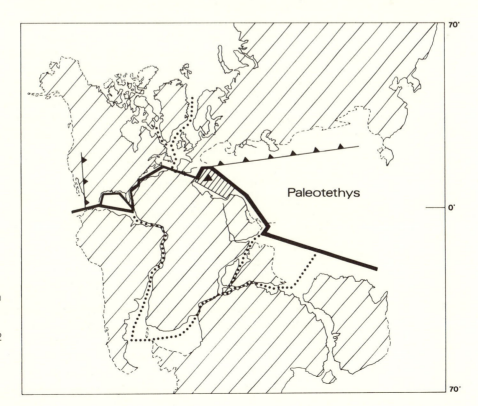

Fig. 16.1. THe Permian Paleotethys (after Laubscher and Bernoulli 1977). The northwestern margin of the Paleotethys, according to an interpretation presented in the text (Chapters 12 and 16), was an active margin, where the Paleotethys plate was subducted under Europe along a northwest-dipping Benioff zone.

early Mesozoic collision of Eurasia and microcontinental fragments detached from Gondwana (see Chapter 15).

(3) There are no relics of the Permian ocean in the Tethyan mountains of Europe. Triassic ocean sediments and submarine volcanics are known in the Cimmerides of Dobrogea, Crimea, and Turkey (Sengör 1984). The Alpine ophiolites are Jurassic; they are relics of the Tethys.

(4) Relics of the Permian Paleotethys have not been positively identified in the Mediterranean region either. The eastern Mediterranean was considered the relic of a Mesozoic ocean (Figure 16.2), called Mesogea by Biju-Duval *et al.* (1977). The western Mediterranean and the Aegean basins are Tertiary back-arc basins.

(5) The shallow marine sedimentary sequences of the Apennines, the Dinarides, and the Hellenides, and possibly also that of the Taurides, were the sedimentary cover of a microcontinent. This is the continent of the Mediterranean plate (see Chapter 12), called Apulia/Anatolia by Biju-Duval *et al.* (Figure 16.2).

(6) The Western Alps owe their origin to the collision of the Apulia microcontinent and Europe, after the consumption of the Tethys Ocean.

(7) The present plate boundary between Africa and Europe is marked by the subduction zone under the inner wall of the Hellenic Trench. This plate boundary is a young feature, probably not more than 10 or 15 Ma in age.

(8) The Mediterranean Ridge, with its two arcuate segments averaging some 150 km in width, is located midway between the Cretan Island Arc and the North African continental mass. The topography is rough, with low hills and small depressions. The highest elevation on the crest of the ridge is 1330 meters below sea level, standing thus some 2500 m above the Ionian and Herodotus abyssal plains (Figure 16.3).

(9) Late Neogene tectonics caused the subduction of the African plate, of which the Mediterranean Ridge is a part, under the European plate at a rate of about 2.5 cm/yr during the late Cenozoic (Le Pichon 1968). The floor of the Hellenic Trench at this plate margin has been depressed in places to a depth of more than 5000 m. Also in the late Neogene, the island of Cyprus near the eastern end of the ridge was uplifted above sea level, and the Troodos Mountains now stand more than 1000 m above sea level.

How is the field of possibilities restrained by this set of boundary conditions and a vast store of data on land geology and marine geophysics? Is there a unique solution for the origin of the submerged mountain chain called the Mediterranean Ridge?

The three possible modes of mountain-building, as we might recall, are extensional, wrench faulting, and compressional (see Chapter 13). The bathymetry of the ridge is similar to that of a mid-ocean ridge, but geophysical evidence has indicated that the feature was not formed by extension during seafloor-spreading (Ryan *et al.* 1969). There is no indication, and no one has ever suggested, that the ridge was formed by wrench faulting. Through this process of elimination, we can conclude that the Mediterranean Ridge is a compressional mountain range, comparable either to the Circum-Pacific or the Tethyan types of orogenic belt.

Ryan *et al.* (1969) noted that "the Mediterranean Ridge is a long and low swell or flexure that superficially resembles the seaward outer ridge of other trenches. Most likely, the swell has been formed in response to the same regional tectonic forces that produced the trenches." They assumed that the Mediterranean Ridge has always been the northern edge of the African plate, which is subducted under the European continent.

If the ridge was formed by an ocean-continent interaction, the eastern Mediterranean should be underlain by ocean crust. However, gravity data indicate that the ridge is underlain by an earth's crust 20–25 km thick. Assuming a 3–4 km sedimentary sequence, one has to postulate "that some thickening has occurred in either the sedimentary layer, the crustal layer, or both" (Ryan *et al.* 1969, p. 437). In plain words, the ridge is not a simple flexure of ocean crust like the outer ridge of ocean trenches; the ridge is underlain by a compressional mountain chain.

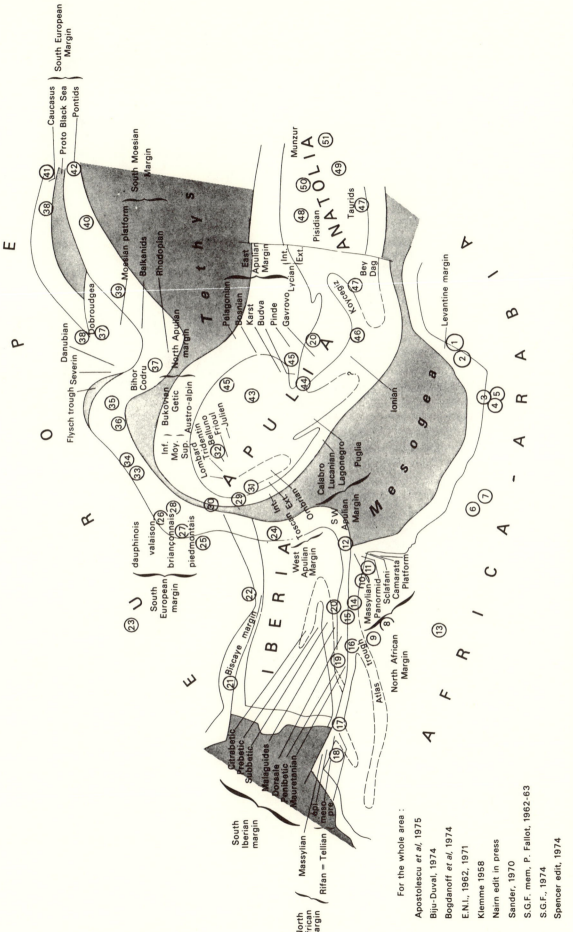

Fig. 16.2. A hypothesis of the Late Mesozoic paleogeography of the Tethys and Mediterranean. The hypothesis representing the current consensus is illustrated in this sketch by Biju-Duval et al. (1977): Europe was separated from Apulia/Anatolia by the Tethys, and the latter from Africa by Mesogea. I am proposing that Apulia/Anatolia shown on this map was not a microcontinent, but a series of relic island arcs and relic back-arc basins, forming the "soft underbelly" of Europe. Relic island-arcs discussed in the text, but not shown on this map, are the Brianconnais, the Austroalpine/Carpathian/Serbo-Macedonian, and the Mediterranean arcs (designated B, C, and E, respectively, in the text).

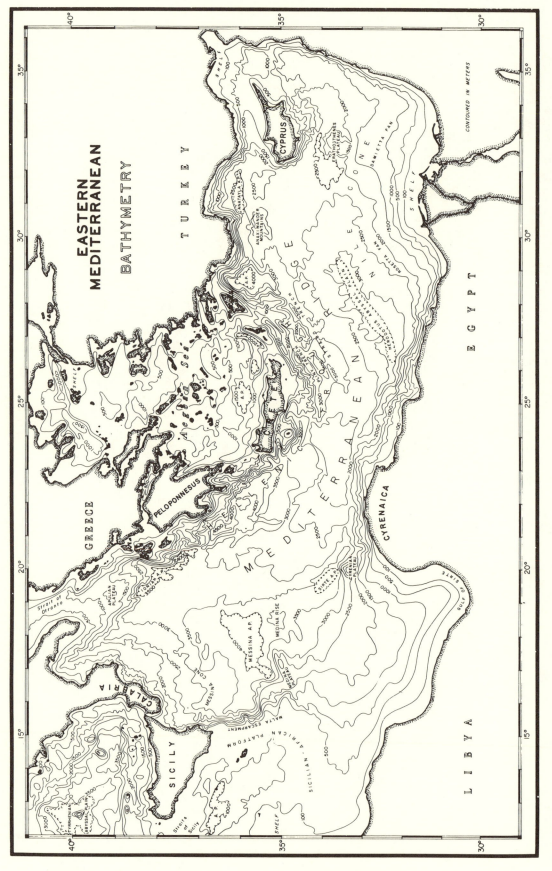

Fig. 16.3. Bathymetry of the Mediterranean Ridge (after Ryan et al. 1969). The seafloor south of the Mediterranean Ridge, according to my interpretation, is a relic of the Paleotethys Ocean (see text for details).

Seismic evidence indicates that the late Neogene deformation of the ridge is manifested by a series of thrusts, and "this ridge is similar to an accretionary prism but it is much wider with respect to its height. . . . In a sense, then, the Hellenic Trench is in the situation of a fore-arc basin" (Le Pichon 1982). However, this late Neogene deformation is caused by the subduction along a north-dipping Benioff zone under Crete, and the deformation did not start until about the Late Miocene. During a previous stage of orogenic deformation that continued until the Middle Miocene, the Hellenides were deformed, when the Vardar Ocean between Apulia and the Serbo-Macedonian massif was consumed (Hsü 1981). The alemanide-facies rocks of the Hellenides crop out on the island of Crete, and they are present under the inner wall of the Hellenic Trench (Ryan and Hsü 1973). The deformation responsible for the relief of the ridge also took place prior to the Late Miocene; the Upper Miocene Messinian is only a thin drape of pelagic sediments on the ridge, but a thick evaporite deposit in eastern Mediterranean basins. The compressional mountains under the ridge had their origin long before the subduction of the African plate under Europe along the present plate margin.

Ryan *et al*. attributed the deformation of the ridge to the counterclockwise rotation of Italy, which "drove the Apulian Ridge eastward. The initiation of this movement during late Eocene created an entirely new paleogeography in this region. . . . By the end of the Middle Miocene time the tectonic wave had reached the Apulian foreland. It is thought that *at that period two continental lithospheric plates collided*" (Ryan *et al*. 1969, p. 480).

I have italicized the last sentence, because their postulate that the Mediterranean Ridge is a pre–Middle Miocene collisional type of orogen has been largely ignored. The idea was not developed into a working hypothesis, because of the prevailing stamp collector's mentality that there can be no theoretical geology. A wealth of marine geophysical and on-land geological data has been accumulated, but has not been interpreted within the framework of this "speculative" model.

Since the ridge is submerged, one has to theorize what the other "continent" was that collided with Apulia/Anatolia to form the Mediterranean Ridge. This southern continent cannot be Africa, because it is still situated south of the relic Mesogea Ocean, as represented by the Messina and Herodotus abyssal plains (see Figure 16.2). To theorize about the unknown, we need at least "two molars."

Geology of Cyprus

Our "two molars" are the island of Cyprus.

Two major tectonic units are present on the island of Cyprus. The Kyrenia Range is underlain mainly by upper Mesozoic and Cenozoic formations, forming disharmonic folds and décollement thrusts. The southern half of the island is underlain mainly by Troodos Ophiolite and by Mamonia Mélange. That the Troodos represents ancient ocean crust is a consensus interpretation. That the Mamonia Mélange represents a tectonic mélange is also a consensus; the Troodos Ophiolite is just a huge slab in the mélange of a suture zone. Assuming that the Cyprus deformation had the same southerly vergence as the late Neogene structures, it is generally thought that the Troodos Ophiolite was thrust upward and southward along a north-dipping thrust fault (see Robertson *et al*. 1991). The continent under the suture zone is called Monia, and the Eratosthenes Seamount was apparently part of "Monia."

The Kyrenia formations have a style of deformation typical of a peeled-off passive-margin sequence; the style of deformation is distinctly different from that of rigid-basement thrusting. I thus suggested in a 1989 memorandum to the JOIDES Planning Committee that the Miocene Benioff zone south of Cyprus dipped to the south, not to the north. According to this model, the Troodos Ophiolite was the ocean floor of a back-arc basin that had its origin in Late Cretaceous back-arc seafloor-spreading. The Upper Cretaceous–Tertiary sediments of the Kyrenia Range constituted a passive-margin sequence north of the back-arc basin. The frontal arc should then be the structural elevation south of Cyprus, where the Eratosthenes Seamount is located! The tectonics of the Cyprus deformation is that of a back-arc basin collapse. The ocean lithosphere (Troodos) under the back-arc basin was subducted under the arc (Eratosthenes), and the Mamonia Mélange was the south-dipping accretionary wedge under the arc, before it was sandwiched in the suture after a collision. The plunger action of the overriding arc and the mélange caused the décollement deformation of the Kyrenia Range sequence.

Assuming that Cyprus is the tip of that part of the Mediterranean Ridge that happened to have risen above sea level, the collision envisioned by Ryan *et al.* in 1969 should have been between an Apulia/Anatolia arc and another island-arc complex. The Eratosthenes Seamount was a volcanic island on the eastern end of the arc, whereas the submarine Cyrene Mountains north of Cyrenaika in Libya may have been another near the western end. Elsewhere, the arc has been sutured onto Apulia and is no longer distinguished by topographic elevations. Only the present arcuate shape of the ridge is a clue to its arc ancestry.

This arc-arc collision model predicts that the Mediterranean Ridge, especially the northern inner belt of the ridge, is underlain by an alemanide similar to that of the Kyrenia Range. The fragmentary evidence gives support to this prediction: hemipelagic Lower Cretaceous, Oligocene, and Lower Miocene clays have been found in the mud diapirs of the ridge (Cita and Camerlenghi 1992). Those sediments, being the offshore deep-water equivalent of the shallow marine carbonates of Apulia, could be the passive-margin sediments on the southern Apulian margin. We sampled Middle Miocene flysch at DSDP Sites 126 and 377; we may have drilled into a flysch nappe. Slabs of wildflysch were encountered at DSDP Site 129, located north of the Strabo Trench (Ryan *et al.* 1973). The petrography of the wildflysch sandstone indicates that the northern carbonate shelf cannot be the source terrane for the flysch sediments. The debris derived from andesitic and metamorphic rocks must have come from an island-arc complex (Ryan *et al.* 1973, p. 743). The only source area then had to be an island arc to the south, in the general area where the Eratosthenes Seamount now stands. That the "seamount" was indeed a volcanic construction is verified by the presence of magnetic anomalies (Kempler and Garfunkel 1992).

The Soft Underbelly of Europe

All Swiss children learn in school that the Alps were formed when Europe and Africa collided. My children were thus surprised on their first trans-Alpine trip when they encountered Italy, not Africa, south of the mountains. I had to tell them, for the sake of simplification, that Italy was a part of Africa before the collision.

The classic paradigm postulates the collision of Europe and Africa to explain the origin of the Alps. The two-plate models relating the Alpine deformation to the relative displacement between the European and African plates predicted a change from an extensional to compressional Tethyan regime at 81 Ma, or Santonian time, when Europe first separated from North America and was displaced eastward at a faster rate than Africa. This Late Cretaceous timing of the compression is, however, contradicted by the evidence of the Alpine geology, because the subduction of the southern Tethyan margin started at the beginning of the Cretaceous, some 50 Ma earlier.

To harmonize the geological and geophysical evidence from the Alps, I proposed a three-plate model to interpret the geological evolution of the Western Alps (see Chapter 12). It was a step in the right direction, but it was again a necessary simplification. The geology of Europe has impressed many that the evolution of the Tethys must be interpreted in terms of numerous microcontinents between Europe and Africa (e.g., Dewey *et al.* 1973). What were those microcontinents? How did they form?

Microcontinents, or small areas underlain by continental crust, are islands, relic island-arcs, or submarine banks of various origins. One hypothesis is to assume that the Paleotethyan microcontinents were split off from Gondwanaland or Africa and that they "drifted" across a wide expanse of ocean before their collision with Europe. The genesis of the Paleotethyan/Tethyan mountain chains is thus interpreted in terms of repeated suturing of Eurasia with "waves" of Gondwana fragments (Sengör 1984).

My work in China during the last decade has, however, led me to a drastically different interpretation for the tectonic evolution of the Paleotethys. The Jurassic Tethys was not, as presumed in the current paradigm, bordered on both sides by passive continental margin; it was not an Atlantic type of ocean. Geological evidence clearly indicates to me that Eurasia was bordered on the south, since the Permo-Triassic time at least, by archipelagos of island arcs, remnant arcs, and back-arc basins. The archipelagos of the West Pacific today are the appropriate analogue model for Paleotethys: Eurasia could be said to have had a "soft underbelly," because the continent was fringed by several chains of active and relic island-arcs. Paleotethys, as it is recognized in the ophiolite mélanges of Indosinides, was not an ocean; it was a name given collectively to the marginal seas south of the Eurasian continent. The microcontinents of the Indosinides were not split off from Gondwanaland; they were remnant arcs of Europe, situated north of an outer arc of the Eurasian plate. The portrait of the Jurassic Tethys as a Mediterranean ocean was perhaps a reflection of European egocentricity; Alpine Tethys was merely a marginal sea comparable to the South China Sea, and the Helvetic sequence was deposited on the northern shore of that back-arc basin.

The West Pacific is a region where microcontinents or arcs abound today. Between the islands or relic arcs are marginal basins, underlain largely by ocean crust. Except for Timor, the deformation of Southeast Asia is not caused by the collision of Australia and Asia. The late Mesozoic and Cenozoic tectonic evolution of the region is a history of back-arc seafloor-spreading and of back-arc-basin collapse. Adopting the West Pacific as an actualistic analogue, I propose that the origin of Tethyan mountains is mainly a process of back-arc basin collapse. The collision of Europe and Africa took place only in Algeria, Tunisia, and Sicily. The last remnant of the ocean between the two continents is now buried under a thick blanket of the eastern Mediterranean sediments. Millions of years have yet to elapse before northeastern Africa can collide with southeastern Europe. The orogenic structures of Europe, according to this viewpoint, have resulted largely from arc-arc or arc-continent collision within the European plate. The Alpine ophiolites were mainly the ocean lithosphere of back-arc or inter-arc basins.

I borrow the expression "soft underbelly of Europe" from Winston Churchill to designate my model for the evolution of the Paleotethys/Tethys, which is similar to the Chinese model presented in Chapter 15. I propose that Mesozoic Europe was flanked on its farthest south side by an outer volcanic arc and by several chains of remnant arcs. The arcs and basins are, from north to south,

(A) Europe
  (1) North Penninic (Valais Trough of Tethys)
(B) Brianconnais/Subbrianconnais
  (2) South Penninic (Piedmont Trough of Tethys)
(C) Austroalpine/Carpathian/Serbo-Macedonian/North Anatolian
  (3) Apusine/Vardar/Anatolian basins (Vardar Ocean)
(D) Apulia/Anatolian
  (4) Antalya/Troodos and their western extension (Mesogea)
(E) Mediterranean Arc (Cyrene Mts./Outer Mediterranean Ridge/Eratosthenes Seamount)
  (5) Messina and Herodotus abyssal plains (Mesogea and/or Paleotethys)
(F) Africa

We should recall that the southeastern Europe of the Permian Period was the site of an Andean type of active margin, where the Paleotethyan ocean lithosphere was thrust under, along a NW-dipping subduction zone, the European margin (see Chapter 12). A slice of continental margin split off from Europe during the Triassic to form an island arc. That arc was later split off into several chains of remnant arcs, while the Mediterranean Arc (E) remained the outer arc. As in the Banda Arc or the Mariana Arc, there should have been active volcanism in the region of the outer arc, throughout the time during which ocean lithosphere was subducted. The other arcs (B, C, D), however, had active volcanism only during the Permian and/or Triassic. After they became remnant arcs, their sedimentary cover should have consisted largely of marine sediments.

Magmatism does not become active again in back-arc regions until the time of back-arc collapse. The active volcanoes of the Philippines, for example, came into existence after the ocean floor beneath the South China Sea had been subducted down a Benioff zone east of the Manila Trench; the partial melting of the subducted lithosphere has given rise to the back-arc volcanism. The Swiss Alps underwent a similar history: the early Paleogene volcanism there (see Chapter 5) was related to the subduction of a back-arc basin that has been called Alpine Tethys.

Applying the West Pacific model to interpret Alpine geology, I postulate that the back-arc basins north of the Mediterranean Arc came into existence in the Mesozoic. The Cretaceous and Tertiary collapses of back-arc basins had led to various orogenic deformations: the subduction of the Piedmont Trough started in the Early Cretaceous, although the final collision of the Austroalpine Arc and the sutured Europe/Brianconnais was Late Eocene; the Dinnarides and Hellenides owe their origin to the Cretaceous collision of the Carpathian/Serbo-Macedonian and Apulian arcs. The Troodos ophiolites mark the suture zone between the Kyrenia remnant arc and the Eratosthenes outer arc. According to this scenario, the Mediterranean Ridge is the youngest of all Alpine-Mediterranean mountains, having been formed by the collision of the Apulia/ Anatolian and the Mediterranean Arc during mid-Tertiary time.

The soft underbelly hypothesis predicts that the relics of the Paleotethys are to be found south of the Mediterranean Arc. In a talk given at Trieste, Jan Makris presented magnetic data for his conclusion that the Mediterranean Sea southeast of the Eratosthenes Seamount is underlain by reversely magnetized ocean crust (see Wang and Makris, 1992). This could well be a relic of the Late Permian Paleotethys, when the ocean crust formed during the Magnetic Epoch of Kiaman Reversal was negatively magnetized.

The adoption of the soft underbelly hypothesis would require a major revision of the plate-tectonic theory of orogeny. Mountain-building is in many instances not a suturing of colliding continents. The "soft underbellies" of colliding plates are not rigid. In these last three chapters, I have presented enough arguments for the hypothesis that the geological evolution of mountains is largely a history of the spreading out and reconsolidation of those soft underbellies, although a Wilsonian cycle is eventually terminated by an intercontinental collision. These were the kind of new ideas, I thought before I went to Trieste, that needed to be tested by such a costly undertaking as the Ocean Drilling Project. I was wrong; the "wild idea" was but a "monkey wrench" that would upset the smooth "mopping-up" operations by holders of MCS.

I hope that the holes proposed by my friends will be drilled. Perhaps a future exploration of the Mediterranean Ridge is possible, now that the project has been extended for five years. If the outer Mediterranean Ridge, including the Eratosthenes Seamount, was a Mesozoic active margin, where the Mesozoic and early Cenozoic Benioff zone dipped to the north, drilling on the southeastern Eratosthenes escarpment should penetrate an accretionary wedge of mélanges, i.e., mixtures of Paleotethyan crust, Mesozoic/lower Tertiary volcanics, and active-margin sediments. I myself shall have retired from active research when and if the hypothesis is tested. That does not matter; truth will prevail.

Theoretical Geology

Kevin Burke gave a talk for me at the 29th International Geological Congress, Kyoto, Japan, on the fractal geometry of catastrophes, and the conclusion was to predict catastrophe on the basis of the relation

$$\log t = -a + b \log M \tag{16.1}$$

which states that the waiting time $\log t$ is inversely proportional to the magnitude of the event M, where a and b are two empirical constants. This is the essence of the theory of actualistic catastrophism: small events occur daily, whereas very big catastrophes occur once every thousand, or million, or billion years (Hsü 1983). After the talk Burke was taken to task by a geologist in the audience: Hsü's generalization must be wrong, because there cannot be one relation shared by such diverse events as storms, volcanic eruptions, earthquakes, and meteorites. The detractor would not be appeased, even after he was told that the fractal geometry as indicated by eqn. (16.1) is a statistical relation, not a theoretical assumption.

I recount this episode to reemphasize the stamp collector's mentality of many geologists, who are convinced that natural phenomena are too diverse to be generalized. With such an attitude, one would have to think Newton was wrong when he tried to use the same rule to explain the planetary motions, the fall of an apple, and the flow of a river. As I wrote in the beginning of this chapter, there is no theoretical geology, because of the prejudice that there can be no theoretical geology. In fact, we do have theories: the theory of mantle convection, the theory of lithospheric plates, the theory of seafloor spreading, the theory of plate tectonics, the theory of evolution, the theory of metamorphic facies, the theory of actualistic catastrophism, and now the theory of tectonic facies. The theories formulated by geophysicists and geochemists are very relevant to geology. We should not forget that we have long passed the day when geology could be a travelogue or a simple narrative of observations.

Geology has undergone two revolutionary changes. The Huttonian/Smithian foundation of classical geology had its root in observations. There were brilliant discoveries, but the mopping up became stale exercises until the process-oriented approach of the plate-tectonic theory took geology out of a dead-end street. Theoreticians have mainly been geophysicists; the stamp collector's mentality is too deeply rooted in geology. I write this last chapter perhaps as my last contribution to geology, seditious pamphleteering for another revolution. I call for a revolutionary change in attitude. Ideas are not cheap, and they are good for more than "bull sessions" and cocktail parties. Ideas are very expensive, and industry is spending billions on research to acquire new ideas for new products. Geologists should cease to insist that one cannot generalize (i.e., geology is not science); they should not continue to hold on to the Murchisonian notion that a good geologist is one who can walk faster than his assistants.

Postscript

I gave a short course on the concept of tectonic facies last September, and my audience was disturbed by my iconoclastic approach to tectonics. I was asked:

"Do you believe in any of the basic principles in tectonics?"

"Yes, I do; I believe in the basic tenet of the Wilsonian cycle."

In the postulate of the Wilsonian cycle, a rift in the continental interior is the first stage in the breakup of a continent. The Tethys, formerly known as the Alpine Geosyncline, has traditionally been considered a Permo-Triassic intracontinental rift. The Tethys became an ocean in the Early Jurassic, and the ocean was fringed by a European passive margin on the north and an African passive margin on the south. The European and African plates began to approach each other in the Cretaceous, and the collision of Europe and Africa caused the orogenic deformation of the Alps. The Wilsonian cycle thus proceeded from intracontinental rifting to mid-ocean seafloor spreading, to consumption of the intervening ocean lithosphere, and finally to intercontinental collision. This was my basic faith, and this *Geology of Switzerland* was written on the basis of this model.

While the manuscript of this opus is being copyedited, I have come to the realization that I have been undergoing, during the last six months, a revolutionary change in my thinking on tectonics. If I were asked the same question now, I would reply:

"None, not even the Wilsonian cycle, nor the theory of plate tectonics as it is taught now!"

One of the greatest hindrances to an understanding of the origin of mountains is the geosynclinal theory of mountain building. The crust of the earth, according to this theory, can be divided into two categories: the stable areas and the mobile belts. Mountains are mobile belts, and a mountain chain of the Tethyan type is supposedly formed by the compression of a mobile belt between two stable blocks. The geosynclinal theory seems to have passed into oblivion since the innovation of the plate tectonic theory. The decline is, however, more apparent than real. The geological nomenclature has been revolutionized, but not the geological thinking. The concept of a Wilsonian cycle of plate displacements replaces the postulate of a geosynclinal cycle, but it now seems clear to me that the Wilsonian cycle is a thinly veiled reincarnation of the discarded geosynclinal theory: The eugeosyncline of the classical theory is now compared to the Atlantic type of spreading ocean. The miogeosynclinal shallow-water sediments of the carbonate-orthoquartzite association are now categorized as passive-margin sequence. The ophiolite "nappes" marking "zones of continental collision" are now called tectonic mélanges at suture zones of plate collision. We now have a plate-tectonic theory of mountain building, but the new theory is, as a matter of fact, not much different from the classical theory, except for the width of the mobile belt prior to the "collision."

More than 20 years ago, shortly after the plate-tectonic theory was first innovated, Dan McKenzie placed the southern boundary of the European plate within the Mediterranean Sea. No, he could not do that, I told him. The plate boundary should be the suture between Europe and Africa, and this suture zone lies within the Alps, the Carpathians, and the Hellenides. McKenzie accepted this solution: he stood at the Hörnlihütte on the northern face of the Matterhorn, and told his BBC audience that he was astride the European and African plates. From the Alps, the plate boundary of the suturing found its way to the Vardar zone of the Dinarides and thence to the southern margin of the Hellenic Arc, where his plate boundary was originally placed. This plate boundary is depicted in standard textbooks and reference books. Looking back, I now believe that McKenzie was on the right track when he placed the plate boundary between Europe and Africa under the Mediterranean; I had made a mistake.

My idea that the plate boundary was the Tethyan suture could be traced to my schooling in the geosynclinal theory of orogeny. The classical theory postulated that the Tethyan mountains owed their origin to the collision of Europe and Africa; the hypothesis was explicitly advanced by Argand in 1911, and has remained a paradigm to the present day. In Argand's scheme the present Mediterranean Sea did not come into existence until the Tertiary, after the Europe-Africa collision. He was wrong, because the eastern Mediterranean is Mesozoic or older; only the Balearic, Tyrrhenian, and Aegean basins are Neogene. Just the same, many of us continue to follow Argand's tectonic synthesis and assume that Italy, or the "Adriatic plate," is a North African promontory.

The two-plate theory of Alpine tectonics has led to numerous contradictions. In our papers on the Alpine-Mediterranean tectonics, both Alan Smith and I had to assume the existence of several microcontinents between Europe and Africa to account for the geology of Mediterranean Europe. Dewey and others went to extremes and assumed some 20 microplates. In this *Geology of Switzerland*, I made a valiant attempt to reduce the number of plates and still preserve the essence of the plate-tectonic doctrine. A three-plate model was formulated, assuming the presence of a Mediterranean plate between Europe and Africa (see Chapter 12).

Many puzzles of the Alpine-Mediterranean tectonics cannot be resolved by assuming this three plate-model. Why was the European active margin not fringed by a magmatic arc, such as that on the active margin of circum-Pacific continents? Where was the eastern extension of the Penninic Ocean? How did the Paleotethys evolve into the Tethys and the Mediterranean? What is the geological history of the eastern Mediterranean? Is it a relic of the Paleotethys, or a relic of the Neotethys? What is the origin of the Mediterranean Ridge, an outer ridge south of a trench or a special kind of fore-arc basin? What is the origin of the Alboran Sea and the continental crustal fragments found in the mountains around the Alboran Sea?

Whereas the geology of the Alps can be explained in terms of interactions between the European and Mediterranean plates, the origin of the Mediterranean remains puzzling, because of uncertainties concerning the plate interaction between the Mediterranean and African plates.

I have been working for the last 15 years in the Paleotethys/Tethys mountains from Indonesia at one end to Spain at the other, but mostly in China and Switzerland. The revelation came to me last October that we were all wrong in assuming that the Western Alps were formed by the continental collision of Europe and Africa. The inspiration was a talk by Jan Makris at Trieste; he told us that the ocean floor of the eastern Mediterranean Sea between the Mediterranean Ridge and North Africa is reversely magnetized. Suddenly, it occurred to me that this is the relic of the Paleotethys Ocean that we have been searching for during the last few decades. North of the relic ocean should be marginal seas on the active margin of the European plate. There is no room for a Mediterranean plate. The southern continental margin of Europe is manifested by the passive-margin sequence in the Helvetic Alps, but the southern plate-margin of Europe should have been an active margin, a frontal volcanic arc, like the Banda-Sunda Arc of the Indonesian Archipelago.

Nowhere in the Mediterranean is such a frontal volcanic arc apparent. It must have been buried, I concluded. The fossil island-arc and ocean trench must be buried under the late Neogene sediments of the Mediterranean Ridge. During the last few months I visited several geophysics institutes in Europe and searched for vestiges of the buried arc. I found an arcuate belt of a positive magnetic anomaly, fringed by an arcuate belt of a negative free-air gravity anomaly, marking the position of an arc-trench complex that is now buried under the upper Neogene sediments on the southern slope of the

Mediterranean Ridge. This was the southern plate-margin of Europe until the Middle Miocene, before the active margin was shifted to its present site under the Hellenic Arc. South of the buried arc-trench complex is the negatively magnetized ocean crust. Europe and Africa are still being separated by the last relic of an open ocean. Its size has been much reduced since Permian time, but the two continents have not yet met in a collision.

If the southern boundary of the European plate is placed south of the summit of the Mediterranean Ridge, the areas of Mesozoic and Cenozoic tectonic activity in southern and central Europe have to be considered back-arc deformations. The simultaneous compressional, extensional, and transcurrent deformations in the back-arc region of a continental plate are exemplified by the geology of the Southwest Pacific. The plate boundary of Asia is defined by the Banda-Sunda Arc of Indonesia and the Mariana Arc of the West Pacific (Figure 17.1). Between the active plate-margin defined by the frontal arc and the Asian mainland are numerous remnant arcs and back-arc basins. Some of the basins are still active centers of seafloor spreading, such as the Mariana Basin behind the Mariana Arc. Some are no longer active, such as the West Philippine Basin or the Malaysian Basin. Some are actively being pressed together, such as the Celebes Basin, the Sulu Basin, and the South China Sea. The lithosphere of the South China Basin is thrust under the Manila Trench (M.T. in Figure 17.1), and the northern tip of the basin is eliminated by the collision of mainland Asia with the Philippine Arc along the Kenting Mélange zone of the Taiwan Central Range (K.M. in the figure). Some back-arc basins have already been completely eliminated by a process called back-arc basin collapse. The back-arc basin between what was once an active volcanic arc of southern Borneo and a remnant nonvolcanic arc of Sarawak was, for example, a Cretaceous and Paleogene basin. The basin was consumed by an arc-arc collision in

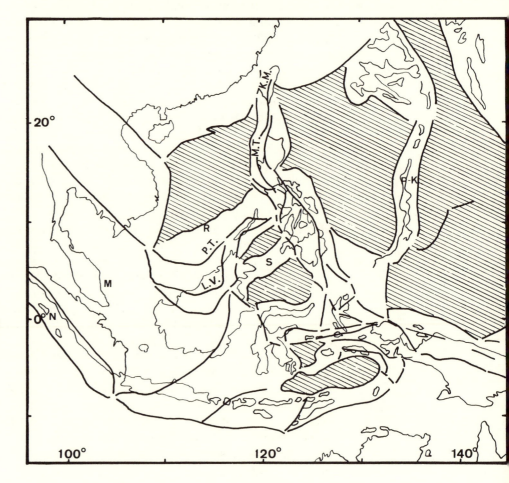

Fig. 17.1. Back-arc basins and remnant arcs of the Southwest Pacific. K.M.: Kenting Mélange; L.V.: Lupar Valley Mélange; M: Malaysian Basin; M.T.: Manila Trench; N: Nios Island; P-K: Palau-Kyushu Ridge; S: Sulu Ridge; P.T.: Palawan Trough; R: Reed Bank.

the Late Eocene, marked by the suture of the Lupar Valley Mélange (L.V. in Figure 17.1). The Palawan Trough (P.T.) is another collapsed back-arc basin. Between the back-arc basins are remnant arcs, such as the Palau-Kyushu Ridge (P-K) between the Western Philippine and Parece-Vela basins, the Sulu Ridge (S) between the Sulu and Celebes basins, and the Reed Bank (R) between the South China Basin and the Palawan Trough. The sedimentary sequences on top of the subsiding remnant arcs are similar to those on passive margins.

The key to understanding the geology of the Alps is the belated recognition that the southeastern European margin was a magmatic arc in Permian time like the Andes today (see Chapter 12). This northern margin of the Paleotethys Ocean was changed during the Mesozoic into an island-arc type of active margin. Between the continent (Europe) and the ocean (Paleotethys) should have been an outer volcanic arc. The existence of such an arc in the Eratosthenes and Medina areas has been verified by geophysical evidence. Elsewhere the arc is deeply buried under the sediments of the Mediterranean Ridge.

The geological history of Europe is a manifestation that the active margin of Mediterranean Europe has been continuously deformed since the early Mesozoic (Figure 17.2). There were numerous small centers of seafloor spreading, resulting in the genesis of the back-arc basins north of the frontal arc, and those basins were floored by ocean crust or very thin continental crust. Various types of deep-sea deposits have been laid down on the deep-sea floor of those basins, including pelagic, hemipelagic, and flysch sediments. There were also numerous remnant arcs, such as the Brianconnais Swell, and realms of platform-carbonate and hemipelagic sedimentation (in Italy and the Balkans), which are traditionally considered African passive margin.

Assuming that the frontal volcanic arc was present where the Mediterranean Ridge is, the lack of magmatic activity in the Tethyan mountains is understandable. The Tethyan back-arc basins, bounded by remnant arcs, were situated far from the frontal arc at the southern plate-boundary. There was magmatic activity during the Permiam and Triassic, when the frontal arc was not yet far distant. Arc magmatism near local subduction zones caused the intrusion of granite at depth and the extrusion of volcanics at the surface.

The Neo-Tethyan basins, according to this scenario of a "soft underbelly of Europe," were located in a nonrigid part of a lithospheric plate near an active margin. The back-arc basins of Mediterranean Europe were not eliminated by intercontinental collision. The back-arc basin collapse and the consequence arc-continent or arc-arc collisions caused the genesis of the Alpine-Mediterranean mountains such as the Pyrenees,

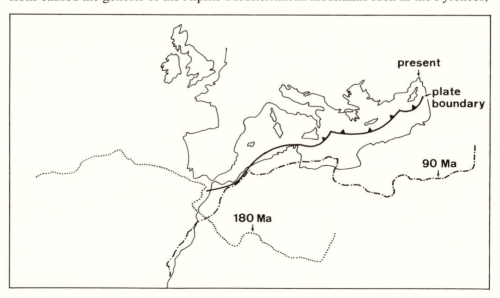

Fig. 17.2. Position of the present northern continental margin of Africa during the last 180 million years. On the basis of the data provided by magnetic lineations of the Atlantic Ocean, the African plate was found to have first moved eastward from 180 to 90 Ma, and to have rotated counterclockwise since then. Subduction of the lithospheric plate under the Paleotethys during the early Mesozoic caused the change from an Andean type to an island-arc of European margin. Subduction of the plate caused the consumption of the Paleotethys; the ocean basin south of the Mediterranean Ridge is the last remnant of the once vast ocean.

the Western Alps, the Apennines, the Carpathians, the Dinarides, the Hellenides, the Taurides, the Troodos, etc. Orogenic deformations were largely a manifestation of intraplate adjustments. Local Benioff zones of subduction were present in back-arc regions, where the lithospheric plate under a back-arc basin was subducted down a back-arc trench. Ophiolite mélanges were formed at shallow depths and high-pressure metamorphic rocks were formed at great depths. Small areas underlain by ocean crust in Mediterranean Europe, such as the Caspian Sea, the Black Sea, the Antalya Basin, the Rhodos Basin, and part of the Ionian Basin, are relic back-arc basins; they were formed during Jurassic or Cretaceous back-arc spreading, and they were separated from one another after the intervening remnant arcs collided.

This southern island-arc margin of Europe extended westward from the eastern Mediterranean and became the transform margin of the western Mediterranean, where the plate boundary is now buried under the shelf sediments of the Mediterranean or the Numidian Flysch of North Africa. My new interpretation of the Alpine-Mediterranean tectonics is that the collision of the European and African continents has not yet taken place. The active margin of Europe is still separated from the passive margin of Africa by the Levantine Sea in the eastern Mediterranean. The passive margin of continental Europe, behind the Balearic and Tyrrhenian back-arc basins, is still separated from the transform margin of Africa by the western Mediterranean.

The postulate of active orogenesis in back-arc regions inside an active plate-margin verifies the geological evidence against the theory of plate tectonics. The severe deformation of orogenic belts of considerable width clearly falsifies the assumption of the rigidity of lithospheric plates. Other mountain chains, such as the Appalachians, the Caledonides, the American Cordilleras, and the various orogenic belts of China discussed in Chapters 13, 14, and 15 of this opus, also give evidence that orogenic deformation was largely ductile deformation in back-arc regions, long before intercontinental collision took place. The model of orogenesis by back-arc collapse is more the rule than the exception.

I have communicated my hypothesis on the origin of mountains to the Planning Committee of the International Ocean Drilling Project. They recognize that the idea contradicts conventional wisdom, but they also recognize that I have seldom erred with my unconventional postulates. I have thus been encouraged to submit a proposal to search, in the area east and south of Cyprus, for the buried Mesozoic/Cenozoic European plate margin. I am hopeful that the ruling paradigm of the Wilsonian cycle can finally be falsified by drilling.

Kenneth J. Hsü
Zürich, May 7, 1993

References

Ackermann, A. 1986. Le Flysch de la nappe du Niesen. *Eclogae Geol. Helv.* 79:641–84.

Allemann, F. 1952. Die "Couches rouges" der Sulzfluh-Decke im Fürstentum Liechtenstein. *Eclogae Geol. Helv.* 45:294–98.

American Commission on Stratigraphic Nomenclature. 1961. Code of stratigraphic nomenclature. *Am. Assoc. Pet. Geol. Bull.* 45:645–65.

Ampferer, O., and W. Hammer. 1911. Geologischer Querschnitt durch die Ostalpen von Allgäu zum Gardasee. *Jahrb. Geol. Reichsanst.* 61:531–710.

Amstutz, A. 1952. Sur l'évolution des structures alpines. *Archives Sci.* 4:323–29.

Amstutz, A. 1971. Formation des Alpes dans le segment Ossola-Tessin. *Eclogae Geol. Helv.* 64:149–50.

Andreasson, P. G., D. G. Gee, and S. Sukotjo. 1985. Seve eclogites in the Norrbotten Caledonides, Sweden. In D. G. Gee and B. A. Sturt (eds.), *The Caledonide Orogen: Scandinavia and Related Areas*, pp. 887–902. Chichester: Wiley.

Aplonov, S., K. J. Hsü, and V. Ustritsky. 1992. Relic back-arc basins of Eurasia and their hydrocarbon potentials. *Island Arc*, 1:71–77.

Arbenz, K. 1947. Geologie des Hornfluhgebietes. *Beitr. Geol. Karte Schweiz, N. F.* 89:1–91.

Arbenz, P. 1919. Probleme der Sedimentation und ihre Beziehung zur Gebirgsbildung in den Alpen. *Naturforsch. Ges. Zürich, Vierteljahresschrift, Jahrg.* 64:246–75.

Argand, E. 1911. Les nappes de recouvrement des Alpes Pennines et leurs prolongements structuraux. *Mat. Carte Géol. Suisse, N. S.* 31.

Argand, E. 1916. Sur l'arc des Alpes Occidentales. *Eclogae Geol. Helv.* 14:145–91.

Argand, E. 1922. La géologie des environs de Zermatt. *Actes Soc. Helv. Sci. Nat.* 8:96–110.

Argand, E. 1924. La tectonique de l'Asie. *Proc. 13th Congr. Géol. Int.* 1:171–372.

Armstrong, T. R., R. J. Tracy, and W. E. Hames. 1992. Contrasting styles of Taconic, East Acadian and West Acadian metamorphism, central and western New England. *J. Metamorph. Geol.* 10:415–26.

Badoux, H. 1945. La géologie de la Zone des cols entre la Sarine et le Hahnenmoos. *Beitr. Geol. Karte Schweiz, N. F.* 84:1–70.

Badoux, H. 1967. Géologie abrégée de la Suisse. In *Guide géologique de la Suisse*, Part 1, pp. 1–44. Basel: Wepf.

Badoux, H., M. Burri, E. Lanterno, and M. Vuagnat. 1967. Excursion No. 6. In *Geologischer Führer der Schweiz*, Part 2, pp. 95–108. Basel: Wepf.

Badoux, H., and G. de Weisse. 1959. Les bauxites siliceuses de Dréveneuse. *Soc. vaudoise Sci. Nat. Bull.* 67:169–77.

Bailey, E. B. 1935. *Tectonic Essays, Mainly Alpine*. Oxford: Clarendon. 200 pp.

Bailey, E. H., W. P. Irwin, and D. L. Jones. 1964. Franciscan and related rocks and their significance in the geology of western California. *Calif. Div. Mines Geol. Bull.* 183:1–177.

Baird, A. W., and J. F. Dewey. 1986. Structural evolution in thrust belts and relative plate motion: the upper Pennine Piedmont Zone of the internal Alps, southwest Switzerland and northwest Italy. *Tectonics* 5:375–87.

Barbier, R. 1948. Les zones ultradauphinoises et subbriançonnaises. *Corte Géol. Fr. Mém.*, 291 pp.

Barth, A. P., and D. J. May. 1992. Mineralogy and pressure-temperature-time path of Cretaceous granulite gneisses, south-eastern San Gabriel Mountains, southern California. *J. Metamorph. Geol.* 10:529–44.

Barth, A. P., J. L. Wooden, and D. J. May. 1992. Small scale heterogeneity of Phanerozoic lower crust: evidence from isotopic and geochemical systematics of mid-Cretaceous granulite gneisses, San Gabriel Mountains, southern California. *Contrib. Mineral. Petrol.* 109:394–407.

Baud, A. 1972. Observations et hypothèses sur la géologie de la partie radicale des Préalps médians. *Eclogae Geol. Helv.* 65:43–55.

Bayer, A. 1982. Untersuchungen im Habkern-Mélange ("Wildflysch") zwischen Aare und Rhein. ETH Zürich, Dissertation no. 6950. 184 pp.

Bearth, P. 1973. Gesteins- und Mineralparagenesen aus den Ophiolithen von Zermatt. *Schweiz. Mineral. Petrogr. Mitt.* 53:299–334.

Bearth, P. 1974: Zur Tektonik der Ossola- und Simplon-Region. *Eclogae Geol. Helv.* 67:509–513.

Bearth, P. 1976. Zur Gliederung der Bündnerschiefer in the Region Zermatt. *Eclogae Geol. Helv.* 69:149–61.

Bearth, P., W. Nabholz, A. Streckeisen, and E. Wenk. 1967. Simplonpass: Brig-Domodossola. In *Geologischer Führer der Schweiz*, Part 5, pp. 336–50. Basel: Wepf.

Beck, P. 1911. Geologie der Gebirge nördlich von Interlaken. *Beitr. Geol. Karte Schweiz, N. F.* 29:1–100.

Beck, P. 1918. Die Niesen-Habkerndecke und ihre Verbreitung im helvetischen Faciesgebiet. *Eclogae Geol. Helv.* 12:65–151.

Bernoulli, D. 1964. Zur Geologie des Monte Generoso. *Beitr. Geol. Karte Schweiz, N. F.* 118:1–134.

Bertrand, M. 1884. Rapports de structure des Alps de Glaris et du bassin houiller du Nord. *Bull. Soc. Géol. Fr.* 3:318–30.

Biju-Duval, B., J. Dercourt, and X. Le Pichon. 1977. From the Tethys Ocean to the Mediterranean Sea: a plate tectonic model of the evolution of the western Alpine system. In B. Biju-Duval, and L. Montadert (eds.), *Structural History of the Mediterranean Basins*, pp. 143–83. Paris: Editions Technip.

Biq, C. 1977. The Kenting Melange and the Manila Trench. *Proc. Geol. Soc. China* 15:119–22.

Bitterli, P. 1945. Geologie der Blauen- und Landskronkette südlich von Basel. *Beitr. Geol. Karte Schweiz, N. F.* 81:1–73.

Blaas, J. 1902. *Geologischer Führer durch die Tiroler und Vorarlberger Alpen.* Innsbruck.

Blau, R. V. 1966. Molasse und Flysch im östlichen Gurnigelgebiet (Kt. Bern). *Beitr. Geol. Karte Schweiz, N. F.* 125:1–151.

Bluck, B. J., A. N. Halliday, M. Aftalion, and R. M. MacIntyre. 1980. Age and origin of Ballantrae ophiolite and its significance to the Caledonian orogeny and Ordovician time scale. *Geology* 8:492–95.

Blüm, W. 1987. Diagenese permischer Schüttfächer-Sandsteine der Nordschweiz. *Eclogae Geol. Helv.* 80:369–82.

Bocquet, J., M. Delaloye, and J. C. Hunziker. 1974. K-Ar and Rb-Sr dating of blue amphiboles, micas and associated minerals from the western Alps. *Contrib. Mineral. Petrol.* 47:7–26.

Bolli, H. 1944. Zur Stratigraphie der Oberen Kreide in den höheren helvetischen Decken. *Eclogae Geol. Helv.* 37:218–328.

Bolli, H., M. Burri, A. Isler, W. Nabholz, N. Pantic, and P. Probst. 1980. Der nordpenninische Raum zwischen Westgraubünden und Brig. *Eclogae Geol. Helv.* 73:779–97.

Briegel, U., and C. Goetze. 1978. Estimates of differential stress recorded in the dislocation structure of Lochseiten limestone (Switzerland). *Tectonophysics* 48:61–76.

Bucher, K. 1977. Die Beziehung zwischen Deformation, Metmorphose und Magmatismus im Gebiet der Bergeller Alpen. *Schweiz. Mineral. Petrogr. Mitt.* 57:414–34.

Bucher, W. 1933. The Deformation of the Earth's Crust. Princeton, N.J.: Princeton University Press. 518 pp.

Bullard, E. C., J. E. Everett, and A. G. Smith. 1965. The fit of the continents around the Atlantic. *R. Soc. London, Philos. Trans. (A)* 258:41–51.

Burchfiel, B. C., and G. A. Davis. 1981. Mohave Desert and environs. In W. G. Ernst (ed.), *The Geotectonic Development of California*, pp. 217–52. Englewood Cliffs, N.J.: Prentice-Hall.

Bürgisser, H. M. 1980. Der "Appenzellergranit"-Leitniveau des Hörnli-Schuttfächers. ETH Zürich, Dissertation no. 6582. 196 pp.

Burke, K., J. F. Dewey, and W.S.F. Kid. 1976. Dominance of horizontal movements, arc and microcontinental collisions during the later permobile regime. In B. Windley (ed.), *The Early History of the Earth*, pp. 113–30. London: Wiley.

Burri, M. 1958. La zone de Sion-Courmayeur au Nord du Rhône. *Beitr. Geol. Karte Schweiz, N. F.* 105:1–45.

Burri, M. 1967. De Pont de la Morge à Sierre. In *Geologischer Führer der Schweiz*, Part 3, pp. 131–35. Basel: Wepf.

Buxtorf, A. 1907. Geologie des Weissensteintunnels. *Beitr. Geol. Karte Schweiz, N. F.* 21.

Buxtorf, A. 1908. Zur Tektonik der Zentralschweizerischen Kalkalpen. *Zentralbl. Dtsch. Geol. Ges.* 60:126–97.

Buxtorf, A. 1918. Ueber die tektonische Stellung der Schlieren und Niesenflyschmasse. *Verh. Naturforsch. Ges. Basel* 29:270–75.

Cadisch, J. 1953. Geologie der Schweizer Alpen. Basel: Wepf. 480 pp.

Cadisch, J., and A. Streckeisen. 1967. Klosters-Dorf—Davos. In *Geologischer Führer der Schweiz*, Part 8, pp. 727–34. Basel: Wepf.

Cadisch, J., P. Bearth, and F. Spaenhauer. 1941. Ardez. *Erläuterungen geol. Atlas Schweiz.* Bern: Francke. 51 pp.

Cadisch, J., H. Eugster, and E. Wenk. 1963. Scuol-Tarasp. *Erläuterungen geol. Atlas Schweiz.* Bern: Francke. 51 pp.

Caminos, R. 1979. Cordillera Frontal. In *Geologia Regional Argentina*, pp. 397–453. Cordoba, Argentina: Acad. Nac. Ciencias de Cordoba.

Caron, C. 1972. La Nappe Supérieure des Prealps. *Eclogae Geol. Helv.* 65:57–73.

Caron, C. 1976. La nappe du Gurnigel dans les Préalpes. *Eclogae Geol. Helv.* 69:297–308.

Caron, C., and A. Escher. 1980. Excursion II. In *Geology of Switzerland: A guide-book*. Basel: Wepf.

Chamberlin, R. T. 1910. The Appalachian folds of central Pennsylvania. *J. Geology* 18:228–51.

Chen, G. D. 1989. *Tectonics of China*. Oxford: Pergamon. 258 pp.

Chen, H. 1975. On the isotopic ages of some granites and metamorphic rocks from Northwest China. *Acta. Geol. Sin.* 49:45–60.

Chen, Pei-ji. 1991. Classification and correlation of Cretaceous in South China (in Chinese). In X. G. Zhu (ed.), *Cretaceous System of South China*, pp. 25–40. Nanjing: Nanjing University Press.

Cheng, Y. 1986. Magmatic and metamorphic rocks of China. In Z. Yang, Y. Cheng, and H. Wang, *The Geology of China*, pp. 187–234. Oxford: Oxford University Press.

Cita, M. B., and A. Camerlenghi. 1992. The mud diapirs of the Mediterranean Ridge. *Proc. Congr. Comm. Int. Expl. Sci. Med.* 33:388.

Cloos, M. 1982. Flow melanges: numerical modelling and geologic constraints on their origin in the Franciscan subduction complex, California. *Geol. Soc. Am. Bull.* 93:330–45.

Cook, F. A., D. S. Albaugh, L. D. Brown, S. Kaufman, J. E. Oliver, and R. D. Hatcher, Jr. 1979. Thin-skinned tectonics in the crystalline southern Appalachians: COCORP seismic reflection profiling of the Blue Ridge and Piedmont. *Geology* 7:563–67.

Cornelius, H. P. 1913. Geologische Beobachtungen im Gebiete des Forno-Gletschers (Engadin). *Zentralbl. Mineral. Geol. Paleontol.* 1913:246–52.

Cornelius, H. P. 1935. Geologie der Err-Julier-Gruppe. *Beitr. Geol. Karte Schweiz, N. F.* 70 (Part 1): 1–321.

Cornelius, H. P. 1940. Zur Auffassung der Ostalpen im Sinne der Deckenlehre. *Z. Dtsch. Geol. Ges.* 92:4–5.

Coward, M. P. 1984. The Moine Thrust structures. In D. J. Fettes, and A. L. Harris, (eds.), *Synthesis of the Caledonian Rocks of Britain*, pp. 259–80. Dordrecht: Reidel.

Coward, M. P., D. Dietrich, and R. G. Parks. 1989. *Alpine Tectonics*. London: Geological Society, Special Publication no. 45. 450 pp.

Cross, T. A., and R. H. Pilger, Jr. 1978. Constraints on absolute motion and plate interaction inferred from Cenozoic igneous activity in the western United States. *Am. J. Sci.* 278:865–902.

Crowell, J. C. 1955. Directional-current structures from the Prealpine Flysch, Switzerland. *Geol. Soc. Am. Bull.* 66:1351–84.

Crowell, J. C. 1981. An outline of the tectonic history of southeastern California. In W. G. Ernst (ed.), *The Geotectonic Development of California*, pp. 583–613. Englewood Cliffs, N.J.: Prentice-Hall.

Dal Piaz, G. V. 1972. Le métamorphisme de haute pression et basse temperature dans l'évolution du bassin ophiolithique alpino-appenninique. *Schweiz. Mineral. Petrogr. Mitt.* 54:399–424.

Dalziel, I.W.D. 1981. Back-arc extension in the southern Andes: A review and critical appraisal. *Philos. Trans. R. Soc. London, Ser. A* 300:319–35.

Dana, J. D. 1873. On some results of the earth's contraction from cooling, including a discussion of the origin of mountains and the nature of the earth's interior: *Am. J. Sci., ser. 3* 5:423–43; 6:6–14, 104–15, 161–72.

de Raaf, M. 1934. La géologie de la nappe du Niesen entre Sarine et la Simme. *Beitr. Geol. Karte Schweiz, N. F.* 68:1–105.

Debelmas, J. 1955. Les zones subbrianconnais et brianconnais occidentale entre Vallouise et Guillestre (Hautres-Alpes). *Carte Géol. Fr. Mém.*, 171 pp.

Deutsch, A. 1973. Datierung an Alkaliamphibolen und Stilpnomelan aus der südlichen Platta-Decke (Graubünden). *Eclogae Geol. Helv.* 76:295–308.

Dewey, J. F. 1988. Lithospheric stress, deformation, and tectonic cycles: The disruption of Pangaea and the closure of Tethys. In M. G. Audley-Charles, and A. Hallam, *Gondwana and Tethys*, pp. 23–40. Oxford: Oxford University Press.

Dewey, J. F., W. C. Pitman, W.B.F., Ryan, and J. Bonin. 1973. Plate-tectonics and the evolution of the alpine system. *Geol. Soc. Am. Bull.* 84:3137–80.

Dickinson, W. R. 1970. Second Penrose conference: new global tectonics. *Geotimes* 15 (no. 4):18–22.

Diebold, P., and W. M. Müller. 1985. Szenarien der geologischen Langzeitsicherheit. *NAGRA Tech. Rep. 84–26.* 110 pp.

Diefenbach, H. L. 1988. Geology of the Wilerhorn Region. ETH Zürich, *Diplomarbeit.* 94 pp.

Dietrich, D., and M. Casey. 1989. A new tectonic model for the Helvetic nappes. In M. P. Corward, D. Dietrich, and R. G. Park (eds.), *Alpine Tectonics*, pp. 47–63. London: Geological Society, Special Publication no. 45.

Dobson, J. P., and F. Heller. 1993. Triassic paleomagnetic results from the Yangtze Block, S. E. China. *Geophys. Res. Lett.* 20:1391–94.

Dössegger, R., and W. Müller. 1976. Die Sedimentserien der Engadiner Dolomiten und ihre lithostratigraphische Gliederung. *Eclogae Geol. Helv.* 69:229–38.

Dronkert, H. 1987. Diagenesis of Triassic evaporites in northern Switzerland. *Eclogae Geol. Helv.* 80:397–414.

Eberli, G. 1987. Die jurassischen Sedimente in den Ostalpinen Decken Graubündens. ETH Zürich, Dissertation no. 7835. 203 pp.

Eberli, G. 1988. The evolution of the southern continental margin of the Jurassic Tethys Ocean as recorded in the Allgäu Formation of the Austroalpine Nappes of Graubünden. *Eclogae Geol. Helv.* 81:175–214.

Ehlig, P. 1958. Geology of the Mount Baldy region of the San Gabriel Mountains, California. UCLA, *Ph.D. Dissertation.* 153 pp.

Ehlig, P. 1981. Origin and tectonic history of the basement terrane of the San Gabriel Mountains, Central Transverse Ranges. In W. G. Ernst (ed.), *The Geotectonic Development of California*, pp. 252–83. Englewood Cliffs, N.J.: Prentice-Hall.

Elie de Beaumont, L. 1830. De l'age relatif des montagnes. *Rev. Fr.* 15:1–58.

Elie de Beaumont, L. 1831. Researches on some of the revolutions on the surface of the Globe. *Philos. Mag., N. S.* 10:241–64.

Ellenberger, F. 1952. Sur l'extension des faciès brianconnais en Suisse, dans les Préalpes médians et le Pennides. *Eclogae Geol. Helv.* 45:285–86.

England, P. C., and T.J.B. Holland. 1979. Archimedes and the Tauern eclogites: the role of buoyancy in the preservation of exotic tectonic blocks. *Earth Planet. Sci. Lett.* 44:287–94.

Ernst, W. G. 1975. Systematics of large-scale tectonics and age progressions in Alpine and circum-Pacific blueschist belts. *Tectonophysics* 26:229–46.

Escher, A. 1846. Gebirgskunde. In O. Heer and J. J. Blumer-Heer (eds.), *Der Canton Glarus*, vol. 7, *Gemälde der Schweiz*. St. Gall: Huter. 41 pp.

Escher, A. 1866. Sur la géologie du Canton de Glaris. *Actes Soc. Helv. Sci. Nat., 50e session, Neuchâtel*, pp. 71–75.

Escher, A. 1988. Sur la géologie du Grand Saint Bernard entre le val de Bagnes et le Mischabel. *BfU Landeshydrologie und Geol. Geol. Ber.* 6:1–26.

Evans, B. W. 1989. Metamorphism under extreme conditions. *Episodes* 12:191–92.

Felber, P. 1984. Der Dogger der zentralschweizerischen Klippen. *Mitt. Geol. Inst. ETH, N. F.* 246:1–256.

Fettes, D. J., A. L. Harris, and L. M. Hall. 1984. The Caledonian geology of the Scottish Highlands. In D. J. Fettes and A. L. Harris (eds.), *Synthesis of the Caledonian Rocks of Britain*, pp. 303–34. Dordrecht: Reidel.

Finger, W. 1978. Die Zone von Samaden (Unterostalpine Decken, Graubünden) und ihre jurassischen Brekzien. *Mitt. Geol. Inst. ETH, N. F.* 224.

Finger, F., and H. P. Steyrer. 1990. I-type granitoids as indicators of a late Paleozoic convergent ocean-continent margin along the southern flank of the central European Variscan orogen. *Geology* 18:1207–10.

Flisch, M. 1986. Die Hebungsgeschichte der oberostalpinen Silvretta-Decke seit der mittleren Kreide. *Bull. Ver. Schweiz. Pet. Geol. Ing.* 53:23–49.

Flores, G. 1955. Discussion. In E. Beneo, Les résultats des études pour la recherche pétrolifere en Sicilie (Italie). *Proc. 4th World Petroleum Congr.* 1:121–22.

Flügel, H. W., F. P. Sassi, and P. Grecula. 1987. Pre-Variscan and Variscan events in the Alpine-Mediterranean Mountain Belts. Bratislava: Alfa. 487 pp.

Franke, W. 1989. Tectonostratigraphic units in the Variscan belt of central Europe. *Geol. Soc. Am. Spec. Paper* 230:67–90.

W. Frei, P. Heitzmann, P. Lehner, and P. Valasek. 1989. Die drei Alpentraversen von NFP 20. *Bull. Ver. Schweiz. Pet. Geol. Ing.* 55:13–43.

Frey, M., V. Trommsdorff, and E. Wenk. 1980. Alpine metamorphism of the Central Alps. Excursion no. VI. In Swiss Geological Committee (ed.), *Geology of Switzerland*, pp. 295–316. Basel: Wepf.

Fricker, P. 1960. Geologie der Gebirge zwischen Val Ferret und Combe d'A, Wallis. *Eclogae Geol. Helv.* 53:33–132.

Frotzheim, N., and G. P. Eberli. 1990. Extensional detachment faulting in the evolution of a Tethys passive continental margin, Eastern Alps, Switzerland. *Geol. Soc. Am. Bull.* 102:1297–1308.

Fuchs, T. 1877. Welche Ablagerungen haben wir als Tiefseebildungen zu betrachten? *N. Jahrb. Mineral. Geol. Paläont. Beil.* 2:487–584.

Fujian Bureau of Geology. 1985. Regional Geology of Fujian Province. Beijing: Geology Publishing House. 673 pp.

Fumasoli, M. W. 1974. Geologie des Gebietes nördlich und südlich der orio-Tonale-Linie, im Westen von Gravedona (Como, Italia). ETH Zürich, *Dissertation.* 230 pp.

Gagnebin, E. 1942. Les idées actuelles sur la formation des Alps. *Acta Soc. Helv. Sci. Nat.* 1942:47–53.

Gansser, A. 1967. Splügen-San Bernardino-Misox-Castione. In *Geologischer Führer der Schweiz*, Part 8, pp. 803–13. Basel: Wepf.

Gautschi, A., and A. Montrasio. 1978. Die andesitisch-basaltischen Gänge des Bergeller Ostrandes und ihre Beziehung zur Regional- und Kontakmetamorphose. *Schweiz. Mineral. Petrogr. Mitt.* 58:329–34.

Gee, D. G., and B. A. Sturt (eds.). 1985. *The Caledonide Orogen: Scandinavia and Related Areas.* Chichester: Wiley. 1266 pp.

Geikie, A. 1905. *The Founders of Geology.* London: MacMillan. 486 pp.

Geological Atlas of China (in Chinese). 1973. Beijing: Chinese Academy of Geological Sciences. 149 pp.

Gerber, E. 1925. Geologie des Gurnigels und der angrenzenden subalpinen Molasse. *Beitr. Geol. Karte Schweiz, N.F.* 50:1–45.

Gerlach, H. 1883. Die Pennischen Alpen. *Beitr. Geol. Karte Schweiz* 27:1–159.

Gignoux, M., and L. Moret. 1933. Déscription géologique du basin supérieur de la Durance. *Trav. Lab. Géol. Univ. Grenoble* 21.

Gilluly, J. 1948. Origin of granite. *Geol. Soc. Am. Mem.* 28:1–139.

Gilluly, J. 1949. The distribution of mountain building in geologic time. *Geol. Soc. Am. Bull.* 60:561–90.

Grabau, A. W. 1924. Migration of geosynclines. *Geol. Soc. China Bull.* 3:154–69.

Grabau, A. W., and M. O'Connell. 1917. Were the graptolite shales, as a rule, deep or shallow water deposits? *Geol. Soc. Am. Bull.* 28:959–64.

Greenly, E. 1919. *The Geology of Anglesey.* London: Mem. Geol. Survey UK. 980 pp.

Gressly, A. 1837. Observations géologiques sur les terrains des chaînes Jurassiques du canton de Soleure, et des contrées limitrophes. *Actes Soc. Helv. Sci. Nat. Solothurn*, pp. 126–32.

Gressly, A. 1838. Observations géologiques sur le Jura Soleurois. *Nouv. Mém. Soc. Helv. Sci. Natur.* 2:1–349.

Griggs, D. T. 1939. A theory of mountain building. *Am. J. Sci.* 237:611–50.

Grünenfelder, M., and T. W. Stern. 1960. Das Zirkon-Alter des Bergeller-Massivs. *Schweiz. Mineral. Petrogr. Mitt.* 40:253–59.

Guillaume, H. 1957. Géologie du Montsalvens. *Beitr. Geol. Karte Schweiz, N. F.* 104:1–170.

Güller, A. 1947. Zur Geologie der südlichen Mischabel- und der Monte-Rosa-Gruppe. Mit Einschluss des Zmutt-Tales westlich Zermatt. *Eclogae Geol. Helv.* 40:39–161.

Gulsen, B. L. 1973. Age relations in the Bergell region of the South-East Swiss Alps, with some geochemical comparisons. *Eclogae Geol. Helv.* 66:293–313.

Gulsen, B. L., and T. E. Krogh. 1973. Old lead components in the young Bergell Massif, South-East Swiss Alps. *Contrib. Mineral. Petrol.* 40:239–52.

Gümbel, W. 1878. Ueber die im stillen Ocean auf dem Meeresgrunde vorkommenden Manganknollen. *Sitzungsber. Bayer. Akad. Wiss.* 8:189–209.

Gygi, R. A. 1969. Zur Stratigraphie der Oxford-Stufe (oberes Jura-System) der Nordschweiz und des süddeutschen Grenzgebietes. *Beitr. Geol. Karte Schweiz, N. F.* 136:1–123.

Haarmann, W. 1930. Die Oszillationstheorie: eine Erklärung der Krustenbewegungen von Erde und Mond. Stuttgart: Enke. 260 pp.

Hall, J. 1859. Paleontology. *Geol. Surv. New York* 3:66–96.

Hamilton, W. B. 1969. Mesozoic California and the underflow of Pacific mantle. *Geol. Soc. Am. Bull.* 67:1319–30.

Hamilton, W. B. 1987. Crustal geological processes of the United States. *Geol. Soc. Am. Mem.* 172:743–81.

Hanson, G. N., M. Grünenfelder, and G. Soptrayanova. 1969. The geochronology of a recrystallized tectonite in Switzerland—the Roffna gneiss. *Earth Planet. Sci. Lett.* 5:413–22.

Hantke, R. 1954. Die fossile Flora der obermiozänen Oehninger-Fundstelle Schrotzburg. *Denkschr. Schweiz. Naturforsch. Ges.* 82/2:1–118.

Harris, A. L., and D. J. Fettes. 1984. In D. J. Fettes and A. L. Harris (eds.), *Synthesis of the Caledonian Rocks of Britain*, pp. 113–85. Dordrecht: Reidel.

Hatcher, R. D., Jr. 1978. Tectonics of the western Piedmont and Blue Ridge, Southern Appalachians. *Am. J. Sci.* 278:276–304.

Hatcher, R. D., Jr. 1984. The Moine Thrust zone: a comparison with Appalachian faults and the structure of orogenic belts. In D. J. Fettes and A. L. Harris (eds.), *Synthesis of the Caledonian Rocks of Britain*, pp. 247–57. Dordrecht: Reidel.

Hatcher, R. D., Jr. 1987. Tectonics of the Southern and Central Appalachian Internides. *Annu. Rev. Earth Planet. Sci.* 15:337–62.

Haug, E. 1925. Contribution à une synthèse stratigraphique des Alpes occidentales. *Bull. Soc. Geol. Fr., 4th Ser.* 25:97.

Haxel, G., and J. Dillon. 1978. The Pelona-Orocopia Schist and Vincent-Chocolate Mountain thrust system, southern California. In D. G. Howell and K. A. McDougall (eds.), Mesozoic Paleogeography of the western United States, *Pacific Section, Soc. Econ. Paleontologists and Mineralogists, Pacific Coast Paleogeography Symp.* 2:453–69.

Heierli, H. 1967. Maloja-St. Moritz-Zernez. *Geologischer Führer der Schweiz*, Part 9, pp. 864–72. Basel: Wepf.

Heim, Alb. 1878. Untersuchung über den Mechanismus der Gebirgsbildung im Anschluss an die geologische Monographie der Tödi-Windgällen-Gruppe. Basel: Schawabe. Vol. 1, 348 pp.

Heim, Alb. 1919–22. *Geologie der Schweiz*. Leipzig: Tauchnitz. Vol. 1, 704 pp; vol. 2/1, 476 pp.; and vol. 2/2, 541 pp.

Heim, Alb. 1922. Die Mythen. *Neujahrsbl. Naturforsch. Ges. Zürich*, 124.

Heim, Alb. 1929. An der Erkenntniswurzel alpiner Tektonik. *Vierteljahrsschr. Naturforsch. Ges. Zürich* 1929:213–33.

Heim, Alb. 1932. *Bergsturz und Menschenleben*. Zürich: Fretz und Wasmuth. 218 pp.

Heim, Alb., and Arn. Heim. 1917: Der Kontakt von Gneiss und Mesozoikum am Nordrand des Aarmassivs bei Erstfeld. *Vierteljahrsschr. Naturforsch. Ges. Zürich* 62:423–51.

Heim, Arn. 1908. Die Nummuliten- und Flyschbildungen der Schweizeralpen. *Abh. Schweiz. Paläontol. Ges.* 35:1–301.

Heim, Arn. 1911. Zur Tektonik des Flysches in den östlichen Schweizeralpen. *Beitr. Geol. Karte Schweiz. N. F.* 31:37–48.

Heim, Arn., 1923. Der Alpenrand zwischen Appenzell und Rheintal (Fähnern-Grüppe) und das Problem der Kreiden-Nummuliten. *Beitr. Geol. Karte Schweiz, N. F.* 53:1–51.

Heinrich, C. A. 1986. Eclogite-facies regional metamorphism of hydrous mafic rocks in the Central Alpine Adula Nappe. *J. Petrol.* 27:123–54.

Herb, R. 1962. Geologie von Amden mit besonderer Berücksichtigung der Flyschbildungen. *Beitr. Geol. Karte Schweiz, N. F.* 114:1–130.

Hofmann, F. 1957. Untersuchungen in der subalpinen und mittelländischen Molasse der Ostschweiz. *Eclogae Geol. Helv.* 50:289–322.

Holmes, A. 1929. Radioactivity and earth movements. *Trans. Geol. Soc. Glasgow* 18:559–606.

Homewood, P. 1974. Le flysch du Meilleret (Préalpes romandes) et ses relations avec les unités l'encadrant. *Eclogae Geol. Helv.* 67:349–401.

Homewood, P. 1977. Ultrahelvetic and North-Penninic Flysch of the Prealps: A general account. *Eclogae Geol. Helv.* 70:627–41.

Hsü, K. J. 1955. Granulites and mylonites of the region about Cucamonga and San Antonio Canyons, San Gabriel Mountains, California. *Univ. Calif. Publ. Geol. Sci.* 30:223–324.

Hsü, K. J. 1958. Isostasy and a theory for the origin of geosynclines. *Am. J. Sci.* 256:305–27.

Hsü, K. J. 1960. Paleocurrent structures and paleogeography of the Ultrahelvetic Flysch basins. *Geol. Soc. Am. Bull.* 71:577–610.

Hsü, K. J. 1968. Principles of melanges and their bearing on the Franciscan-Knoxville paradox. *Geol. Soc. Am. Bull.* 79:1063–74.

Hsü, K. J. 1969a. Role of cohesive strength in the mechanics of overthrust faulting and of landsliding. *Geol. Soc. Am. Bull.* 80:927–53.

Hsü, K. J. 1969b. A preliminary analysis of the statics and kinetics of the Glarus overthrust. *Eclogae Geol. Helv.* 62:143–54.

Hsü, K. J. 1970. The meaning of the word flysch: A short historical search. *Geol. Assoc. Canada, Spec. Paper* 7:1–11.

Hsü, K. J. 1971a. Franciscan mélanges as a model for eugeosynclinal sedimentation and underthrusting tectonics. *J. Geophys. Res.* 76:1172–70.

Hsü, K. J. 1971b. Origin of Alps and Western Mediterranean. *Nature (London)* 233:44–48.

Hsü, K. J. 1972. Origin of saline giants. *Earth Sci. Rev.* 8:371–96.

Hsü, K. J. 1973. The Odyssey of Geosyncline. In R. N. Ginsburg (ed.), *Evolving Concepts in Sedimentology*, pp. 66–92. The Johns Hopkins University Studies in Geology, no. 21.

Hsü, K. J. 1975. Catastrophic debris streams generated by rockfalls. *Geol. Soc. Am. Bull.* 86:129–40.

Hsü, K. J. 1979. Thin-skinned plate tectonics during Neoalpine orogenesis. *Am. J. Sci.* 279:353–66.

Hsü, K. J. 1981. Thin-skinned plate-tectonic model for collision-type orogenesis. *Sci. Sin.* 24:100–10.

Hsü, K. J. 1982. *Ein Schiff revolutioniert die Wissenschaft.* Hamburg: Hofmann und Campe. 303 pp.

Hsü, K. J. 1983. Actualistic catastrophism. *Sedimentology* 30:3–9.

Hsü, K. J. 1986. *The Great Dying.* San Diego: Harcourt Brace Jovanovich. 292 pp.

Hsü, K. J. 1988. Relic back-arc basins. In K. L. Kleinspehn and C. Paola (eds.), *New Perspectives in Basin Analysis*, pp. 245–64. New York: Springer.

Hsü, K. J. 1989. Time and place in Alpine orogenesis—the Fermor Lecture. In M. P. Coward, D. Dietrich, and R. G. Parks (eds.), *Alpine Tectonics*, pp. 421–43. London: Geological Society, Special Publication no. 45.

Hsü, K. J. 1991a. Exhumation of high-pressure metamorphic rocks. *Geology* 19:107–10.

Hsü, K. J. 1991b. The concept of tectonic facies. *Tech. Univ. Istanbul, Bull.* 44:25–42.

Hsü, K. J. 1992. *Challenger at Sea.* Princeton, N.J.: Princeton University Press.

Hsü, K. J., and U. Briegel. 1991. *Geologie der Schweiz.* Basel: Birkhäuser. 219 pp.

Hsü, K. J., and K. Kelts. 1971. Seismic investigation of Lake Zurich. Part II: Geology. *Eclogae Geol. Helv.* 63:525–38.

Hsü, K. J., and K. Kelts. (eds.). 1984. Quaternary geology of Lake Zurich: An interdisciplinary investigation by deep-lake drilling. *Contrib. Sedimentol.* 13:1–210.

Hsü, K. J., J. Li, H. Chen, Q. Wang, S. Sun, and A.M.C. Sengör. 1990. Tectonics of South China: Key to understanding West Pacific geology. *Tectonophysics* 183:9–39.

Hsü, K. J., J. Li, Q. Wang, and S. Sun. 1988. Melanges around the Junggar Basin: A case history of orogenic deformation of a relic back-arc basin. *Proc. 1st Int. Conf. Asian Marine Geol.*, pp. 25–38.

Hsü, K. J., and W.B.F. Ryan. 1973. Summary of the evidence for extensional and compressional tectonics in the Mediterranean. *Initial Reports of the Deep Sea Drilling Project*, vol. 13, pp. 1011–19. Washington, D.C.: U.S. Government Printing Office.

Hsü, K. J., and S. O. Schlanger. 1971. Ultrahelvetic flysch sedimentation and deformation related to plate tectonics. *Geol. Soc. Am. Bull.* 82:1207–18.

Hsü, K. J., S. Sun, J. Li, H. Chen, and H. Pen. 1989. Mesozoic overthrust tectonics in China: Reply to a discussion. *Geology* 17:386–87; 17:669–93.

Hsü, K. J., Q. Wang, J. Li, and J. Hao. 1991. Geologic evolution of the Neimonides: A working hypothesis. *Eclogae Geol. Helv.* 84:1–30.

Hsü, K. J., Q. Wang, J. Li, D. Zhou, and S. Sun. 1987. Tectonic evolution of Qinling Mountains, China. *Eclogae Geol. Helv.* 80:735–52.

Hsü, K. J., Y. Yao, J. Li, and Q. Wang. 1992. Geology of the Beishan Mountains and the tectonic evolution of Northwest China. *Eclogae Geol. Helv.* 85:213–25.

Huang, T. K. 1935. Etude géologique de la région Weissmies-Portjengrat (Valais). *Bull. Soc. Neuchâteloise Sci. Nat.* 60.

Hubbert, M. K., and W. W. Rubey. 1959. Role of fluid pressure in mechanics of overthrust faulting. *Geol. Soc. Am. Bull.* 70:115–206.

Hug, J. 1917. Die letzte Eiszeit der Umgebung von Zürich. *Vierteljahresschr. Naturforsch. Ges. Zürich* 62:125–42.

Hunziker, J. 1974. Rb-Sr and K-Ar age determination and the Alpine tectonic history of the Western Alps. *Ist. Geol. Min. Univ. Padova, Mem.* 31:1–54.

Hunziker, J. C., J. Desmons, and G. Martinotti. 1989. Alpine thermal evolution in the central and western Alps. In M. P. Coward, D. Dietrich, and R. G. Park (eds.), *Alpine Tectonics*, pp. 353–67. London: Geological Society, Special Publication no. 45.

Hurford, A. J., M. Flisch, and E. Jäger. 1989. Unravelling the thermo-tectonic evolution of the Alps: a contribution from fission track analysis and mica dating. In M. P. Coward, D. Dietrich, and R. G. Park. (eds.), *Alpine Tectonics*, pp. 369–98. London: Geological Society, Special Publication no. 45.

Hutton, D.H.W. 1989. Pre-Alleghanian terrane tectonics in the British and Irish Caledonides. *Geol. Soc. Am., Spec. Paper* 230:47–65.

Iten, W. B. 1948: Zur Stratigraphie und Tektonik der Zone du Combin. *Eclogae Geol. Helv.* 41:144–246.

Jäckli, H. 1941. Geologische Untersuchungen im nördlichen Westschams (Graubünden). *Eclogae Geol. Helv.* 34:17–105.

Jäger, E. 1973. Die alpine Orogenese im Lichte der radiometrischen Altersbestimmung. *Eclogae Geol. Helv.* 66:11–21.

Jeannet, A., W. Leupold, and D. Beck. 1935. Stratigraphische Profile des Nummulithikums von Einsiedeln-Iberg. *Ber. Schweiz. Naturforsch. Ges.* 1:35–51.

Joly, J. J. 1926. *The Surface History of the Earth.* Oxford: Oxford University Press.

Jones, O. T. 1938. On the evolution of a geosyncline. *Q. J. Geol. Soc. London* 94:60–110.

Karig, D., and J. C. Ingle *et al.* 1975. *Initial Reports of the Deep Sea Drilling Project*, vol. 31. Washington, D.C.: U.S. Government Printing Office. 927 pp.

Katz, H. R. 1973. Contrasts in tectonic evolution of orogenic belts in the Southeast Pacific. *J. R. Soc. N. Z.* 3:333–62.

Kaufmann, F. J. 1877. Geologische Beschreibung der Kalk und Schiefergebirge der Kantone Schwyz und Zug und des Bürgenstocks bei Stans. *Beitr. Geol. Karte Schweiz* 14:1–180.

Kaufmann, F. J. 1886. Emmen- und Schlierengegenden nebst Umgebungen bis zur Brünig-strasse und Linie Lungern-Grafenort. *Beitr. Geol. Karte Schweiz* 24:1–607.

Kelling, G., W. S. McKerrow, and P. Stone. 1984. Field guide to the Lake District and Southern Uplands. In D. J. Fettes and A. L. Harris. (eds.), *Synthesis of the Caledonian Rocks of Britain*, pp. 77–111. Dordrecht: Reidel.

Kelts, K. 1978. Geological and sedimentary evolution of Lakes Zürich and Zug. ETH Zürich, Dissertation no. 6146. 250 pp.

Kelts, K. 1981. A comparison of some aspects of sedimentation and translational tectonics and the Mesozoic Tethys and Northern Penninic Margin. *Eclogae Geol. Helv.* 74:317–38.

Kempler, D., and Z. Garfunkel. 1992. The Eratosthenes Seamount: a fossil superstructure in the eastern Mediterranean. *Proc. Congr. Comm. Int. Expl. Sci. Med.* 33:391.

King, P. B. 1977. *The Evolution of North America.* Princeton, N.J.: Princeton University Press. 197 pp.

Kober, L. 1921. *Der Bau der Erde.* Berlin: Bornträger.

Kossmat, F. 1927. Gliederung des varistischen Gebirgsbaues. *Abh. Sächs. Geol. Landesamtes* 1:1–39.

Kuhn, J. A. 1972. Stratigraphisch-mikropaläontologische Untersuchungen in der äusseren Einsiedler Schuppenzone und im Wägitaler Flysch. *Eclogae Geol. Helv.* 65:483-553.

Lancelot, Y., and E. Seibold. 1977. The evolution of the central northeastern Atlantic: Summary of the results of DSDP Leg 41. *Initial Reports of the Deep Sea Drilling Project* vol. 41, pp. 1215–45. Washington, D.C.: U.S. Government Printing Office.

Laubscher, H. 1965. Ein kinematisches Modell der Jurafaltung. *Eclogae Geol. Helv.* 58:231–318.

Laubscher, H. 1987. Die tektonische Entwicklung der Nordschweiz. *Eclogae Geol. Helv.* 80:287–303.

Laubscher, H., and D. Bernoulli. 1977. Mediterranean and Tethys. In A.E.M. Nairn, W. H. Kanes, and F. G. Stehli (eds.), *The Ocean Basins and Martins*, vol. 4A, *The Eastern Mediterranean*, pp. 1–28. New York: Plenum.

Laubscher, H., and D. Bernoulli. 1980. Cross-section from the Rhine Graben to the Po Plain. Excursion no. III. In Swiss Geological Committee (ed.), *Geology of Switzerland*, pp. 183–209. Basel: Wepf.

Laubscher, H., and D. Bernoulli. 1982. History and deformation of the Alps. In K. J. Hsü (ed.), *Mountain Building Processes*, pp. 169–80. London: Academic Press.

Leemann, A. 1988. Geologie im südpenninisch-unterostalpinen Grenzbereich auf der nördlichen Alp Flix (Oberhalbstein). ETH Zürich, *Diplomarbeit*. 88 pp.

Leggett, J. K., W. S. McKerrow, and M. H. Eales. 1979. The Southern Upland of Scotland: A Lower Paleozoic accretionary prism. *J. Geol. Soc. London* 138:167–76.

Lemoine, M. 1953. Remarques sur les caractères et l'évolution de la paléogéographie de la zone brianconnais au Secondaire et au Tertiare. *Soc. Géol. Fr. Bull., Sér. 6* 3:105–20.

Le Pichon, X. 1968. Seafloor spreading and continental drift. *J. Geophys. Res.* 73:3661–97.

Le Pichon, X. 1982. Land-locked ocean basins and continental collision: the eastern Mediterranean as a case-example. In K. J. Hsü. (ed.), *Mountain Building Processes*, pp. 201–12. London: Academic Press.

Le Pichon, X., and J. C. Sibuet. 1981. Passive margins: a model of formation. *J. Geophys. Res.*, 86:3708–20.

Leupold, W. 1937. Zur Stratigraphie der Flyschbildungen zwischen Linth und Rhein. *Eclogae Geol. Helv.* 30:1–23.

Leupold, W. 1942. Neue Beobachtungen zur Gliederung der Flyschbildungen der Alpen zwischen Reuss und Rhein. *Eclogae Geol. Helv.* 35:247–91.

Leupold, W. 1966. Südhelvetischer Flysch (Obereocän-Flysch). *Lex. Stratigr. Int. 7c, Schweizer Alpen und Südtessin*, pp. 1112–16.

Leupold, W. 1967a. Einsiedler Flysch. *Lex. Stratigr. Int. 1, Europe, Fasc. Suisse*, pp. 322–59.

Leupold, W. 1967b. Excursion no. 31. In *Geologischer Führer der Schweiz*. Basel: Wepf.

Li, S., *et al.* 1989. The timing of collision between the North China Block and the South China Block (in Chinese). *Sci. Sin., Ser. B.*, 3:312–19.

Lombard, A. 1971. La nappe du Niesen et son Flysch. *Beitr. Geol. Karte Schweiz, N. F.* 141:1–252.

Löw, S. 1987. Die tektono-metamorphe Entwicklung der nördlichen Adula-Decke. *Beitr. Geol. Karte Schweiz, N. F.* 161:1–83.

Lu, Y. 1981. Provincialism, dispersal, development and phylogeny of trilobites. *Geol. Soc. Am., Spec. Paper* 187:143–50.

Lu, C., and K. J. Hsü. 1992. Tectonic evolution of Taiwan Mountain Belt. *Pet. Geol. Taiwan* 27:21–26.

Lüdin, P. 1987. Flysche und tektonische Melanges im Südpenninisch/Unterostalpinen Grenzbereich (Arosa-Zone, Mittelbünden und Rätikon, Schweiz). Univ. Basel, *Dissertation*. 281 pp.

Lugeon, M. 1902. Sur la coupe géologique massif du Simplon. *Bull. Soc. Vaudoise. Sci. Nat.* 38:39–41.

Lugeon, M. 1938. Quelques faits nouveaus dans les Préalpes internes Vaudoises. *Eclogae Geol. Helv.* 31:1–20.

Lugeon, M. 1943. Une nouvelle hypothèse tectonique: la Diverticulation. *Bull. Soc. Vaudoise Sci. Nat.* 62:260.

Lugeon, M., and E. Argand. 1905. Sur les grandes nappes de recouvrement de la zone Piémont. *C. R. Acad. Sci., Paris* 140:1491.

McConnell, R. B. 1951. La nappe du Niesen et ses abords entre les Ormonts et la Sarine. *Beitr. Geol. Karte Schweiz, N. F.* 95.

McKenzie, D. P. 1978. Some remarks on the development of sedimentary basins. *Earth Planet. Sci. Lett.* 40:25–32.

Massey, N.W.D. 1986. Metchosin Igneous Complex, southern Vancouver Island: Ophiolite stratigraphy developed in an emergent island setting. *Geology* 14:602–5.

Mattauer, M. S., *et al.* 1985. Tectonics of the Qinling Belt: build-up and evolution of eastern Asia. *Nature (London)* 317:496–500.

Matter, A. 1987. Faciesanalyse und Ablagerungsmilieus des Permokarbons im Nordschweizer Trog. *Eclogae Geol. Helv.* 80:345–67.

Matter, A., P. Homewood, C. Caron, D. Rigassi, J. van Stuivenberg, M. Weidmann, and W. Winkler. 1980. Excursion no. V: Flysch and Molasse of western and central Switzerland. In Swiss Geological Committee (ed.), *Geology of Switzerland: A Guidebook*, pp. 261–94. Basel: Wepf.

May, D. J., and N. W. Walker. 1989. Late Cretaceous juxtaposition of metamorphic terranes in the southeastern San Gabriel Mountains, California. *Geol. Soc. Am. Bull.* 101:1246–67.

Mayer-Eymar, K. 1898. Systematisches Verzeichnis der Versteinerungen des Parisian der Umgegend von Einsiedeln. *Beitr. Geol. Karte Schweiz* 30:1–100.

Menard, H. W. 1986. *The Ocean of Truth.* Princeton, N.J.: Princeton University Press. 353 pp.

Milnes, A. G. 1974a. Structure of the Pennine Zone (Central Alps): A new working hypothesis. *Geol. Soc. Am. Bull.* 85:1727–32.

Milnes, A. G. 1974b. Post-nappe folding in the western Lepontine Alps. *Eclogae Geol. Helv.* 67:333–48.

Milnes, A. G. 1987. Tectonic evolution of the southern Andes, Tierra del Fuego: A summary. In J. Schaer and J. Rodgers (eds.), *The Anatomy of Mountain Ranges*, pp. 174–77. Princeton, N.J.: Princeton University Press.

Milnes, A. G., M. Greller, and R. Müller. 1981. Sequence and style of major post-nappe structures, Simplon-Pennine Alps. *J. Struct. Geol.* 3:411–20.

Milnes, A. G., and H.-U. Schmutz. 1978. Structure and history of the Suretta nappe (Pennine zone, Central Alps): a field study. *Eclogae Geol. Helv.* 71:19–34.

Misra, K. C., and F. B. Keller. 1978. Ultramafic bodies in the southern Appalachians: a review. *Am. J. Sci.* 278:389–418.

Mitchell, A.H.G., and W. S. McKerrow. 1975. Analogous evolution of the Burma orogen and the Scottish Caledonides. *Geol. Soc. Am. Bull.* 86:305–15.

Moore, R. C. 1949. Meaning of facies. *Am. Assoc. Pet. Geol. Bull.* 42:2718–44.

Murchison, R. I. 1850. On the geological structure of the Alps, Apennines and Carpathians, more especially to prove a transition from Secondary to Tertiary rocks, and the development of Eocene deposits in South Europe. *Q. J. Geol. Soc. London* 5:157–312.

Murray, J., and A. F. Renard. 1891. Report on the deep-sea deposits based upon specimens collected during the voyage of H.M.S. "Challenger" in the years 1873–1876. In *"Challenger" Reports.* Edinburgh: Her Majesty's Stationery Office. 525 pp.

Nabholz, W. 1945. Geologie der Bündnerschiefergebirge zwischen Rheinwald, Valser- und Safiental. *Eclogae Geol. Helv.* 38:1–120.

Nabholz, W. 1967. Chur-Vals. In *Geologischer Führer der Schweiz*, Part 8, pp. 743–75. Basel: Wepf.

Nänny, P. 1948. Zur Geologie der Prättigauschiefer zwischen Rhätikon und Plessur. Zurich: Fretz. 127 pp.

Neumayr, M. 1887. *Erdgeschichte*, Vol. 1. Leipzig: Bibliographisches Institut. 653 pp.

Nicholson, H. A. 1890. Address on recent progress in paleontology as regards invertebrate animals. *Trans. Edinburgh Geol. Soc.* 6:53–69.

Nievergelt, P., and V. Dietrich. 1977. Die andesitisch-basaltischen Gänge des Piz Lizun (Bergell). *Schweiz. Mineral. Petrogr. Mitt.* 57:267–80.

Niggli, E. 1953. Die metamorphen Bildung. In J. Cadisch, *Geologie der Schweizer Alpen*, pp. 80–119. Basel: Wepf.

Niggli, E., and W. Nabholz. 1967. San Gions-Passhöhe—Acquacalda. *Geologischer Führer der Schweiz*, Part 5, pp. 404–11. Basel: Wepf.

Oberhänsli-Langenegger, H. 1978. Mikropaläontologische und sedimentologische Untersuchungen in der Amdener Formation. *Beitr. Geol. Karte Schweiz, N. F.* 150:1–83.

Oberhauser, R. 1983. Mikrofossilfunde in Nordwestteil des Unterengadiner Fensters sowie im Verspalaflysch des Rätikon. *Jahrb. Geol. Bundesanst.* 126:71–93.

Oberholzer, J. 1933. Geologie der Glarneralpen. *Beitr. Geol. Karte Schweiz, N. F.* 28:1–626.

Opdyke, N. D., *et al.* 1987. Florida as an exotic terrane: Paleomagnetic and geochronologic investigation of lower Paleozoic rocks from the subsurface of Florida. *Geology* 15:900–903.

Oxburgh, R. 1982. Heterogeneous lithospheric stretching in early history of orogenic belts. In K. J. Hsü (ed.), *Mountain Building Processes*, pp. 85–94. London: Academic Press.

Pantic, N., and A. Gansser. 1977. Palynologisch Untersuchungen in Bündnerschiefern (I). *Eclogae Geol. Helv.* 70:59–81.

Pantic, N., and A. Isler. 1978. Palynologisch Untersuchungen in Bündnerschiefern (II). *Eclogae Geol. Helv.* 71:447–65.

Park, R. G. 1988. *Geological Structures and Moving Plates.* Glasgow: Blackie. 337 pp.

Pavoni, N. 1967. Falätsche-Uetliberg bei Zürich. In *Geologischer Führer der Schweiz*, Part 7, pp. 538–42. Basel: Wepf.

Peach, B. N., J. Horne, W. Gunn, C. T. Clough, L. W. Hinxman, and J. M. Gadell. 1888. Report on recent work of the Geological Survey in the northwest Highlands of Scotland. *Q. J. Geol. Soc. London* 44:378–441.

Penck, A., and E. Brückner. 1909. *Die Alpen in Eiszeitalter*, Vol. 2. Leipzig: Tauchnitz. 716 pp.

Pfiffner, O. A. 1978. Der Faltens- und Kleindeckenbau im infrahelvetikum der Ostschweiz. *Eclogae Geol. Helv.* 71:61–84.

Phillip, R. 1982. Die Alkaliamphibole der Platta-Decke zwischen Silsersee und Lunghinpass (Graubünden). *Schweiz. Mineral. Petrogr. Mitt.* 62:437–55.

Pitcher, W. 1982. Granite type and tectonic environment. In K. J. Hsü (ed.), *Mountain Building Processes*, pp. 19–40. London: Academic Press.

Platt, J. P. 1987. The uplift of high-pressure–low-temperature metamorphic rocks. *Philos. Trans. R. Soc. London, Series A* 321:87–102.

Playfair, J. 1802. *Illustrations of the Huttonian Theory of the Earth.* Edinburgh: William Creech. 528 pp.

Radomski, A. 1961. On some sedimentological problems of the Swiss Flysch Series. *Eclogae Geol. Helv.* 54:451–58.

Ramsay, J. G. 1989. Fold and fault geometry in the western Helvetic nappes of Switzerland and France and its implications for the evolution of the arc of the western Alps. In M. P. Coward, D. Dietrich, and R. G. Park (eds.), *Alpine Tectonics*, pp. 33–46. London: Geological Society, Special Publication no. 45.

Ramseyer, K. 1987. Diagenese des Buntsandsteins und ihre Beziehung zur tektonischen Entwicklung der Nordschweiz. *Eclogae Geol. Helv.* 80:383–96.

Rast, N., B. H. O'Brien, and R. J. Wardle. 1976. Relationships between Precambrian and lower Paleozoic rocks of the "Avalon Platform" in New Brunswick, the northeast Appalachians and the British Isles. *Tectonophysics* 30:315–38.

Rast, N., B. A. Sturt, and A. L. Harris. 1984. Early deformation in the Caledonian-Appalachian orogen. In A. L. Harris and D. J. Fettes (eds.), *The Caledonian-Appalachian Orogen*, pp. 111–22. London: Geological Society, Special Publication no. 13.

Reading, H. G. (ed.). 1986. *Sedimentary Environments and Facies*, 2nd ed. Oxford: Blackwell. 615 pp.

Reed, R. D. 1933. *Geology of California.* Tulsa, Okla.: American Association of Petroleum Geologists. 355 pp.

Reinhard, M., and H. Preiswerk. 1934. Excursion no. 30: Nufenengebiet. In *Geologischer Führer der Schweiz*, pp. 495–99. Basel: Wepf.

Richter, M. 1957. Die Allgäu-Vorarlberger Flyschzone und ihre Fortsetzung nach Westen und Osten. *Z. Deutsch. Geol. Ges.* 108:156–74.

Roberts, D., and D. G. Gee. 1985. An introduction to the structure of the Scandinavian Caledonides. In D. G. Gee and B. A. Sturt (eds.), *The Caledonide Orogen: Scandinavia and Related Areas.* pp. 887–902. Chichester: Wiley.

Robertson, A.H.F., P. D. Clift, P. J. Degnan, and G. Jones. 1991. Paleogeographic and paleotectonic evolution of the Eastern Mediterranean Neotethys. *Palaeogeogr. Palaeoclimatol. Palaeoecol.* 87:289–343.

Rodgers, J. 1983. The life history of a mountain range—the Appalachians. In K. J. Hsü (ed.), *Mountain Building Processes*, pp. 229–42. London: Academic Press.

Rodgers, J. 1987. The Appalachian geosyncline. In J. Schaer and J. Rodgers (eds.), *The Anatomy of Mountain Ranges*, pp. 241–58. Princeton, N.J.: Princeton University Press.

Rodgers, J., and K. J. Hsü. 1990. Letter to the Editor. *Geology* 18:192.

Roessli, F. 1967. Luzern-Alpnachstad. *Geologischer Führer der Schweiz*, Part 7, pp. 583–89. Basel: Wepf.

Rollier, L. 1918. Ueber alpine Kreide und Nummuliten-Formation. *Eclogae Geol. Helv.* 14:669–74.

Rothpletz, A. 1900. *Geologische Alpenforschungen. I. Das Grenzgebiet zwischen den Ost- und West-Alpen und die Rhaetische Ueberschibung.* Munich: Lindauer. 176 pp.

Rück, P. 1990. Stratigraphisch-sedimentologische Untersuchung der Schamser Decken. ETH Zürich, Dissertation no. 9185. 196 pp.

Ryan, W.B.F., D. J. Stanley, J. B. Hersey, D. A. Farhlquist, and T. D. Allan. 1969. The tectonics and geology of the Mediterranean Sea. In A. Maxwell (ed.), *The Sea* vol. 4, pp. 387–492. New York: Wiley.

Ryan, W.B.F., and K. J. Hsü. 1973. *Initial Reports of the Deep Sea Drilling Project*, vol. 13. Washington, D.C.: U.S. Government Printing Office. 1447 pp.

Ryan, W.B.F., K. Venkatarathnam, and F. Wezel. 1973. In W.B.F. Ryan, and K. J. Hsü, *Initial Reports of the Deep Sea Drilling Project*, vol. 13, pp. 731–46. Washington, D.C.: U.S. Government Printing Office.

Sartori, M. 1987. Structure de la zone du Combin entre les Diablons et Zermatt (Valais). *Eclogae Geol. Helv.* 80:789–814.

Satir, M. 1975. Die Entwicklungsgeschichte der westlichen Hohen Tauern und der südlichen Oetztalmasse aufgrund radiometrischer Altersbestimmungen. *Ist. Geol. Min. Univ. Padova, Mem.* 30:1–84.

Schaer, J. 1987. Introduction: Comparative anatomy in geology. In J. Schaer, and J. Rodgers (eds.), *The Anatomy of Mountain Ranges*, pp. 3–10. Princeton, N.J.: Princeton University Press.

Schaer, J., and J. Rodgers (eds.). 1987. *The Anatomy of Mountain Ranges.* Princeton, N.J.: Princeton University Press. 298 pp.

Schardt, H. 1893. Sur l'origin des Préalpes romandes. *Eclogae Geol. Helv.* 4:129–42.

Schardt, H. 1898. Les régions exotiques du versant Nord des Alpes Suisses (Préalpes du Chablais et du Stockhorn et les Klippes). *Bull. Soc. Vaudoise Sci. Nat.* 34:113–219.

Schardt, H. 1904. Note sur le profil géologique et la tectonique du massif du Simplon, *Eclogae Geol. Helv.* 8:173–200.

Schere, F. 1966. Geologisch-paläontologische Untersuchungen im Flysch und in der Molasse zwischen Thunersee und Eris. *Beitr. Geol. Karte Schweiz, N. F.* 127:1–115.

Schmid, S. M. 1975. The Glarus overthrust: field evidence and mechanical model. *Eclogae Geol. Helv.* 68:247–80.

Schmid, S. M. 1982. Microfabric studies as indicators of deformation mechanisms and flow laws operative in mountain building. In K. J. Hsü (ed.), *Mountain Building Processes*, pp. 95–110. London: Academic Press.

Schmid, S. M., J. N. Boland, and M. S. Patterson. 1977. Superplastic flow in fine-grained limestone. *Tectonophysics* 43:257–91.

Schmid, S. M., P. Rück, and G. Scheurs. 1990. The significance of the Schams nappes for the reconstruction of the paleotectonic and orogenic evolution of the Penninic zone along the NFP 20 East traverse (Grisons, eastern Switzerland). *Soc. Géol. Fr., Mem.* 156:263–87.

Schmidt, C., and H. Preiswerk. 1908. Geologische Karte der Simplon-Gruppe 1:50 000 (mit Erläuterungen). Special Map no. 48. Swiss Geological Committee.

Schreyer, W. 1988. Experimental studies on metamorphism of crustal rocks under mantle pressures. *Mineral. Mag.* 52:1–26.

Sclater, J. G., R. N. Anderson, M. I. Bell. 1971. The elevation of ridges and the evolution of the central eastern Pacific. *J. Geophys. Res.* 76:7888–915.

Seilacher, A. 1958. Zur ökologischen Charakteristik von Flysch und Molasse. *Eclogae Geol. Helv.* 51:1962–2078.

Sengör, A.M.C. 1984. The Cimmeride orogenic system and the tectonics of Eurasia. *Geol. Soc. Am., Spec. Paper* 195:1–82.

Sengör, A.M.C. 1991. Timing of orogenic events: a persistent controversy. In D. W. Müller, J. A. McKenzie, and H. Weissert (eds.), *Controversies in Modern Geology*, pp. 405–73. London: Academic Press.

Sengör, A.M.C., B. A. Natalin, and V. S. Burtman. 1993. Evolution of the Altaid tectonic collage and Paleozoic crustal growth in Eurasia. *Nature* 364:299–308.

Sengör, A.M.C., and A. H. Okurogullari. 1991. The role of accretionary wedges in the growth of continents: Asiatic examples from Argand to plate tectonics. *Eclogae Geol. Helv.* 84:535–97.

Shackleton, N. J., and N. D. Opdyke. 1976. Oxygen-isotope and paleomagnetic stratigraphy of Pacific Core V-28–239 late Pliocene to latest Pleistocene. *Geol. Soc. Am. Mem.* 145:449–64.

Skekan, J. W., D. P. Murray, A. R. Palmer, A. T. Smith, and E. S. Belt. 1978. Significance of fossiliferous Middle Cambrian rocks of Rhode Island to the history of the Avalonian microcontinent. *Geology* 6:694–98.

Smith, A. G. 1971. Alpine deformation and the oceanic areas of the Tethys, Mediterranean and Atlantic. *Geol. Soc. Am. Bull.* 82:2039–70.

Smith, D. C. 1984. Coesite in clinopyroxene in the Caledonides and its implications for geodynamics. *Nature (London)* 310:641–44.

Smoluchowski, M. S. 1909. Some remarks on the mechanics of overthrusts. *Geol. Mag., N. S., Decade V* 6:204–5.

Smythe, D. K., A. Dobinson, R. McQuillin, J. A. Brewer, D. H. Matthews, D. J. Blunder, and B. Kelk. 1982. Deep structure of the Scottish Caledonides revealed by the MOIST reflection profile. *Nature (London)* 299:338–40.

Soper, N. J., and A. J. Barber. 1982. A model for the deep structure of the Moine thrust zone. *Q. J. Geol. Soc. London* 139:127–38.

Stacher, P. 1980. Stratigraphie, Mikrofazies, und Mikropaläontologie der Wang Formation. *Beitr. Geol. Karte Schweiz, N. F.* 152:1–105.

Staeger, D. 1944. Geologie der Wilerhorngruppe zwischen Brienz und Lungern. *Eclogae Geol. Helv.* 37:99–188.

Stanley, R. S., and N. M. Ratcliffe. 1985. Tectonic synthesis of the Taconian orogeny in western New England. *Geol. Soc. Am. Bull.* 96:1227–50.

Staub, R. 1917. Über Faciesverteilung und Orogenese in den südöstlichen Schweizer Alpen. *Beitr. Geol. Karte Schweiz, N. F.* 46:165–98.

Staub, R. 1920. Neuere Ergebnisse der geologischen Erforschung Graubündens. *Eclogae Geol. Helv.* 16:1–26.

Staub, R. 1924. Der Bau der Alpen. *Beitr. Geol. Karte Schweiz, N. F.* 52:1–272.

Staub, R. 1928. *Der Bewegungsmechanismus der Erde.* Berlin: Bornträger. 270 pp.

Staub, R. 1937. Gedanken zum Bau der Westalpen zwischen Bernina und Mittelmeer. I. Teil. *Vierteljahrsschr. Naturforsch. Ges. Zürich* 82:197–336.

Staub, R. 1942. Gedanken zum Bau der Westalpen zwischen Bernina und Mittelmeer. II. Teil. *Vierteljahrsschr. Naturforsch. Ges. Zürich* 87:1–138.

Staub, R. 1958. Klippendecke und Zentralalpenbau: Beziehungen und Probleme. *Beitr. Geol. Karte Schweiz, N. F.* 103:1–184.

Steinmann, G. 1905. Gibt es fossile Tiefseeablagerungen von erdgeschichtlicher Bedeutung? *Geol. Rundsch.* 16:435–68.

Stephens, M. B., and D. G. Gee. 1985. A tectonic model for the evolution of the eugeosynclinal terranes in the central Scandinavian Caledonides. In D. G. Gee and B. A. Sturt (eds.), *The Caledonian Orogen*, pp. 954–78. Chichester: Wiley.

Stille, H. 1924. *Grundfragen der vergleichenden Tektonik.* Berlin: Bornträger. 443 pp.

Streckeisen, A. 1948. Der Gabbrozug Klosters-Davos-Arosa. *Schweiz. Mineral. Petrogr. Mitt.* 28:195–214.

Streiff, V. 1962. Zur östlichen Beheimatung der Klippendecken. *Eclogae Geol. Helv.* 55:79–134.

Studer, B. 1825. *Beiträge zu einer Monographie der Molasse.* Bern: Jenni.

Studer, B. 1827. Remarques géognostiques sur quelques parties de la chaine septentrionale des Alpes. *Ann. Sci. Nat. Paris* 11:1–47.

Studer, B. 1848. Sur la véritable signification du nom de Flysch. *Verh. Schweiz. Naturforsch. Ges.* 33:32–35.

Studer, B. 1853. *Geologie der Schweiz*, vol. 2. Bern: Stämpfli. 497 pp.

Studer, B. 1872. *Index der Petrographie und Stratigraphie der Schweiz und ihrer Umgebung.* Bern: Dalp.

Suess, E. 1875. *Die Entstehung der Alpen.* Vienna: W. Braunmüller. 168 pp.

Suess, E. 1904. Ueber das Inntal bei Nauders. *Sitzungsber. Kai. Akad. Wien* 114:699.

Suess, E. 1909. *Das Antlitz der Erde*, vol. 3. Leipzig: Freitag. 789 pp.

Suess, F. E. 1937. Baustein zur einer Synthese Tektogenese. *Fortschr. Geol. Paläont.* 13:1–86.

Sujkowski, Z. L. 1957. Flysch sedimentation. *Geol. Soc. Am. Bull.* 68:543–54.

Tercier, J. 1925. Sur la géologie de la Berra et l'emplacement originel du Flysch des Préalpes externes. *Bull. Soc. Fribourg. Sci. Nat.* 28:1–14.

Termier, P. 1903. Les Nappes des Alpes Orientales et la Synthèse des Alpes. *Bull. Soc. Géol. Fr., 4th Ser.* 3:711–66.

Thöni, M. 1983. The thermal climax of the early alpine metamorphism in the Austroalpine thrust sheet. *Ist. Geol. Min. Univ. Padova, Mem.* 36:211–39.

Tollmann, A. 1977. *Geologie von Oesterreich*, vol. 1. Vienna: Deuticke. 766 pp.

Trommsdorff, V. 1980. Alpine metamorphism and Alpine intrusions. In *Geology of Switzerland: A Guidebook*, Part A, pp. 82–87. Basel: Wepf.

Trommsdorff, V., and P. Nievergelt. 1983. The Bregaglia (Bergell) Iorio Intrusive and its field relations. *Mem. Soc. Geol. Ital.* 26:55–68.

Trümpy, R. 1954. La zone de Sion-Courmayeur dans le haut Val Ferret valaisain. *Eclogae Geol. Helv.* 47:315–59.

Trümpy, R. 1955. Remarques sur la corrélation des unités Penniques externes entre la Savoie et le Valais et sur l'origine des nappes Préalpines. *Bull. Soc. Géol. Fr., 6th Ser.* 5:217–29.

Trümpy, R. 1959. Hypothesen über die Ausbildung von Trias, Lias und Dogger im Untergrund des schweizerischen Molassebeckens. *Eclogae Geol. Helv.* 52:435–48.

Trümpy, R. 1960. Paleotectonic evolution of the central and western Alps. *Geol. Soc. Am. Bull.* 71:843–908.

Trümpy, R. 1969. Die helvetischen Decken der Ostschweiz: Versuch einer palinspastischen Korrelation und Ansätze zu einer kinematischen Analyse. *Eclogae Geol. Helv.* 62:105–42.

Trümpy, R. 1973. The timing of orogenic events in the central Alps. In K. A. De Jong and R. Scholton (eds.), *Gravity and Tectonics*, pp. 229–52. New York: Wiley.

Trümpy, R. 1980. An outline of geology of Switzerland. In *Geology of Switzerland: A Guidebook*, Part A. Basel: Wepf. 104 pp.

Trümpy, R. 1991. The Glarus nappes: a controversy of a century ago. In D. W. Müller, J. A. McKenzie, and H. Weissert (eds.), *Controversies in Modern Geology*, pp. 385–404. London: Academic Press.

van Bemmelen, R. W. 1933. The undation theory of the development of the earth's crust. *Proc. 16th Int. Geol. Congr., Washington, D.C.* 2:965–82.

Vening-Meinesz, F. A. 1930. Maritime gravity surveys in the Netherlands East Indies. *Ned. Akad. Wetensch. Proc. Ser. B* 33:566–77.

Vogler, W., and G. Voll. 1976. Fabrics and metamorphism from tonalite, granitic augengneiss and Tonale Series at the southern margin of the Swiss Alps. *Schweiz. Mineral. Petrogr. Mitt.* 56:635–40.

von Buch, L. 1805. Geognostische Uebersicht von Neu-Schlesien. In J. Ewald, J. Roth, and H. Eck (eds.), *Leopold von Buch's gesammelte Schriften*, vol. 1, pp. 719–39. Berlin: G. Reimer Verlag.

Walther, J. 1897. Ueber Lebensweise fossiler Meeresthiere. *Z. Dtsch. Geol. Ges.* 49:209–73.

Wang, H. 1986. Geotectonic development of China. In Z. Yang, Y. Cheng, and H. Wang *The Geology of China*, pp. 235–78. Oxford: Oxford University Press.

Wang, J., and J. Makris. 1992. New magnetic maps and the tectonic implications of the eastern Mediterranean. *Proc. Congr. Comm. Int. Expl. Sci. Med.* 33:155.

Wang, X., J. G. Liou, and H. K. Mao. 1988. Coesite-bearing eclogite from the Dabie Mountains in central China. *Geology* 17:1085–88.

Wegener, R. 1922. *Paläogeographische Darstellung der Theorie der Kontinentalverschiebungen.* Leipzig: Deuticke. 27 pp.

Wegmann, R. 1962. Zur Geologie der Flyschgebiete südlich Elm. University of Zürich, *Dissertation.* 259 pp.

Weissert, H. 1974. Die Geologie der Casanna bei Klosters. ETH Zürich, *Diplomarbeit.* 93 pp.

Weissert, H., and D. Bernoulli. 1985. A transform margin in the Mesozoic Tethys: evidence from the Swiss Alps. *Geol. Rundsch.* 74:665–79.

Wenk, E. 1973. The structure of the Bergell Alps. *Eclogae Geol. Helv.* 66:255–91.

Wernicke, B. 1985. Uniform-sense normal simple shear of the continental lithosphere. *Can. J. Earth Sci.* 22:108–25.

White, D. A., D. H. Roedder, T. H. Nelson, and J. C. Crowell. 1970. Subduction. *Geol. Soc. Am. Bull.* 81:3431–32.

Wildi, W. 1985. Heavy minerals in Alpine flysch. *G. Geol.* 47:77–100.

Williams, H. R. 1964. The Appalachians in northeastern Newfoundland: A two-sided symmetrical system. *Am. J. Sci.* 262:1137–59.

Wilson, J. T. 1966. Did the Atlantic close and then re-open? *Nature (London)* 211:676–81.

Winkler, W. 1981. Aspekte der Sedimentation des Schlieren-Flysches. *Eclogae Geol. Helv.* 73:311–18.

Winkler, W., and D. Bernoulli. 1986. Detrital high-pressure/low-temperature minerals in a late Turonian flysch sequence of the eastern Alps (western Austria): Implications for early Alpine tectonics. *Geology* 14:596–601.

Yeats, R. S. 1968. Southern California structure, seafloor spreading, and history of the Pacific basin. *Geol. Soc. Am. Bull.* 79:1693–1702.

Yu, Z., Y. Shi, and S. Yang. 1988. Hainan Island—an element of Gondwana. In D. G. Howell and T. J. Wiley (eds.), *Proc. Int. Tectonostratigr. Terrane Conf., Nanking Univ.* pp. 131–32.

Zhang, Q. S. 1980. *Metamorphic Geology of East Qinling* (in Chinese). Changchuan, Jilin: People's Publishing House.

Zhu, X. 1983. On the geodynamics of Chinese Mesozoic and Cenozoic petroliferous basins. In X. Zhu (ed.), *Tectonic Evolution of Mesozoic and Cenozoic Basins of China*, pp. 1–10. Beijing: Scientific Publishing House.

Ziegler, P. 1956. *Zur Stratigraphie des Séquanien im zentralen Schweizer Jura*, pp. 37–102. Bern: Druck Stämpfli.

Ziegler, P. 1985. Caledonian, Acadian-Ligerian, Bretonian, and Variscan Orogens: is a clear distinction justified? In D. G. Gee and B. A. Sturt (eds.), *The Caledonide Orogen: Scandinavia and Related Areas*, pp. 887–902. Chichester: Wiley.

Ziegler, W. H. 1956: Geologische Studien in den Flyschgebieten des Oberhalbsteins (Graubünden). *Eclogae Geol. Helv.* 49:1–78.

Index